The IMA Volumes in Mathematics and Its Applications

Volume 28

Series Editors
Avner Friedman   Willard Miller, Jr.

Institute for Mathematics and
its Applications
IMA

The **Institute for Mathematics and its Applications** was established by
a grant from the National Science Foundation to the University of Minnesota in
1982. The IMA seeks to encourage the development and study of fresh mathematical concepts and questions of concern to the other sciences by bringing together
mathematicians and scientists from diverse fields in an atmosphere that will stim-
ulate discussion and collaboration.

The IMA Volumes are intended to involve the broader scientific community in
this process.

Avner Friedman, Director
Willard Miller, Jr., Associate Director

* * * * * * * * * *

IMA PROGRAMS

1982-1983	**Statistical and Continuum Approaches to Phase Transition**
1983-1984	**Mathematical Models for the Economics of**
	Decentralized Resource Allocation
1984-1985	**Continuum Physics and Partial Differential Equations**
1985-1986	**Stochastic Differential Equations and Their Applications**
1986-1987	**Scientific Computation**
1987-1988	**Applied Combinatorics**
1988-1989	**Nonlinear Waves**
1989-1990	**Dynamical Systems and Their Applications**
1990-1991	**Phase Transitions and Free Boundaries**

* * * * * * * * * *

SPRINGER LECTURE NOTES FROM THE IMA:

The Mathematics and Physics of Disordered Media
> Editors: Barry Hughes and Barry Ninham
> (Lecture Notes in Math., Volume 1035, 1983)

Orienting Polymers
> Editor: J.L. Ericksen
> (Lecture Notes in Math., Volume 1063, 1984)

New Perspectives in Thermodynamics
> Editor: James Serrin
> (Springer-Verlag, 1986)

Models of Economic Dynamics
> Editor: Hugo Sonnenschein
> (Lecture Notes in Econ., Volume 264, 1986)

Kenneth R. Meyer   Dieter S. Schmidt
Editors

Computer Aided
Proofs in Analysis

Springer-Verlag
New York Berlin Heidelberg London
Paris Tokyo Hong Kong Barcelona

Kenneth R. Meyer
Departments of Mathematics
and Computer Science
University of Cincinnati
Cincinnati, OH 45221
USA

Dieter S. Schmidt
Department of Computer Science
University of Cincinnati
Cincinnati, OH 45221
USA

Series Editors

Avner Friedman
Willard Miller, Jr.
Institute for Mathematics and Its Applications
University of Minnesota
Minneapolis, MN 55455
USA

Library of Congress Cataloging-in-Publication Data
Computer aided proofs in analysis / Kenneth R. Meyer, Dieter Schmidt,
  editors.
      p.   cm. — (The IMA volumes in mathematics and its
  applications ; v. 28)
    1. Numerical analysis—Data processing—Congresses. I. Meyer,
  Kenneth R. (Kenneth Ray), 1937–   II. Schmidt, Dieter S.
  III. Series.
  QA297.C638   1990
  519.4'0285—dc20   90-45342

Printed on acid-free paper.

© 1991 by Springer-Verlag New York Inc.
Softcover reprint of the hardcover 1st edition 1991

Camera-ready copy prepared by the IMA.

9 8 7 6 5 4 3 2 1

ISBN-13: 978-1-4613-9094-7   e-ISBN-13: 978-1-4613-9092-3
DOI: 10.1007/978-1-4613-9092-3

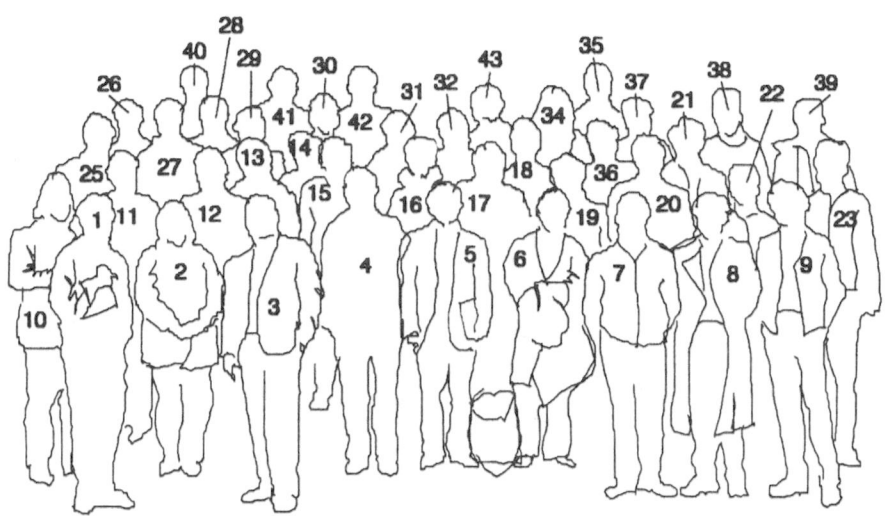

1. Mohamed Elbialy 2. Alessandra Celletti 3. Ken Meyer 4. Marc Jacobs 5. Jyun Fu 6. Rafael de la Llave
7. Alkis Akritas 8. Richard McGehee 9. Chris McCord 10. Bruce Miller 11. Bruno Buchberger 12. Tony Hearn
13. Liam Healy 14. Anne Noonburg 15. Govindan Rangarajan 16. Jong Juang 17. Carmen Chicone 18. Luis Seco
19. Shui-Nee Chow 20. Richard Rand 21. Richard Aron 22. Vincent Coppola 23. Konstantin Mischaikow
25. Andre Deprit 26. Jerry Paul 27. Etienne Deprit 28. Dave Richards 29. Ramon Moore 30. George Corliss
31. Louis Rall 32. Shannon Coffey 34. Mark Muldoon 35. David Hart 36. Johnnie Baker 37. Phil Korman
38. J.P. Eckmann 39. Barry Cipra 40. Dieter Schmidt 41. Oliver Aberth 42. C.Y. Han 43. Bruce Char

The IMA Volumes
in Mathematics and its Applications

Current Volumes:

Forthcoming Volumes:

1988-1989: *Nonlinear Waves* Multidimensional Hyperbolic Problems and Computations (2 Volumes)

   Microlocal Analysis and Nonlinear Waves

Summer Program 1989: *Robustness, Diagnostics, Computing and Graphics in Statistics*

   Robustness, Diagnostics in Statistics (2 Volumes)

   Computing and Graphics in Statistics

1989-1990: *Dynamical Systems and Their Applications*

   An Introduction to Dynamical Systems

   Patterns and Dynamics in Reactive Media

   Dynamical Issues in Combustion Theory

   Twist Mappings and Their Applications

   Dynamical Theories of Turbulence in Fluid Flows

   Nonlinear Phenomena in Atmospheric and Oceanic Sciences

   Chaotic Processes in the Geological Sciences

FOREWORD

This IMA Volume in Mathematics and its Applications

COMPUTER AIDED PROOFS IN ANALYSIS

is based on the proceedings of an IMA Participating Institutions (PI) Conference held at the University of Cincinnati in April 1989. Each year the 19 Participating Institutions select, through a competitive process, several conferences proposals from the PIs, for partial funding. This conference brought together leading figures in a number of fields who were interested in finding exact answers to problems in analysis through computer methods. We thank Kenneth Meyer and Dieter Schmidt for organizing the meeting and editing the proceedings.

Avner Friedman

Willard Miller, Jr.

PREFACE

Since the dawn of the computer revolution the vast majority of scientific computation has dealt with finding approximate solutions of equations. However, during this time there has been a small cadre seeking precise solutions of equations and rigorous proofs of mathematical results. For example, number theory and combinatorics have a long history of computer-assisted proofs; such methods are now well established in these fields. In analysis the use of computers to obtain exact results has been fragmented into several schools.

This volume is the proceedings of a conference which brought together people in symbolic algebra and in interval arithmetic with some independent entrepreneurs who where interested in obtaining precise answers to questions in analysis by computer methods. There were mathematical physicists interested in the stability of matter, functional analysts computing norms in strange function spaces, celestial mechanists analyzing bifurcations, symbolic algebraists interested in exact integration, numerical analysts who had developed interval arithmetic, plus much more. The mix included developers and end users. The papers within reflect the heterogeneous background of the participants.

Barry Cipra, an independent mathematics reporter, attended the conference to research his article "Do mathematicians still do math?" which appeared in the May 19, 1989 issue of *Science*. This article contains long quotes from several of the participants at the conference. It gives a little of the flavor of the conference.

We thank Drs. Charles Groetsch and Jerome Paul for their help in organizing this conference. Patricia Brick, Steven Skogerboe, Kaye Smith and Marise Widmer are to be praised for the excellent TEXing of the manuscripts. The conference was funded by grants from the Departments of Mathematical Sciences and Computer Science of the University of Cincinnati, the National Science Foundation and the Institute for Mathematics and its Applications.

Kenneth R. Meyer

Dieter S. Schmidt

CONTENTS

THE CONVERSION OF A HIGH ORDER PROGRAMMING LANGUAGE FROM FLOATING-POINT ARITHMETIC TO RANGE ARITHMETIC

OLIVER ABERTH†

Floating-point arithmetic is the computation arithmetic that most high-level programming languages provide. A well-known drawback of this arithmetic is that when extended calculations are performed, the final answers have an error that varies widely. Under certain conditions the error is so large as to completely invalidate the results, and the person doing the computation is often unaware that this has occurred.

Computer arithmetic more complicated than floating-point arithmetic can help in controlling the errors that are made in computation. In recent years we have gained experience solving a variety of numerical problems exactly by using the programming language PRECISION BASIC or PBASIC [1],[2], which employs range arithmetic.

Range arithmetic is essentially a variety of interval arithmetic [3],[4], so simplified that the computation for one of the four rational operations $+$, $-$, $\times$, or $\div$ requires a minimal number of extra steps beyond those needed by ordinary floating-point arithmetic. A possible form of range arithmetic is as follows: Suppose that every number is manipulated by the computer in the form

$$(1) \qquad .d_1 d_2 \ldots d_n \pm \epsilon \cdot 10^e$$

Here the decimal digits $d_1, \ldots, d_n$ form the mantissa, with the decimal point preceding the first digit d_1, which must be non-zero if n is larger than 1. The single decimal digit ϵ is the range (or maximum error) of the mantissa, and the integer e defines the number's power of 10 exponent. The range digit ϵ is understood as always associated with the last digit d_n of the mantissa, so that the representation of line (1) is an abbreviation for the interval

$$\left\{ \begin{array}{ccccc} .d_1 & d_2 & \ldots & d_{n-1} & d_n \\ \pm\ .0 & 0 & \ldots & 0 & \epsilon \end{array} \right\} \cdot 10^e$$

Here the number n of mantissa digits is *variable* instead of fixed as is the case with ordinary floating-point arithmetic. Thus constants such as 10/4 or 0 can be represented in range arithmetic as

$$.25 \pm 0 \cdot 10^1 \qquad \text{or} \qquad .0 \pm 0 \cdot 10^0$$

A long computation, such as the solution of a set of k linear equations in k unknowns x_i, normally begins by manipulating input constants, such as those above, which are error free or exact (their range is 0). The successive results of additions,

---

†Mathematics Department, Texas A & M University, College Station, Texas 77843

subtractions, and multiplications of these exact numbers then create other exact numbers which tend to have longer and longer mantissas, such as

$$(2) \qquad\qquad .351275896553 \pm 0 \cdot 10^1$$

To control the length of the mantissas of numbers generated by arithmetic operations, it is useful to employ the parameter *precision*. This parameter is under the control of an executing program, and equals the maximum number of mantissa digits permitted for a result of any rational operation $+$, $-$, $\times$, or $\div$. Thus if the precision is currently set at 10 digits by an executing program, then any generated number is allowed to have a mantissa of no more than 10 digits, and its range must be non-zero if this leads to truncation. With this precision the number of line (2), being a result of an arithmetic operation, would appear as shown below.

$$.3512758965 \pm 1 \cdot 10^1$$

A division of one exact number by another, which does not yield an exact decimal result (such as 1 divided by 3) also gives a result with a 10 digit mantissa and a non-zero range. Thus any arithmetic operation with exact operands may or may not yield an exact result, depending on the operation and the current precision. Arithmetic operations with one or both operands non-exact are executed as interval arithmetic operations and yield a non-exact result. As computation proceeds with these numbers, the mantissa length can be expected to decrease slowly, with the number of mantissa digits dropping steadily from 10 toward 1, so that numbers such as

$$.72176 \pm 2 \cdot 10^5 \qquad \text{or} \qquad .4913 \pm 4 \cdot 10^2$$

are obtained, or even numbers such as

$$.0 \pm 1 \cdot 10^8$$

if the computation is too extensive for the prescribed precision.

A typical computation in range arithmetic can now be described. The problem of solving k non-singular linear equations in k unknowns x_i again can serve as an example. Suppose the goal of the calculation is 7 correct decimal places for the unknowns x_i. The precision is set at some value comfortably above 7 depending on how extensive the computation is. For our example it would be useful to let the precision depend in some fashion on k, the number of equations. To be specific, assume here a choice of 15 for the precision. After the computation in range arithmetic is complete, the values obtained for x_i either indicate that 7 correct decimal places can be printed, or else indicate the need for a higher precision. In the latter case, the amount by which the precision should be increased can be estimated from the least accurate x_i value. The entire computation is then repeated at this higher precision, and the desired answers correct to 7 decimals are eventually produced. The cases where the calculation must be repeated are likely to be cases which result in large errors if the calculations are done in ordinary floating-point arithmetic.

3

With range arithmetic, because the interval width is represented by so few digits (only one in our examples), it serves mainly to bound off the reliable mantissa digits and to cause the automatic discard of questionable mantissa digits as new arithmetic results are generated. In an actual computer implementation, it is efficient to use as many decimal digits for the range as will fit into a unit of memory storage. This is two digits if the unit of storage is a byte, four digits if the unit of storage is a word (2 bytes), and so forth.

Before making a comparison of floating-point arithmetic with range arithmetic, we list some characteristics of each. With floating-point arithmetic different sizes are usually allowed for the mantissa (with names like single precision, double precision, extended precision). These sizes must be specified in advance in a program before it is run. A fixed amount of memory space is allotted for each floating-point number of a particular size, and the mantissa representation may be in either binary or decimal form.

With range arithmetic, the amount of memory space needed for a number is variable instead of fixed, so that the number of digits n in the mantissa must be stored with the number as an element of its memory representation, along with its mantissa, exponent, and range. For an input constant only as many mantissa digits are used as are necessary to correctly express the number, and the range is set to zero. This variability in memory space needed for a number entails a level of indirection in storing and retrieving numbers. A pointer to each number is allocated instead of the memory space to hold the number as with floating-point arithmetic, and this pointer then points to the number created in a section of memory given over to this purpose. Thus management of the memory storage of numbers must be done somewhat like the way it is done with a LISP type language. The mantissa representation must be decimal instead of binary, for otherwise input constants, such as 0.1, would require recomputation every time the precision were increased.

The disadvantages of range arithmetic relative to floating-point arithmetic are easy to discern. These are the increased complexity in referencing a ranged number, and the more complicated rational operations $+$, $-$, $\times$, $\div$. The extra complication of the arithmetic operations is not major, however, since it consists mainly in having to make a preliminary computation of the result range; most of the computer time is spent forming the result mantissa, just as with floating-point. Actually, since the mantissas become shorter and shorter in length as an extended computation proceeds, the later arithmetic operations may execute faster than the equivalent floating-point operations with full-length numbers.

There are a number of significant advantages to range arithmetic. The foremost is that it provides a method for determining the appropriate precision to use in a computation. The number of significant digits that are obtained with a preliminary, or sample, computation is a guide as to whether the precision is sufficient or should be increased. A second major advantage is that when the final results of a computation are printed, or displayed at the computer console, these results can often be obtained correct to the last decimal place. Here we need to differentiate between *straightforward* and *non-straightforward* calculations. A straightforward calcula-

4

tion is one that proceeds from mathematically specified input constants through to end results, by means of a finite length sequence of arithmetic operations or function evaluations. Otherwise the computation is non-straightforward. Thus the calculation by the elimination method of the solutions to a set of linear equations is straightforward, whereas the computation of the zeros of a high degree polynomial by Bairstow's method is non-straightforward, since the method is iterative and the zeros are only approximated.

With a straightforward computation, when the end results are displayed with proper attention paid to the size of the range so that suspect decimals are held back, then these results are automatically obtained correct to the last decimal place. If too few correct decimal places are obtained, it is a simple matter to increase the precision appropriately and repeat the computation. For a non-straightforward calculation, there must be available some method of obtaining a rigorous upper bound E on the error of each result R. Then after the calculation of E is made in range arithmetic, the range of R can be incremented by an amount reflecting the size of E, and then displayed just as if it were the result of a straightforward computation. (In PBASIC a special operation $+/-$ is used to achieve this, so that the displayed value would be $R +/- E$.) If too few correct decimal places are obtained, then the iteration procedure yielding R must be continued.

Two minor advantages of range arithmetic should also be mentioned. It is often convenient to calculate in rational arithmetic. Here a pair of integers p, q are used to represent each rational number r, with r equal to p/q. Rational arithmetic can not easily be programmed in terms of floating-point arithmetic, since no indication is obtained when digits are discarded when doing arithmetic operations on integer pairs. On the other hand, programming rational arithmetic in terms of range arithmetic is easy since the integer pairs yielded by arithmetic operations have zero ranges as long as digits are not discarded.

In numerical programming a popular method of bounding the error is to use interval arithmetic in various forms, with a pair of floating-point numbers to represent each variable. As is well known, this requires certain modifications of the floating-point arithmetic operations because the direction of rounding must be modified in certain situations. Often it is difficult to make the appropriate modifications, especially if the floating-point operations are done by hardware instead of software. This type of interval arithmetic can also be carried out with a pair of ranged numbers representing each variable. Here no modification of the range arithmetic operations is necessary, and the method mentioned of determining what precision of computation to use is applicable here too.

REFERENCES

[1] O. ABERTH, *Precise scientific computation with a microprocessor*, IEEE Transactions on Computers, C-33 (1984), pp. 685–690.
[2] O. ABERTH, *Precise Numerical Analysis*, Wm. C. Brown Publishers, Dubuque, Iowa, 1988.
[3] G. ALEFELD AND J. HERZBERGER, *Introduction to Interval Computation, translated by Jon Rokne*, Academic Press, New York, 1983.
[4] R. E. MOORE, *Methods and Applications of Interval Analysis*, SIAM, Philadelphia, 1979.

SYLVESTER'S FORM OF THE RESULTANT
AND THE MATRIX-TRIANGULARIZATION SUBRESULTANT PRS METHOD

ALKIVIADIS G. AKRITAS*

Summary. Sylvester's form of the resultant is often encountered in the literature but is completely different from the one discussed in this paper; the form described here can be found in Sylvester's paper of 1853 [12], and has been previously used only once, by Van Vleck [13] in the last century. Triangularizing this "rediscovered" form of the resultant we obtain a new method for computing a greatest common divisor (gcd) of two polynomials in $\mathbf{Z}[x]$, along with their polynomial remainder sequence (prs); since we are interested in exact integer arithmetic computations we make use of Bareiss's [4] integer-preserving transformation algorithm for Gaussian elimination. This new method uniformly treats both complete and incomplete prs's and, for the polynomials of the prs's, it provides the smallest coefficients that can be expected *without* coefficient gcd computations.

1. Introduction. In this note we restrict our discussion to univariate polynomials with integer coefficients and to computations in $\mathbf{Z}[x]$, a unique factorization domain. Given the polynomial $p(x) = c_n x^n + c_{n-1} x^{n-1} + \cdots + c_0$, its degree is denoted by $\deg(p(x))$ and c_n, its leading coefficient, by $lc(p)$; moreover, $p(x)$ is called *primitive* it its coefficients are relatively prime.

Consider now $p_1(x)$ and $p_2(x)$, two primitive, nonzero polynomials in $\mathbf{Z}[x]$, $\deg(p_1(x)) = n$ and $\deg(p_2(x)) = m$, $n \geq m$. Clearly, the polynomial division (with remainder) algorithm, call it **PD**, that works over a field, cannot be used in $\mathbf{Z}[x]$ since it requires exact divisibility by $lc(p_2)$. So we use *pseudo-division*, which always yields a pseudo-quotient and pseudo-remainder; in this process we have to premultiply $p_1(x)$ by $lc(p_2)^{n-m+1}$ and then apply algorithm **PD**. Therefore we have:

$$(1) \qquad lc(p_2)^{n-m+1} p_1(x) = q(x)p_2(x) + p_3(x), \quad \deg(p_3(x)) < \deg(p_2(x)).$$

Applying the same process to $p_2(x)$ and $p_3(x)$, and then to $p_3(x)$ and $p_4(x)$, etc. (Euclid's algorithm), we obtain a *polynomial remainder sequence* (prs)

$$p_1(x), p_2(x), p_3(x), \ldots p_h(x), p_{h+1}(x) = 0,$$

where $p_h(x) \neq 0$ is a greatest common divisor of $p_1(x)$ and $p_2(x)$, $gcd(p_1(x), p_2(x))$. If $n_i = \deg(p_i(x))$ and we have $n_i - n_{i+1} = 1$, for all i, the prs is called *complete*, otherwise, it is called *incomplete*. The problem with the above approach is that the coefficients of the polynomials in the prs grow exponentially and hence slow

---

*University of Kansas, Department of Computer Science, Lawrence, Kansas 66045

down the computations. We wish to control this coefficient growth. We observe that equation (1) can also be written more generally as

$$(2) \quad lc(p_{i+1})^{n_i-n_{i+1}+1} p_i(x) = q_i(x)p_{i+1}(x) + \beta_i p_{i+2}(x), \quad \deg(p_{i+2}(x)) < \deg(p_{i+1}(x)),$$

$i = 1, 2, \ldots, h-1$. That is, if a method for choosing β_i is given, the above equation provides an algorithm for constructing a prs. The obvious choice $\beta_i = 1$, for all i, is called the *Euclidean prs*; it was described above and leads to exponential growth of coefficients. Choosing β_i to be the greatest common divisor of the coefficients of $p_{i+2}(x)$ results in the *primitive prs*, and it is the best that can be done to control the coefficient growth. (Notice that here we are dividing $p_{i+2}(x)$ by the greatest common divisor of its coefficients before we use it again). However, computing the greatest common divisor of the coefficients for each member of the prs (after the first two, of course) is an expensive operation and should be avoided. So far, in order to control the coefficient growth and to avoid the coefficient gcd computations, either the *reduced* or the (improved) *subresultant* prs have been used. In the reduced prs we choose

$$(3) \qquad \beta_1 = 1 \quad \text{and} \quad \beta_i = lc(p_i)^{n_{i-1}-n_i+1}, \quad i = 2, 3, \ldots, h-1,$$

whereas, in the subresultant prs we have

$$(4) \quad \beta_1 = (-1)^{n_1-n_2+1} \quad \text{and} \quad \beta_i = (-1)^{n_i-n_{i+1}+1} lc(p_i)H_i^{n_i-n_{i+1}}, \quad i = 2, 3, \ldots, h-1,$$

where

$$H_2 = lc(p_2)^{n_1-n_2} \quad \text{and} \quad H_i = lc(p_i)^{n_{i-1}-n_i} H_{i-1}^{1-(n_{i-1}-n_i)}, \quad i = 3, 4, \ldots, h-1.$$

That is, in both cases above we divide $p_{i+2}(x)$ by the corresponding β_i before we use it again. The reduced prs algorithm is recommended if the prs is complete, whereas if the prs is incomplete the subresultant prs algorithm is to be preferred. The proofs that the β_i's shown in (3) and (4) exactly divide $p_{i+2}(x)$ are very complicated [7] and have up to now obscured simple divisibility properties [11], (see also [5] and [6]). For a simple proof of the validity of the reduced prs see [1]; analogous proof for the subresultant prs can be found in [8].

In contrast with the above prs algorithms, the matrix-triangularization subresultant prs method avoids explicit polynomial divisions. In what follows we present this method. We also present an example where bubble pivot is needed.

2. Sylvester's form of the resultant. Consider the two polynomials in $\mathbb{Z}[x]$, $p(x) = c_n x^n + c_{n-1} x^{n-1} + \cdots + c_0$ and $p_2(x) = d_m x^m + d_{m-1} x^{m-1} + \cdots + d_0$, $c_n \neq 0$, $d_m \neq 0$, $n \geq m$. In the literature the most commonly encountered forms of the resultant of $p_1(x)$ and $p_2(x)$ (both known as "Sylvester's" forms) are:

$$\text{res}_B(p_1, p_2) = \begin{vmatrix} c_n & c_{n-1} & \cdots & & c_0 & 0 & \cdots & 0 \\ 0 & c_n & c_{n-1} & \cdots & & c_0 & \cdots & 0 \\ & & & & & \vdots & & \\ 0 & 0 & \cdots & & c_n & c_{n-1} & \cdots & c_0 \\ d_m & d_{m-1} & \cdots & d_0 & 0 & 0 & \cdots & 0 \\ 0 & d_m & d_{m-1} & \cdots & d_0 & 0 & \cdots & 0 \\ & & & & \vdots & & & \\ 0 & 0 & \cdots & d_m & d_{m-1} & & \cdots & d_0 \end{vmatrix}$$

or

$$\text{res}_T(p_1, p_2) = \begin{vmatrix} c_n & c_{n-1} & \cdots & & c_0 & 0 & \cdots & 0 \\ 0 & c_n & c_{n-1} & \cdots & & c_0 & \cdots & 0 \\ & & & & & \vdots & & \\ 0 & 0 & \cdots & & c_n & c_{n-1} & \cdots & c_0 \\ 0 & 0 & \cdots & d_m & d_{m-1} & & \cdots & d_0 \\ & & & & \vdots & & & \\ 0 & d_m & d_{m-1} & \cdots & d_0 & 0 & \cdots & 0 \\ d_m & d_{m-1} & \cdots & d_0 & 0 & 0 & \cdots & 0 \end{vmatrix}$$

where for both cases we have m rows of c's and n rows of d's; that is, the determinant is of order $m + n$. Contrary to established practice, we call the first Bruno's and the second Trudi's form of the resultant [3]. Notice that $\text{res}_B(p_1, p_2) = (-1)^{n(n-1)/2} \text{res}_T(p, p_2)$. We choose to call Sylvester's form the one described below; this form was "buried" in Sylvester's 1853 paper [12]and is only once mentioned in the literature in a paper by Van Vleck [13]. Sylvester indicates ([12]), p. 426 that he had produced this form in 1839 or 1840 and some years later Cayley unconsciously reproduced it as well. It is Sylvester's form of the resultant that forms the foundation of our new method for computing polynomial remainder sequences; however ,we first present the following theorem concerning Bruno's form of the resultant:

THEOREM 1 (Laidacker [10]). *If we transform the matrix corresponding to* $\text{res}_B(p_1(x), p_2(x))$ *into its upper triangular form* $T_B(R)$, *using row transformations only, then the last nonzero row of* $T_B(R)$ *gives the coefficients of a greatest common divisor of* $p_1(x)$ *and* $p_2(x)$.

The above theorem indicates that we can obtain only a greatest common divisor of $p_1(x)$ and $p_2(x)$ but none of the remainder polynomials. In order to compute both a $gcd(p_1(x), p_2(x))$ and all the polynomial remainders we have to use Sylvester's form of the resultant; this is of order $2n$ (as opposed to $n + m$ for the other forms) and of the following form ($p_2(x)$ has been transformed into a polynomial of degree n by introducing zero coefficients):

$$
\mathrm{res}_S(p, q) =
\begin{vmatrix}
c_n & c_{n-1} & \ldots & c_0 & 0 & 0\ldots 0 \\
d_n & d_{n-1} & \ldots & d_0 & 0 & 0\ldots 0 \\
0 & c_n & \ldots & & c_0 & 0\ldots 0 \\
0 & d_n & \ldots & & d_0 & 0\ldots 0 \\
& & \ldots\ldots\ldots & & & \\
0 & \ldots & 0 & c_n & c_{n-1} & \ldots \; c_0 \\
0 & \ldots & 0 & d_n & d_{n-1} & \ldots \; d_0
\end{vmatrix}
\qquad (S)
$$

Sylvester obtains this form from the system of equations ([12]) pp. 427–428)

$$
\begin{aligned}
p(x) &= 0 \\
q(x) &= 0 \\
x \cdot p(x) &= 0 \\
x \cdot q(x) &= 0 \\
x^2 \cdot p(x) &= 0 \\
x^2 \cdot q(x) &= 0 \\
&\;\cdots \\
x^{n-1} \cdot p(x) &= 0 \\
x^{n-1} \cdot q(x) &= 0
\end{aligned}
$$

and he indicates that if we take k pairs of the above equations, the highest power of x appearing in any of them will be x^{n+k-1}. Therefore, we shall be able to eliminate so many powers of x, that x^{n-k} will be the highest power uneliminated and $n-k$ will be the degree of a member of the Sturmian polynomial remainder sequence generated by $p(x)$ and $q(x)$. Moreover, Sylvester showed that the polynomial remainders thus obtained are what he terms *simplified residues*; that is, the coefficients are the smallest possible obtained without integer gcd computations and without introducing rationals. Stated in other words, the polynomial remainders have been freed from their corresponding *allotrious factors*.

It has been proved [13] that if we want to compute the polynomial remainder sequence $p_1(x)$, $p_2(x)$, $p_3(x), \ldots, p_h(x)$, $\deg(p_1(x)) = n$, $\deg(p_2(x)) = m$, $n \geq m$, we can obtain the (negated) coefficients of the $(i+1)$th member of the prs, $i = 0, 1, 2, \ldots, h-1$, as minors formed from the first $2i$ rows of (S) by successively associating with the first $2i-1$ columns (of the $(2i)$ by $(2n)$ matrix) each succeeding column in turn.

Instead of proceeding as above, we transform the matrix corresponding to the resultant (S) into its upper triangular form and obtain the members of the prs with the help of Theorem 2 below. We also use Bareiss's integer-preserving transformation algorithm [4]; that is:

let $r_{00}^{(-1)} = 1$, and $r_{ij}^{(0)} = r_{ij}$, $i, j = 1, \ldots, n$; then for $k < i, j \leq n$,

$$(5) \qquad\qquad r_{ij}^{(k)} := (1/r_{k-1,k-1}^{(k-2)}) \cdot \begin{vmatrix} r_{kk}^{(k-1)} & r_{kj}^{(k-1)} \\ r_{ik}^{(k-1)} & r_{ij}^{(k-1)} \end{vmatrix}$$

Of particular importance in Bareiss's algorithm is the fact that the determinant of order 2 is divided *exactly* by $r_{k-1,k-1}^{(k-2)}$ (the proof is very short and clear and is described in Bareiss's paper [4]) and that the resulting coefficients are the smallest that can be expected without coefficient *gcd* computations and without introducing rationals. Notice how all the complicated expressions for β_i in the reduced and subresultant *prs* algorithms are mapped to the simple factor $r_{k-1,k-1}^{(k-2)}$ of this method.

It should be pointed out that using Bareiss's algorithm we will have to perform pivots (interchange two rows) which will result in a change of signs. We also define the term *bubble* pivot as follows: if the diagonal element in row i is zero and the next nonzero element down the column is in row $i + j$, $j > 1$, then row $i + j$ will become row i after pairwise interchanging it with the rows above it. Bubble pivot preserves the symmetry of the determinant.

We have the following theorem.

THEOREM 2 ([2]). *Let $p_1(x)$ and $p_2(x)$ be two polynomials of degrees n and m respectively, $n \geq m$. Using Bareiss's algorithm transform the matrix corresponding to $res_S(p_1(x), p_2(x))$ into its upper triangular form $T_S(R)$; let n_i be the degree of the polynomial corresponding to the i-th row of $T_S(R)$, $i = 1, 2, \ldots, 2n$, and let $p_k(x), k \geq 2$, be the kth member of the (complete or incomplete) polynomial remainder sequence of $p_1(x)$ and $p_2(x)$. Then if $p_k(x)$ is in row i of $T_S(R)$, the coefficients of $p_{k+1}(x)$ (within sign) are obtained from row $i + j$ of $T_S(R)$, where j is the smallest integer such that $n_{i+j} < n_i$. (If $n = m$ associate both $p_1(x)$ and $p_2(x)$ with the first row of $T_S(R)$.)*

We see, therefore, that based on Theorem 2, we have a new method to compute the polynomial remainder sequence and a greatest common divisor of two polynomials. This new method uniformly treats both complete and incomplete *prs*'s and provides the smallest coefficients that can be expected without coefficient *gcd* computation.

3. The matrix-triangularization subresultant prs method. The inputs are two (primitive) polynomials in $\mathbf{Z}[x]$, $p_1(x) = c_n x^n + c_{n-1} + \cdots + c_0$ and $p_2(x) = d_m x^m + d_{m-1}x^{m-1} + \cdots + d_0$, $c_n \neq 0$, $d_m \neq 0$, $n \geq m$.

Step 1: Form the resultant $(S), res_S(p_1(x), p_2(x))$, of the two polynomials $p_1(x)$ and $p_2(x)$.

Step 2: Using Bareiss's algorithm (and bubble pivot) transform the resultant (S) into its upper triangular form $T_S(R)$; then the coefficients of all the members of the polynomial remainder sequence of $p_1(x)$ and $p_2(x)$ are obtained from the rows of $T_S(R)$ with the help of Theorem 2.

For this method we have proved [2] that its computing time is:

THEOREM 3. *Let* $p_1(x) = c_n x^n + c_{n-1} x^{n-1} + \cdots + c_0$ *and* $p_2(x) = d_m x^m + d_{m-1} x^{m-1} + \cdots + d_0$, $c_n \neq 0$, $d_m \neq 0$, $n \geq m$ *be two (primitive) polynomials in* $\mathbf{Z}[x]$ *and for some polynomial* $P(x)$ *in* $\mathbf{Z}[x]$ *let* $|P|_\infty$ *represent its maximum coefficient in absolute value. Then the method described above computes a greatest common divisor of* $p_1(x)$ *and* $p_2(x)$ *along with all the polynomial remainders in time*

$$0(n^5 L(|p|_\infty)^2)$$

where $|p|_\infty = \max(|p_1|_\infty, |p_2|_\infty).$

Below we present an incomplete example where bubble pivoting is needed [3]; note that there is a difference of 3 in the degrees of the members of the *prs*, as opposed to a difference of 2 in Knuth's "classic" incomplete example [2].

Example. Let us find the polynomial remainder sequence of the polynomials $p_1(x) = 3x^9 + 5x^8 + 7x^7 - 3x^6 - 5x^5 - 7x^4 + 3x^3 + 5x^2 + 7x - 2$ and $p_2(x) = x^8 - x^5 - x^2 - x - 1$. This incomplete *prs* example presents a variation of three in the degrees of its members (from 7 to 4) and it requires a bubble pivot in the matrix-triangularization method; that is, the special kind of pivot described above will take place between rows that are not adjacent (the pivoted rows are marked by "#").

The matrix-triangularization subresultant prs method

row		degree
1 >	3 5 7 −3 −7 3 5 7 −2 0 0 0 0 0 0 0	(9)
2 >	0 1 0 0 −1 0 0 −1 −1 −1 0 0 0 0 0 0	(8)
3)	0 0 5 7 0 −5 −7 6 8 10 −2 0 0 0 0 0 0	(8)
4 >	0 0 0 −7 0 0 7 −6 −13 −15 −3 0 0 0 0 0	(7)
5)	0 0 0 0 −49 0 0 79 23 19 −55 14 0 0 0 0	(7)
#6)	0 0 0 0 0 −343 0 −24 501 73 93 −413 98 0 0 0 0	(7)
#7)	0 0 0 0 0 0 −2401 −510 −1273 1637 −339 56 −2891 686 0 0 0 0	(7)
8 >	0 0 0 0 0 0 0 2058 4459 7546 3430 2401 0 0 0 0 0	(4)
9)	0 0 0 0 0 0 0 0 −1764 −3822 −6468 −2940 −2058 0 0 0 0	(4)
10)	0 0 0 0 0 0 0 0 0 1512 3276 5544 2520 1764 0 0 0	(4)
11)	0 0 0 0 0 0 0 0 0 0 25811 −18982 4520 −811 −3024 0 0 0	(4)
12 >	0 0 0 0 0 0 0 0 0 0 0 −64205 −77246 −37568 −28403 0 0 0	(3)
13)	0 0 0 0 0 0 0 0 0 0 0 0 2124693 449379 519299 128410 0 0	(3)
14 >	0 0 0 0 0 0 0 0 0 0 0 0 0 −5240853 −1800739 −2018639 0 0	(2)
15)	0 0 0 0 0 0 0 0 0 0 0 0 0 0 −22909248 −24412716 10481706 0	(2)
16 >	0 0 0 0 0 0 0 0 0 0 0 0 0 0 0 −40801132 47620330 0	(1)
17)	0 0 0 0 0 0 0 0 0 0 0 0 0 0 0 0 −398219984 81602264	(1)
18 >	0 0 0 0 0 0 0 0 0 0 0 0 0 0 0 0 0 682427564	(0)

The members of the prs are obtained from the rows whose numbers are followed by ">", except for row 8 in which case the smaller coefficients shown below, in 6 >, are taken as the coefficients of the polynomial. The largest integer generated is 27843817119202448 [17 digits].

Pivoted row 6 during transformation 6. Stored row is:

$$6 > \qquad 0\ 0\ 0\ 0\ 0\ 0\ 0\ 42\ 91\ 154\ 70\ 49\ 0\ 0\ 0\ 0\ 0\ 0$$

Pivoted row 7 during transformation 7. Stored is:

$$7) \qquad 0\ 0\ 0\ 0\ 0\ 0\ 0\ 294\ 637\ 1078\ 490\ 343\ 0\ 0\ 0\ 0\ 0\ 0$$

REFERENCES

[1] AKRITAS, A.G., *A simple validity proof of the reduced prs algorithm*, Computing 38 (1987), 369–372.

[2] AKRITAS, A.G., *A new method for computing greatest common divisors and polynomials remainder sequences*, Numerische Mathematik 52 (1988), 119–127.

[3] AKRITAS, A.G., *Elements of Computer Algebra with Applications*, John Wiley Interscience, New York, 1989.

[4] BAREISS, E.H., *Sylvester's identity and multistep integer-preserving Gaussian elimination*, Mathematics of Computation 22 (1968), 565–578.

[5] BROWN, W.S., *On Euclid's algorithm and the computation of polynomial greatest common divisors*, JACM 18, (1971) 476–504.

[6] BROWN, W.S., *The subresultant prs algorithm*, ACM Transactions On Mathematical Software 4 (1978), 237–249.

[7] COLLINS, G.E., *Subresultants and reduced polynomial remainder sequences*, JACM 14 (1967), 128–142.

[8] HABICHT, W., *Eine Verallgemeinerung des Sturmschen Wurzelzaehlverfahrens*, Commentarii Mathematici Helvetici 21 (1948), 99–116.

[9] KNUTH, D.E., *The art of computer Programming*, Vol. II, 2nd ed.: Seminumeral Algorithms. Addison-Wesley. Reading MA, 1981.

[10] LAIDACKER, M.A., *Another theorem relating Sylvester's matrix and the greatest common divisor*, Mathematics Magazine 42 (1969), 126–128.

[11] LOOS, R., *Generalized polynomial remainder sequences*. In: Computer Algebra Symbolic and Algebraic Computations. Ed. by B. Buchberger, G.E. Collins and R. Loos, Springer Verlag, Wien, New York, 1982, Computing Supplement 4, 115–137.

[12] SYLVESTER, J.J., *On a theory of the syzegetic relations of two rational integral functions, comprising an application to the theory of Sturm's functions, and that of the greatest algebraical common measure*, Philosophical Transactions 143 (1853), 407–584.

[13] VAN VLECK, E.B., *On the determination of a series of Sturm's functions by the calculation of a single determinant*, Annals of Mathematics, Second Series, Vol. 1, (1899–1900) 1–13.

COMPUTING THE TSIRELSON SPACE NORM

JOHNNIE W. BAKER*, OBERTA A. SLOTTERBECK†, and RICHARD ARON‡

Abstract. After a review of Tsirelson space T, a reflexive Banach space containing no isomorphic copies of any ℓ_p space, the authors develop an efficient algorithm for computing the norm of T. Properties of the algorithm, timings, and space considerations are discussed.

AMS(MOS) Subject Classification. 46-04, 46B25.

Keywords. Tsirelson space, norm algorithm.

I. INTRODUCTION. Ever since the introduction of Banach spaces in the early 1930's, a problem of continuing interest for functional analysts has been the structure theory of these spaces. Specifically, a topic of basic research for the last fifty years has been the study of the type and structure of finite and infinite dimensional subspaces of a given Banach space. For example, a well-known, useful result of J. Lindenstrauss and L. Tzafriri [5] is that if every closed subspace of a Banach space is complemented (i.e. if it admits a projection onto the subspace), then the Banach space is in fact a Hilbert space, up to isomorphism.

A natural hope of mathematicians for much of this time has been that the "brickwork" making up the structure of an arbitrary infinite dimensional Banach space consist of "reasonable" pieces. Specifically, it was hoped that Conjecture I below is true:

Conjecture I: **Every infinite dimensional Banach space contains a subspace which is isomorphic to some ℓ_p or to c_0.**

Here, for $1 \leq p < \infty$, ℓ_p consists of those sequences of scalars $(\lambda_j)_j$ for which the norm $\|(\lambda_j)\|_p = [\sum_{j=0}^{\infty} |\lambda_j|^p]^{1/p} < \infty$, and $c_0 = \{(\lambda_j) : \|(\lambda_j)\|_\infty = max|\lambda_j| < \infty\}$. The ℓ_p spaces and c_0 are called the classical Banach spaces. Since the ℓ_p spaces are reflexive for $1 < p < \infty$, Conjecture II is substantially weaker than Conjecture I, and is still open.

Conjecture II: **Every infinite dimensional Banach space contains either an infinite dimensional reflexive subspace or else a subspace isomorphic to c_0 or ℓ_1.**

In 1973, the Soviet mathematician B. S. Tsirelson [6] disproved Conjecture I by

---

* Johnnie W. Baker, Department of Mathematical Sciences, Kent State University, Kent, Ohio 44242, CSNET address: jbaker@kent.edu

†Oberta A. Slotterbeck, Department of Mathematical Sciences, Hiram College, Hiram, Ohio 44234, CSNET address: slotter@kent.edu

‡Richard Aron, Department of Mathematical Sciences, Kent State University, Kent, Ohio 44242, CSNET address: aron@kent.edu

producing a Banach space containing no isomorphic copy of any of the classical Banach spaces. Within a year, T. Figiel and W. B. Johnson [4] had produced another such example, a space which we will call T. In fact, T is the dual of Tsirelson's original space. It is reflexive and has an unconditional basis. Nevertheless, and not surprisingly, the structure of T is extremely complicated and remains far from being well-understood. As will be seen below, the norm on T is defined as a limit of an infinite recursion and, consequently, computations involving this norm have been extremely slow and difficult. In this paper, we describe an algorithm for computing the norm on T.

The plan of this article is the following. After a brief description of the space $(T, \| \ \|)$, we will describe our algorithm. Finally, we will discuss some computational results as well as storage and time considerations.

II. PRELIMINARIES AND EXAMPLES. Let T_0 be the vector space of all sequences $X = (x_1, x_2, \ldots, x_n)$ of real scalars, such that $x_j = 0$ for all big j. We'll write $X = (x_1, x_2, \ldots, x_n)$ for $(x_1, x_2, \ldots, x_n, 0, 0, \ldots)$. For X as above and for $0 \le lo < hi \le n$, write $(lo, hi]$ for the "subvector" $(0, \ldots, 0, x_{lo+1}, \ldots, x_{hi}, 0, 0, \ldots)$. The definition of the norm for Tsirelson space which is used here is the one described below: An admissible partition P of $\{1, 2, \ldots, n\}$ is a set

$$P = \{p_1, p_2, \ldots, p_{k+1}\}$$

of integers with

$$k \le p_1 < p_2 < \ldots < p_{k+1} \le n.$$

For example, the only admissible partition of $\{1, 2, \ldots, 6\}$ having 4 elements is $P = \{3, 4, 5, 6\}$. For $X \in T_0$, define $\|X\|_0 = max_j |x_j|$, and for $m \ge 1$, $\|X\|_m$ is defined as

$$\|X\|_m = \max \begin{cases} \|X\|_{m-1} \\ \frac{1}{2} \max_P \sum_{j=1}^{k} \|(p_j, p_{j+1}]\|_{m-1} \end{cases}$$

where $\max_P$ is the maximum over all admissible partitions P of $\{1, 2, \ldots, n\}$. Finally, we let $\|X\| = \lim_m \|X\|_m$, and we let T be the completion of T_0 with this norm.

We give two simple examples to illustrate the computations involved, and also to provide some motivation for the algorithm which follows:

Example 1. Let $X = (0, 0, 0, 1, 1, 1)$. So $\|X\|_0 = 1$ and, taking the partition $P = \{3, 4, 5, 6\}$ which was mentioned above, we see that $\|X\|_1 \ge \frac{1}{2}[\|(3, 4]\|_0 + \|(4, 5]\|_0 + \|(5, 6]\|_0] = \frac{3}{2}$. In fact, it is easy to see that $\|X\| = \frac{3}{2}$.

The next example, although still quite trivial, illustrates both the computations involved and our method for finding the Tsirelson norm.

Example 2. Let $X = (4, 1, 2, 6, 5, 3)$. For each subvector $Y = (lo, hi]$ of X, let's calculate $\|Y\|_0$, storing our answer in the (lo, hi) position of a 6 by 6 matrix. Thus, we get the following half-matrix: ·

lo \ hi	1	2	3	4	5	6
0	4	4	4	6	6	6
1		1	2	6	6	6
2			2	6	6	6
3				6	6	6
4					5	5
5						3

To find $\|X\|_1$, we will use the $\|Y\|_0$'s from this matrix. Similarly, to find $\|X\|_2$, we will use the $\|Y\|_1$'s which will be displayed in a similar half-matrix. Observe that once the half-matrix of $\|Y\|_1$ norms has been found, the preceding half- matrix, of $\|Y\|_0$ norms can be discarded.

It is not difficult to see that in our calculation of $\|Y\|_m$, we can omit the first ($lo = 0$) row and $hi = 1$ column, substituting instead $\|Y\|_{m-1}$ in the (n, n) position. Thus, in this example, the half-matrix we will be dealing with is the more efficient, albeit less aesthetic, one given below:

lo \ hi	2	3	4	5	6
1	1	2	6	6	6
2		2	6	6	6
3			6	6	6
4				5	5
5					3
6					6

III. THE ALGORITHM. The algorithm presented here works only on the dense subspace T_0 of T. In fact, there are two computer programs, NORM and TRACE, which have been developed based on this algorithm. The TRACE algorithm includes the code for NORM, but also allows the user the choice of tracing intermediate results as they are generated. Both programs were written in standard Pascal and tested on a VAX 11/780 running under UNIX (version 4.2 UCB). Details concerning these programs, including the Pascal code and instructions on how to use the programs, are given in the appendix of [1]. As in Example 2, to calculate $\|X\|_m$, we will first calculate and store $\|Y\|_{m-1}$ for all subvectors Y of X. present, we assume that these values are all stored in an $n \times n$ matrix called OLDNORMS, indexed by i, j, where $1 \leq i \leq n$, $2 \leq j \leq n$. The norm $\|(\ell, h)\|_{m-1}$ is stored at position (ℓ, h) of this matrix. Note that approximately half of the storage locations are used. A more efficient storage is actually employed in our algorithm. However, it is convenient to visualize storage in the above form in the algorithm description. The storage scheme for an additional matrix, NEWNORMS, described below, is the same as for OLDNORMS.

In the actual algorithm described below, the number k of subvectors in a particular partition P involved in the calculation of $\|Y\|_m$ is called PARTS, and m is called LEVEL. The amount of work required for the calculation of $\|Y\|$ can be significantly reduced by saving the values

$$Y(k, lo, hi) = \max_{P_k} \sum_{j=1}^{k} \|(p_j, p_{j+1})\|_{m-1}$$

for successive values of k $(k = 1, 2, \ldots, \lfloor \frac{n}{2} - 1 \rfloor)$. Here $0 \leq lo < hi \leq n$, $\lfloor j \rfloor$ is the greatest integer $\leq j$, and $P_k = \{p_1, p_2, \ldots, p_{k+1}\}$ ranges over all subsets of integers satisfying

$$max\{k, lo\} \leq p_1 < p_2 < \ldots < p_{k+1} \leq hi.$$

It is convenient to assume these values will be stored in the three dimensional $(\lfloor n/2 \rfloor - 1) \times n \times (n-1)$ matrix NORMSUMS with $Y(k, lo, hi)$ stored at location NORMSUMS (k, lo, hi). For each fixed k, the values $Y(k, lo, hi)$ are stored in the $n \times (n-1)$ submatrix NORMSUMS$(k, ., .)$ in the same manner as used for OLDNORMS earlier in this section. These values can then be used to calculate the corresponding values for the $n \times (n-1)$ submatrix NORMSUMS$(k+1, ., .)$, as indicated in step 13 of the following algorithm.

Again, at most half the storage locations are used. A more efficient storage scheme will be introduced after the algorithm is described.

In describing the algorithm, it is convenient to introduce some procedures which accomplish specific tasks. A brief description of these procedures together with their nested structure is indicated below (i.e., indented procedures are called by the preceding procedure which is less indented):

```
program NORM(input,output);

    procedure EnterVector: Handles input of the vector.

    procedure ProcessVector: Calculates the m-norms of the current
                             vector entered and reports the value
                             of each m-norm as it is calculated.
                             If the m-norm values stabilize,
                             terminate.

        procedure NextLevel: Calculates the next level of m-norms for
                             the subvectors of the vector entered.

        procedure SumsOfNorms: Calculates values for appropriate sums
                               of norms of subvectors and stores the
                               results in NORMSUMS matrix.

        procedure UpdateNorm: Updates the current calculation of the
```

norms of subvectors using values in the
NORMSUMS x matrix. Results are stored
in the NEWNORMS matrix.

End of NORM.

The following is a step-by-step outline of the control flow during execution of
the algorithm. The name of the procedure containing each step is stated prior to
the step. Variables and named constants in the code are introduced as needed and
are given in capitals.

1. (EnterVector) Prompt for vector, read it, and set DIM = length of vector
   read.

2. (Norm) Set LEVEL to 0.

3. (ProcessVector) Prompt user for highest level M permitted for m-norm and
   assign this value to TOPLEVEL.

4. (ProcessVector) Set STABILIZED to FALSE.

5. (ProcessVector) { Begin loop to calculate m-norm for each level M successively, $M = 0, 1, \ldots$, TOPLEVEL.} While LEVEL $\leq$ TOPLEVEL and STA-
   BILIZED is false, repeat steps (6) - (26).

6. (ProcessVector) Set STABILIZED to TRUE.

7. (NextLevel) { Initialize NEWNORMS matrix for zero level.} If LEVEL is
   zero, execute steps (8) - (10), else go to (11).

8. (NextLevel) Calculate the sup-norm of the subvector from position (LO+1)
   to position HI and store in the array location NEWNORMS[LO,HI] for all
   integers LO and HI with $0 \leq$ LO $\leq$ HI $\leq$ DIM.

9. (NextLevel) { Store the 0-norm of original vector.} Set NEWNORMS[DIM,DIM]
   to sup-norm of the original vector.

10. (NextLevel) Copy values in NEWNORMS into OLDNORMS.

11. (NextLevel) { Begin calculations for the three dimensional matrix NORM-
    SUMS for this level.} If LEVEL > 0, execute steps (12) - (25) , else go to
    (26).

12. (NextLevel) Copy the values of the two dimensional array NEWNORMS into
    the two dimensional submatrix NORMSUMS [1,.,.].

13. (SumsOfNorms) { Next, start calculating values for the two dimensional sub-
    matrix NORMSUMS [PARTS+1,.,.] from the two dimensional submatrix
    NORMSUMS [PARTS,.,.].} For PARTS starting at 1 until $\lfloor$DIM/2$\rfloor$ do steps
    (14) - (24). { $\lfloor$DIM/2$\rfloor$ denotes the greatest integer of DIM/2.}

14. (SumsOfNorms) For LO starting at PARTS+1 until DIM-1-PARTS do steps (15) - (16).

15. (SumsOfNorms) For HI starting at (PARTS+1) + LO until DIM do step (16).

16. (SumsOfNorms) Set NORMSUMS[PARTS+1,LO,HI] to the maximum of the values OLDNORMS[LO,MID] + NORMSUMS[PARTS,MID,HI] for all integers MID satisfying LO + 1 ≤ MID ≤ HI - PARTS.

17. (UpdateNorm) { Steps (17) - (24) update the NEWNORMS matrix from m-norms to $(m + 1)$-norms, using the values calculated for the two-dimensional submatrix NORMSUMS [PARTS+1,.,.].} For LO starting at 1 until DIM-1 do block (18) - (23).

18. (UpdateNorm) For HI starting at LO+1 until DIM do steps (19) - (23).

19. (UpdateNorm) Initialize X to zero.

20. (UpdateNorm) { Calculate ASSIGNMENT A value for X.} If PARTS + 1 ≤ LO and (PARTS + 1) + LO ≤ HI, let X be NORMSUMS[PARTS+1,LO,HI].

21. (UpdateNorm) { Calculate ASSIGNMENT B value for X.} If LO < PARTS + 1 and 2(PARTS+1)≤ HI, let X be NORMSUMS[PARTS+1,PARTS+1,HI].

22. (UpdateNorm) { Update norm of vector (LO,HI) using contributions from NORMSUMS[PARTS+1, .,.].} Replace NEWNORMS[LO,HI] with the larger of its present value and $X/2$.

23. (UpdateNorm) Set STABILIZED to FALSE if the value of NEWNORMS[LO,HI] was increased in step (22).

24. (UpdateNorm) { Update the norm of the original vector for this level.} Set NEWNORMS[DIM,DIM] to be the larger of its present value and NEWNORMS[1,DIM].

25. (NextLevel) Copy the values of the two dimensional array NEWNORMS into the two dimensional array OLDNORMS.

26. (ProcessVector) Set LEVEL to LEVEL + 1 and return to (5).

IV. SPACE REQUIREMENTS. As mentioned earlier, the storage scheme we presented for NEWNORMS and OLDNORMS is inefficient. Instead of using two-dimensional arrays for NEWNORMS and OLDNORMS, we use one-dimensional arrays. The norm information for the subvector (LO,HI) is stored at location (LO-1)N - LO(LO+1)/2 + HI in the linear arrays. For the vector $X = (x_1, x_2, \ldots, x_n)$, this leads to the following storage locations for the subvectors of X:

| NUMBER OF SUBVECTORS | | | |
ARRAY INDEX	SUBVECTOR	MATRIX INDEX	NUMBER OF VECTORS
1	$(0, x_2)$	$(1,2)$	
2	$(0, x_2, x_3)$	$(1,3)$	$n-1$
$\vdots$			
$n-1$	$(0, x_2, x_3, \ldots, x_n)$	$(1,n)$	
n	$(0, 0, x_3)$	$(2,3)$	
$n+1$	$(0, 0, x_3, x_4)$	$(2,4)$	n-2
$\vdots$			
$2n-3$	$(0, 0, x_3, i \ldots, x_n)$	$(2,n)$	
$2n-2$	$(0, 0, 0, x_4)$	$(3,4)$	
$2n-1$	$(0, 0, 0, x_4, x_5)$	$(3,5)$	$n-3$
$\vdots$			
$3n-6$	$(0, 0, 0, x_4, \ldots x_n)$	$(3,n)$	
$\vdots$			
$n(n-1)/2$	$(0, \ldots, 0, x_n)$	$(n-1,n)$	1
$n(n-1)/2 + 1$	$(x_1, x_2, \ldots x_n)$	(n,n)	1

The storage of the norm of the vector $(x_1, x_2, \ldots, x_n)$ was at (n,n) in the matrix scheme. Under the new approach, we store it at the next location after the storage location for $(0, 0, \ldots x_n)$, i.e. at index $n(n-1)/2 + 1$.

The storage of NORMSUMS is also inefficient. For a fixed index PARTS, NORMSUMS[PARTS,LO,HI] is stored in a linear array with

$$\text{INDEX} = \text{(LO-1)N - LO(LO+1)/2 + HI}.$$

Thus, NORMSUMS can be regarded as a two-dimensional array and (PARTS,LO,HI) can be stored at NORMSUMS[PARTS,INDEX]. However, one additional space saving scheme is used. After NORMSUMS[PARTS,.] is calculated, it is needed only in the calculation of NORMSUMS[PARTS+1,.]. After this calculation is completed, NORMSUMS[PARTS,.] can have its storage space used again. As a result, we need only two values for the first component in the index for NORMSUMS. This is accomplished by replacing NORMSUMS[PARTS,INDEX] with NORMSUMS[PARTS mod 2, INDEX].

The total number of storage locations in each array NEWNORMS and OLD-NORMS is $n(n-1)/2 + 1$. The number of storage locations for NORMSUMS is $2[n(n-1)/2 + 1] = n(n-1)/2$. Also, n locations are required to store the vector X entered by the user. Consequently, the total storage for arrays is $2n^2 - n + 4$. Thus, the storage required is $O(n^2)$, where n is the maximum length that is permitted for the vector entered by the user.

V. TWO SWITCHES IN THE ALGORITHM. The variable STABI-
LIZED is a switch. It has an initial value of TRUE for each norm level calculated
(see step 6). As the norm values for subvectors are updated in step (22), the value
of STABILIZED is changed to FALSE if the norm of any subvector is increased. If
STABILIZED is still TRUE after the calculations of the NEWNORMS matrix, then
the NEWNORMS and OLDNORMS matrices are identical and the calculation for
the next level will also be identical. As a result, if STABILIZED is TRUE after the
calculation of a given level, the value of all m-norms for all m greater than this level
for each subvector will be the same as they are for this level. The norm value for
this level for the original vector entered by the user can be returned as the Tsirelson
space norm of the vector.

A variable SUMSGROW is used as another switch in the procedure SumsOfNorm
(but is not exhibited in our simplified description of the algorithm in section III).
It has an initial value of FALSE for each pass through the outer "parts" loop
(i.e., steps 13 to 24). If a larger value is found for the two-dimensional submatrix NORMSUMS[(PARTS+1)mod 2, .,.], i.e. with PARTS fixed, than for
the corresponding position in NORMSUMS[PARTS mod 2,., .] then SUMSGROW
has its value changed to TRUE. Otherwise, these two-dimensional submatrices of
NORMSUMS are equal and no values in the matrix NEWNORMS will be changed
by the procedure UpdateNorms. The calculations for NORMSUMS[(PARTS+2)
mod 2, ., .] will be the same as for NORMSUMS[(PARTS+1) mod 2, ., .]. Consequently, NORMSUMS[k mod 2, ., .] need not be calculated for $k \geq$ PARTS + 1
and the calculation of the norm for this level is complete.

VI. TIMINGS. Using the definition of the Tsirelson space norm, one can
use recursion to develop a more natural algorithm than the preceding algorithm.
Unfortunately, the recursive version runs much slower than the one presented here.
The timings chart given below includes timings for an implementation of the natural
recursive algorithm which was developed earlier by the authors.

The CPU timings given below were obtained on a VAX 11/780 running UNIX
(Version 4.2 UCB). The notation

$$3 * \overline{1, 2, \ldots, 9, 0}$$

means that the block 1,2,3,4,5,6,7,8,9,0, is repeated three times. The columns represent the following:

A   The number of m-norm levels calculated using the recursive algorithm.

B   The timing in CPU seconds for the recursive algorithm.

C   The timing in CPU seconds for NORM.

D   The level at which the m-norm stabilized.

E   The norm value.

VECTOR	A	B	C	D	E
7,7,7,7,4,4,4	4	24.45	.050	2	7.5
15,14,13,...,2,1	4	KILLED after one hour of connect time.	.534	3	23.75
$3*\overline{1,2,\ldots,0}$			6.400	3	40.5
$10*\overline{1,2,\ldots,0}$			651.817	3	128.75

Observe that as the length n of the vector increases, the time required by NORM to calculate the norm of the vector increases rapidly. (See column C.) In fact, it is not difficult to show that the time complexity of NORM is exponential in n. However, the time complexity of the natural recursive algorithm is exponential in both n and m, the number of levels of the m-norm calculated. On the other hand, with NORM, the time required to calculate the $(m + 1)$-norm after the m-norm has been calculated is essentially the same as the time required to calculate the $(m + 2)$-norm after the $(m + 1)$-norm has been calculated. That is, the amount of work required to calculate the m-norm of a vector using NORM is linear with respect to m.

As the recursive algorithm did not provide an easy method of determining when the m-norms stabilized, Column A gives the actual number of m-norm levels that were calculated. As a result, this algorithm provided only information about the m-norms of a vector. When the same value was obtained for the m-norm of a vector for several successive values of m, it was natural to assume that the norm of the vector equaled the repeated m-norm. However, this was only a guess, and the recursive algorithm did not seem to lend itself to a method for calculating the actual norm of a vector.

VII. STOPPING TIME QUESTION. Based on the problem cited in the preceding paragraph, it might seem reasonable to believe that if a vector had the same m-norm for two successive values of m, this m-norm value would be the norm value of the vector. However, it is not difficult to find vectors with finitely many nonzero terms which have an m-norm equal to an $(m + 1)$-norm, but with this m-norm value unequal to the norm value. It appears reasonable to believe that for every pair of positive integers m and k, there exists a vector $X = (x_1, x_2, \ldots, x_m)$ with $\|X\|_m = \|X\|_{m+1}$ for $1 \leq i \leq k$, but $\|X\|_m < \|X\|$. Therefore, if the same value is obtained for two or more successive m-norms of a vector, one cannot automatically assume this value is also the norm value of that vector. This leads to the following question:

Problem 1. If $X = (x_1, x_2, \ldots, x_n)$ and k is a positive integer, find the minimal value of k (as a function of n alone) such that if m is a positive integer with $m+k \leq n$ and $\|x\|_m = \|X\|_{m+1}$ for $1 \leq i \leq k$, then $\|X\|_m = \|X\|$.

The following easy to prove fact provides a partial answer to the preceding problem.

Theorem. If $X = (x_1, x_2, \ldots, x_n)$, then $\|X\|_m = \|X\|$ for $m \geq \lfloor (n-1)/2 \rfloor$.

Based on this result, a sufficient condition on k in the preceding problem is to take $k = \lfloor (n-1)/2 \rfloor$. However, this is possibly not a minimal value for k.

A consequence of the preceding theorem is that there exists a positive integer t such that $\|X\|_0, \|X\|_1, \ldots \|X\|_j, \ldots$ stabilizes at level $j = t$ for all vectors X of length n. Let $j(n)$ be the minimal value of t above.

Problem 2. In the above setting,

(a) Find a reasonably tight upper bound for $j(n)$ for each positive integer n.

(b) Determine a formula for $j(n)$.

An answer to either part of Problem 2 would allow a user to estimate the time required in the worst case to evaluate the norm of a vector. Recall, the time required to calculate the $(m+1)$-norm after the m-norm has been calculated is essentially the same as the time required to calculate the $(m+2)$-norm after the $(m+1)$-norm has been calculated for all $m \geq 0$. Thus, if an upper bound for the value of $j(n)$ is k and r is the CPU time required to calculate the 1-norm of a vector after the 0-norm is calculated using the NORM program, then kr is an approximate upper bound for the CPU time needed to calculate the norm.

VIII. SOFTWARE.

The authors are interested in making a copy of the software available to potential users, either by shipping a copy of the software to the users net address or by sending a floppy disk or tape to the user. To make arrangements to secure a copy of the software, please contact one of the authors.

REFERENCES

(1) P. G. CASAZZA AND T. J. SHURA, *Tsirelson's Space* (with an Appendix by J. Baker, O. Slotterbeck, and R. Aron), Lecture Notes in Mathematics, Springer-Verlag, 1989.

(2) P. G. CASAZZA, *Tsirelson's space*, Proceedings of Research Workshop on Banach Space Theory (Iowa City, Iowa, 1981), Univ. Iowa, 9 - 22.

(3) P.G. CASAZZA, W.B. JOHNSON, AND L. TZAFRIRI, *On Tsirelson's space*, Isr. J. Math., 47 (1984) 81-98.

(4) T. FIGIEL AND W. B. JOHNSON, *A uniformly convex Banach space which contains no ℓ_p*, Comp. Math., Vol. 29, Fasc. 2, 1974, 179-190.

(5) J.LINDENSTRAUSS AND L.TZAFRIRI, *On the complemented subspaces problem*, Isr. J. Math. 9, 263-269 (1971).

(6) B. S. TSIRELSON, *Not every Banach space contains an embedding of ℓ_p or c_0*, Functional Anal. Appl., 8, 1974, 138-141. (Russian translation).

FLOATING-POINT SYSTEMS FOR THEOREM PROVING*

G. BOHLENDER†, J. WOLFF VON GUDENBERG†
AND W.L. MIRANKER‡

Abstract. There are a number of existing floating-point systems (programming languages and libraries) which allow the computer user to produce computation with guarantees. These systems may be conveniently used as a component in theorem proving. Following a brief description of the methodology which underlies these systems, we tabulate them, indicating some properties and noting where they may be obtained for use. A brief appendix supplies an introduction to the formalism of the methodology.

§1. INTRODUCTION

Is the modern digital floating-point computer actually being used as a component in theorem proving? Certainly many of the contributions in these proceedings show that this is the case. Yet anyone familiar with these computers knows that it is very difficult to learn what floating-point computers actually do. Their arithmetic operations are usually not completely specified. Certainly not at a level of care which a proof would require. These computers and their systems are sprinkled with argument dependent exceptions which are often concealed. Moreover, even ideal floating-point computation produces well-known pathological errors, some of which are spectacular. (In fact, the spectacular errors are the more benign, since they more readily show themselves than do the subtle errors.) These errors are due, of course, to cancellations, i.e., a loss of information in the floating-point arithmetic process.

We must make the distinction between fixed point and floating point and their respective theorem proving capacities clear. As a fixed point or data (i.e., bit) processor, the digital computer is a simpler device, usually rather completely specified. (Indeed, a fixed point computer is, more or less, a Turing machine with a finite tape.) In this form, the digital computer has a history of use as a theorem prover.

However, here, when we speak of a computer aided proof, we mean a floating-point calculation. By means, which we can only view as heroic, these theorem provers have actually penetrated into the system with which they compute (computer and software) with a rigorous mathematical thoroughness. In this way they make sure that their computation is as solid as any other logical part of their proof apparatus. Shall we say that they tame their own computer, by the sweat of their intellectual brows.

Independently of these roll-your-own floating-point aided theorem proving efforts, a systematic body of work has developed which addresses the "floating-point as mathematics" issue [17]. Theory, applications, systems, packages, and hardware;

---

*This manuscript was prepared in October, 1989 in Obsteig, Austria, at a Workshop on Computer Arithmetic organized by U. Kulisch.

†University of Karlsruhe, Institut für Angewandte Mathematik, D-7500 Karlsruhe, West Germany

‡Department of Mathematical Sciences, IBM T.J. Watson Research Center, P.O. Box 218, Yorktown Heights, NY 10598

even commercial products comprise this body of work. It is a vigorous and growing subject [21, 23], but as it is set in the numerical analysis/systems area, it is more or less unknown to mathematicians, in particular, to theorem provers. Indeed the orientation of this subject has been toward scientific computation. As a result of this work, the floating-point computer is now provided with a concise axiomatic specification. It may be employed by theorem provers with the reliability they require. Moreover, the associated interfaces, being user friendly, make exploitation quite straight forward. Let's refer to these systems as *validating floating-point systems*.

In §2 we give a concise and informal excursion through the development and state of floating-point as it relates both to theorem proving and to the validating floating-point systems. In §3 we tabulate the validating floating-point systems, describing generic properties. We give references to where, when they are available, copies of these systems may be obtained. Some of these systems are developmental , but others are commercial products. Some of the development systems are available from University sources. A brief appendix giving some technical details is also included.

§2. FLOATING-POINT COMPUTATION

2.1 Historical Development. Floating-point arithmetic proceeds by replacing the reals by a finite set (the floating-point numbers, usually normalized) and by replacing the arithmetic operations by corresponding approximating operations defined on this finite set. There is an immediate loss of information when floating-point arithmetic is used. There is also a loss of a number of familiar critical laws and properties of rings and fields, such as associativity of repeated addition, uniqueness of the additive inverse, and the like. Cancellations adversely impact the accuracy with which algorithms, especially iterative algorithms may be computed.

Over the years, a collection of ad hoc approximate methods and constructs were devised to regain, in part, some of these basic lost properties. Constituents of this collection are denormalized numbers/gradual underflow, symbols for representing and computing with infinity, signed zeros, etc. The method of residual correction was introduced to floating-point (this is often attributed to Wilkinson). With residual correction came the need to compute ever more accurate residuals, and this contributed to the introduction of double precision and higher precision formats.

In the modern theory of computer arithmetic, most of the ad hoc fixes, just referred to, are seen to be irrelevant. This theory simplifies and makes rigorous the specification of floating-point arithmetic. It enables the digital computer to provide floating-point computation with guarantees! This theory traces its origin to two fundamental contributions. The first is interval arithmetic, started by R. Moore [24]. The second is Kulisch/Miranker (K/M)-arithmetic [20,22].

Moore's interval arithmetic which is based on directed roundings (rounding upward/downward) broke the ground for floating-point computation with guaranteed results. It allowed for representation of the continuum on a computer, and so, opens the way to extend the computer (a tool of finite mathematics) to a tool of real analysis.

At first, interval arithmetic quickly became discredited, because its naive use led directly to rapidly expanding intervals of little value or interest. Today somewhat more subtle techniques generate potent contracting interval methods.

The K/M arithmetic, which includes directed roundings and intervals, provides an axiomatic specification of floating-point numbers and operations. The operations are defined by an algebraic mapping principle called semimorphism. Semimorphism is used to define floating-point arithmetic in all common spaces of computation (reals, matrices, vectors, complex versions, interval versions, etc.) The definition is mathematically simple and always the same. For convenience to the reader, we include an appendix which describes this process for the data types, Real and Real Matrix. (We shall presently come to the use of semimorphism to define arithmetic and other operations in computer versions of function spaces.)

An informal notion of semimorphism: A mapping (rounding) from the ideal space (e.g. the reals) into its computer representable subspace (e.g. the normalized floating-point numbers) is applied to the exact result of an operation between elements in the subspace. This defines each computer operation with a loss of information limited to the effect of a single rounding [20]. In the case of matrix multiplication this means that an exact dot product with only one rounding is necessary and must be provided by the computing system.

2.2 Implementations of the theory. In 1980, the first implementation of these ideas was embodied in a compiler for a PASCAL extension called PASCAL-SC [3]. The first commercial implementation was made by IBM in a software package for 370 machines called ACRITH and released in 1983 [10]. Standard functions (sin, exp, cosh, etc.) are implemented with high accuracy and in interval versions. With intervals, the precision of standard function evaluation is controllable. The exact evaluation of expressions containing standard functions requires standard functions with dynamic precision, since precision decreases with arithmetic combining. This process makes use of a so-called staggered correction format proposed by H. Stetter [26].

ACRITH also includes E-methods, the name of a methodology for validating the solution of numerical problems. These interval methods are an efficient computer implementation devised by E. Kaucher and S. Rump [15] of computer versions of contraction mapping processes initiated by R. Krawczyk [16] and others in the late '60's. When handled properly, E-methods provide existence and uniqueness statements of mathematical problems as well as the numerically generated bounds.

Extension of all of these computational ideas (floating-point numbers, roundings, semimorphism, computer operations, interval methods, contracting mappings for validations, etc.) to function spaces and function space problems is a development made by E. Kaucher and W. L. Miranker [12, 13, 14]. Among other capabilities this computational function space methodology provides a framework for extending E-methods to IVP/BVP for differential equations and to integral equations. In the Kaucher/Miranker development, approximation theoretic methods are recast into computer arithmetic/data-type form called ultra-arithmetic [6]. This makes the correspondence between function space problems and methods, on the one hand

and algebraic problems and methods, on the other hand transparent. It is this correspondence which guides the development of the new computational function space methods, including E-methods.

2.3 Floating-point in theorem proving. The collection of all these ideas and methods provides an effective framework for computation. Since it is built on an axiomatic basis, when *faithfully* implemented, the corresponding computer system becomes a part of an analyst's mathematical repertoire. He may use the computations with an absolute certainty, in the mathematical sense. A computation so produced is the same as any other mathematical deduction. It may be used at face value in a proof or derivation.

2.4 IEEE-norm. A collection of workers, including W. Cody and W. Kahan [5], have proposed a standard for floating point formats, roundings, arithmetic precision etc. which is commonly called the IEEE-norm. It provides many of the special ad hoc devices previously referred to, and this provision is found useful to many computer users. It likewise provides a strong basis for residual correction methods by means of several layers of data format precision, some of enormous range. The IEEE-norm is restricted to a special data format which is not required by the K/M-theory. These special operations, however, fulfill the requirements of a semimorphism of the K/M-theory. The IEEE-norm falls short of including semimorphic operations for the operations in the vector spaces and their interval correspondents which are required by the K/M-theory. Specifically the IEEE-norm falls short of including a semimorpic dot product. A computer equipped with K/M arithmetic is a vector processor in a mathematical sense, providing all vector and matrix operations and their interval correspondents to full machine accuracy (by semimorphism) while the IEEE-norm provides scalar operations only. Thus K/M arithmetic provides a powerful tool endowing a computer with mathematical properties usable in theorem proving.

The major and indeed quite significant achievement of this norm is its support of floating-point portability. For this reason it has achieved a wide acceptance and constitutes a valuable contribution to computation.

While the IEEE-norm and the K/M theory are definitely not the same (neither in category nor in detail), there are no conflicts between them.

2.5 Programming languages/libraries. The K/M theory and its associated methodologies, thus far so briefly described, contains a large number of constituents. How are they to be handled in a congenial (user-friendly) way. This is the role of a programming language (such as FORTRAN-SC [8]) or a library (such as ACRITH). These systems include user-friendly concepts and capabilities. Examples of such relevant to a programming language are an operator concept and function overloading. Roughly speaking, these terms are programming jargon for conventional mathematical notation. For instance, an operator concept means the use of the same symbol, say the times sign, to denote the product $a \times b$, no matter what a and b are. So long as a and b are compatible in the expression $a \times b$ (say when a is a scalar and b is a matrix) the expression is acceptable to the system which parses it correctly.

(This routine mathematical convention is a quite advanced programming language construct!)

§3. VALIDATING FLOATING-POINT SYSTEMS

We summarize the existing software packages for floating-point verification methods in a table. The last column in the table, headed: availability, contains some information for obtaining the package (as a book, a diskette, a commercial product). The table heading: kind, specifies whether the package is a programming language or a library. Programming languages generally provide operations with user specifiable roundings (from among a specific set of roundings) and interval operations as well by means of a user friendly operator notation. Libraries implement these constructs as subroutines, but they are equivalently user friendly.

All of the packages listed provide an exact dot product (equivalently an exact scalar product) which allows for the implementation of vector and matrix operations with maximum accuracy. In some implementations, a special data type for scalar products is provided which facilitates the implementation of scalar product expressions. (An example of the latter is $Ax - b$ for vectors x and b and a matrix A.) This feature is indicated in the column headed: "special properties" and with a table entry: "data type for scalar products".

Some of the package compilers provide for the execution of such scalar product expressions with a result equal to exact evaluation followed by a single rounding. This is indicated in the column headed: "special properties" and with a table entry "scalar product expressions".

The accuracy of the standard functions in these packages is usually better than one unit in the last place (ulp). By usually, we mean that in exceptional cases only 2 ulp accuracy is delivered.

Floating Point Verification Packages

Name/kind	problem solving routines	special properties	availability
PASCAL-SC A programming language, extension of PASCAL, full compiler and runtime system	linear systems, eigenvalues, nonlinear systems, zeros of polynomials, arithmetic expressions	extended set of standard functions (including complex and interval versions), user defined operators, in version 2 additionally: scalar product expressions, multiple precision, dynamic arrays, modules	**Version 1:** available as book including a floppy for IBM-PC [18], and Atari ST [19]; **Version 2:** book [4], system on floppy: in preparation on Atari ST, *planned* on MacIntosh II
ACRITH A FORTRAN 77 library, contains an online training component	linear systems, eigenvalues, sparse systems, nonlinear systems, zeros of polynomials, arithmetic expressions, linear programming	extended set of standard functions (including complex and interval versions), datatype for scalar products, microcode/ hardware assists on several computers	IBM program product for IBM/370 machines under VM/CMS and MVS [10]
FORTRAN-SC A programming language, extension of FORTRAN 77, full compiler and runtime system, based on ACRITH	same as ACRITH	same as ACRITH, dynamic arrays, user defined operators, scalar product expressions	IBM/370 VM/CMS available from IBM Germany [2, 8, 29]
ARITHMOS A FORTRAN 77 library,	linear systems, eigenvalues, zeros of polynomials, arithmetic expressions	datatype for scalar products, microcode assist on several computers	Siemens program product for mainframes under BS 2000 [1]

Name/kind	problem solving routines	special properties	availability
Abacus An interactive programming environment with programming capability	linear systems, etc.	extended set of standard functions (including complex, vector and matrix versions), generic functions/operators	planned on IBM-PC (DOS), Unix [25], from University of Hamburg-Harburg, Prof. S. Rump
Hificomp A FORTRAN 77 library	linear systems, zeros of polynomials, interpolation, Poisson equation, statistical computation	hardware/microcode assist on several computers	product for IBM/370 and IBM-PC from Bulgarian Academy of Sciences [25], Prof. S. Markov
Modula-SC A programming language, extension of Modula-2, precompiler to Modula-2, runtime support in Modula-2	linear systems, eigenvalues, zeros of polynomials, arithmetic expressions	portable to any computer with a Modula-2 compiler	for IBM-PC and Macintosh II from University of Basel, Prof. C. Ullrich [25]
NAG library An Ada library	?	portable to Ada systems	planned for late 1989 [7]
Turbo Pascal SC A precompiler to Turbo Pascal	linear systems, etc.		planned on IBM-PC, from University of Hamburg-Harburg, Prof. S. Rump [25]
APL/PCXA A programming language, APL extension	linear systems	multiple precision arithmetic	book and floppy [9] for IBM-PC

Appendix. K/M arithmetic for the data-types, Real and Real Matrix.

In this appendix we sketch the K/M methodology for defining floating-point arithmetic through use of rounding mappings and semimorphism. As an example, we deal only with the reals and the real matrices. The extension to other data types and intervals as well is entirely analogous. However, we omit this extension, and refer to [20,22] for details.

A floating-point system $R = R(b, \ell, e1, e2)$ with base b, ℓ digits in the mantissa and exponent range $e1..e2$ consists of a finite number of elements. They are equally spaced between successive powers of b and their negatives. This spacing changes at every power of b. A floating-point system has the appearance of a screen placed over the real numbers. Indeed, the expression floating-point screen is often used.

Next we turn to the arithmetic operations $+, -, \times, /$. These operations for real numbers are approximated by floating-point operations. If x and y are floating-point numbers, the exact product $x \times y$ itself is not usually a floating-point number of $R(b, \ell, e1, e2)$ since the mantissa of $x \times y$ has 2ℓ digits. For related reasons, the exact sum of $x + y$ is also not usually a floating-point number. Since a computer must be able to represent the results of its own operations, the result of a floating-point operation must be a floating-point number. The best we can do is to round the exact result into the floating-point screen and take the rounded version as the definition of the floating-point operation.

If $*$ is one of the exact operations, $+, -, \times, /$, let $\boxdot$ denote the corresponding floating-point operation. Then our choice of floating-point operations is expressed by the following mathematical formula.

(RG) $x \boxdot y := \Box (x * y)$ for all $x, y \in R$ and all $* \in \{+, -, \times, /\}$.

In (RG), $\Box$ is a mapping $\Box: \mathbf{R} \to R$. $\Box$ is called a rounding if it has the following properties $(R1)$ and $(R2)$.

(R1) $\Box x = x$ for all $x \in R$,

that is, the screen R is invariant under the mapping $\Box$.

(R2) $x \leq y \Rightarrow \Box x \leq \Box y$ for all $x, y \in \mathbf{R}$,

that is, $\Box$ is monotonic on the real numbers.

The three familiar roundings: to the nearest floating-point number, toward zero or away from zero have properties $(R1)$ and $(R2)$ and the following additional property.

(R4) $\Box (-x) = -\Box x$ for all $x \in \mathbf{R}$.

The mapping property expressed in (RG) for mappings which satisfy $(R1)$, $(R2)$, and $(R4)$ is called, *semimorphism*.

The monotone upwardly and the monotone downwardly directed roundings are denoted by $\triangle$ and ∇. These two roundings, which are used to define interval arithmetic, are characterized by $(R1), (R2)$ and the additional property

(R3) $\nabla x \leq x$ and $x \leq \triangle x$ for all $x \in$ **R**.

Thus, ∇ rounds to the left and $\triangle$ rounds to the right. However, the roundings ∇ and $\triangle$ do not have the antisymmetry property $(R4)$.

All operations defined by (RG) and a rounding with the properties $(R1)-(R2)$ produce results of maximum accuracy in a certain sense which is rounding dependent. In particular, between the correct result (in the sense of real numbers) and the approximate result $x \boxdot y$ (in the sense of the screen of floating-point numbers) no other floating-point number in the screen can be found.

For convenience, we shall refer to the class of roundings which satisfy $(R1), (R2)$, and $(R4)$ along with the special roundings $\triangle$ and ∇ as admissible roundings. We may summarize this discussion by saying that admissible roundings generate maximally accurate floating-point arithmetic through use of (RG).

The same formulas $(RG), (R1)-(R4)$ can be used to define arithmetic for other spaces occuring in numerical computations. For example, operations for floating-point matrices MR can be defined by applying the laws of semimorphism to the real matrices M**R**, i.e. in the above formulas **R** and R have to be replaced with M**R** and MR respectively. Rounded addition and subtraction are performed componentwise, whereas (RG) for matrix multiplication reads

$$X \boxdot Y := \square\left(\sum_{k=1}^{n} x_{ik} * y_{kj} \right)$$

for all matrices $X, Y, \in MR$. Floating-point matrix multiplication is implemented as if an exact scalar product were computed which is then rounded only once onto the screen of floating-point matrices.

At first sight it seems to be doubtful that formula (RG) can be implemented on computers at all. In order to determine the approximation $x \boxdot y$, the exact but unknown result $x * y$ which is, in general, neither computer specifiable nor computer representable seems to be required in (RG). It can be shown, however, that whenever $x * y$ is not representable on the computer, it is sufficient to replace it by an appropriate and representable value $x \overset{\sim}{*} y$. The latter has the property $\square(x * y) = \square(x \overset{\sim}{*} y)$ for all roundings in question. Then $x \overset{\sim}{*} y$ can be used to define $x \boxdot y$ by means of the relations

$$x \boxdot y = \square(x * y) = \square(x \overset{\sim}{*} y)$$

for all $x, y \in R$, respectively for all $x, y \in MR$.

REFERENCES

[1] **ARITHMOS (BS 2000)**, *Kurzbeschreibung*, SIEMENS AG, Bereich Datentechnik, Postfach 83 09 51, D-8000 München 83.

[2] BLEHER, J.H., KULISCH, U., METZGER, M., RUMP, S.M., ULLRICH, CH., WALTER, W., *FORTRAN-SC: A Study of a FORTRAN Extension for Engineering / Scientific Computation with Access to ACRITH*, Computing 39 (November 1987) pp. 93–110.

[3] BOHLENDER, G., ET AL, *PASCAL-SC: A PASCAL for Contemporary Scientific Computation*, RC9009, IBM Research Center, Yorktown Heights, NY (1981).

[4] BOHLENDER, G., RALL, L.B., ULLRICH, CH., WOLFF V. GUDENBERG, J., *PASCAL-SC: A Computer Language for Scientific Computation*, Academic Press (Perspectives in Computing, vol. 17), Orlando (1987) (ISBN 0-12-111155-5).

[5] COONAN, J., ET.AL., *A Proposed Standard for Floating-Point Arithmetic*, SIGNUM Newsletter (1979).

[6] EPSTEIN, C., MIRANKER, W.L., RIVLIN, T.J., *Ultra-Arthmetic* Part 1: *Function Data Types* pp. 1–18, Part 2: *Intervals of Polynomials* pp. 19–29, Mathematics and Computers in Simulation, Vol. 24 (1982).

[7] ERL, M., HODGSON, G., KOK, J., WINTER, D., ZOELLNER, A., *Design and Implementation of Accurate Operators in Ada*, ESPRIT-DIAMOND, Del 1-2/4 (1988).

[8] **FORTRAN-SC:** *FORTRAN for Scientific Computation. Language Description and Sample Programs*, Institute for Applied Mathematics, University of Karlsruhe, P.O. Box 6980, D-7500 Karlsruhe, West Germany (1988).

[9] HAHN, W., MOHR, K., *APL/PCXA: Erweiterung der IEEE Arithmetik für Technisch Wissenschaftlisches Rechnen. Hanser Verlag, München (1988) (ISBN 3-446-15264-4).

[10] **IBM High-Accuracy Arithmetic Subroutine Library (ACRITH)**, *General Information Manual*, GC 33-6163-02, 3rd Edition (April 1986).

[11] KAUCHER, E., KULISCH, U., ULLRICH, CH. (EDS), *Computer Arithmetic, Scientific Computation and Programming Languages*, B.G. Teubner Verlag, Stuttgart (1987) (ISBN 3-519-02448-9).

[12] KAUCHER, E., MIRANKER, W.L., *Self-Validating Numerics for Function Space Problems*, Academic Press (1984) (ISBN 0-12-402020-8).

[13] KAUCHER, E., MIRANKER, W.L., *Residual Correction and Validation in Functoids*, Computing Supplementum 5, *Defect Correction Methods-Theory and Application*, eds. K. Boehmer and H. Stetter, Springer-Verlag (1984).

[14] KAUCHER, E., MIRANKER, W.L., *Validating Computation in a Function Space*, Proceedings of the Conference, International Computing–the Role of Interval Methods in Scientific Computing, R. Moore, ed., Academic Press (1987).

[15] KAUCHER, E., RUMP, S.M., *E-Methods for Fixed Point f(x)=x*, Computing Vol. 28 (1982) pp. 31–42.

[16] KRAWCZYK, R., *Newton-Algorithmen zur Bestimmung von Nullstellen mit Fehlerschranken*, Computing Vol. 4 (1969) pp. 187–201.

[17] KULISCH, U., *Grundlagen des Numerischen Rechnens*, Bibliographisches Institut. (Reihe Informatik, Nr. 19), Mannheim/Wien/Zürich (1976) (ISBN) 3-411-01517-9).

[18] KULISCH, U. (ED.), *PASCAL-SC: A PASCAL Extension for Scientific Computation; Information Manual and Floppy Disks; Version IBM PC/AT; Operating System DOS*, B.G. Teubner Verlag (Wiley-Teubner series in computer science), Stuttgart (1987) (ISBN 3-519-02106-4 / 0-471-91514-9).

[19] KULISCH, U. (ED.), *PASCAL-SC: A PASCAL Extension for Scientific Computation; Information Manual and Floppy Disks; Version ATARI ST.*, B.G. Teubner Verlag Stuttgart (1987) (ISBN 3-519-02108-0).

[20] KULISCH, U., MIRANKER, W.L., *Computer Arithmetic in Theory and Practice*, Academic Press, New York (1981) (ISBN 0-12-428650-x).

[21] KULISCH, U., MIRANKER, W.L.(EDS.), *A New Approach to Scientific Computation*, Academic Press, New York (1983) (ISBN 0-12-428660-7).

[22] KULISCH, U., MIRANKER, W.L., *The Arithmetic of the Digital Computer: A New Approach*, SIAM Review, Vol. 28, No. 1 (March 1986) pp. 1-40.

[23] KULISCH, U., STETTER, H.J. (EDS.), *Scientific Computation with Automatic Result Verification*, Computing Supplementum 6. Springer Verlag, Wien / New York (1988).

[24] MOORE, R.E., *Interval analysis*, Prentice Hall (1966).

[25] *SCAN 89: Abstracts of the International Symposium on Computer Arithmetic and Self-Validating Numerical Methods, Basel (1989)*, proceedings to be published by Academic Press.

[26] STETTER, H.J., *Sequential Defect Correction for High-Accuracy Floating-Point Algorithms*, Lect. Notes Math., vol. 1006 (1984) pp. 186-202.

[27] ULLRICH, CH., WOLFF V. GUDENBERG, J. (EDS.), *Accurate Numerical Algorithms, A Collection of Research Papers*, ESPRIT Series, Springer Verlag (1989).

[28] WALLIS, P.J.L., *Improving Floating-Point Programming*, J. Wiley (to appear 1989).

[29] WALTER, W., *FORTRAN-SC, A FORTRAN Extension for Engineering / Scientific Computation with Access to ACRITH: Language Description with Examples*, In: Moore, R.E, *Reliability in Computing. The Role of Interval Methods in Scientific Computing*, Academic Press, New York (1988) pp. r43–62.

COMPUTER ALGEBRA AND INDEFINITE INTEGRALS

MANUEL BRONSTEIN*

Abstract. We give an overview, from an analytical point of view, of decision procedures for determining whether an elementary function has an elementary antiderivative. We give examples of algebraic functions which are integrable and non-integrable in closed form, and mention the current implementation status of various computer algebra systems.

1. Introduction. Consider the following two indefinite integrals:

$$I_1 = \int \frac{dx}{x\sqrt{1-x^3}}.$$

$$I_2 = \int \frac{x\,dx}{\sqrt{1-x^3}}$$

Although the integrands look similar, it turns out that

$$I_1 = \log(x) - \frac{1}{3}\log(x^3 - 2\sqrt{1-x^3} - 2)$$

while I_2 cannot be expressed in closed form using the elementary functions of calculus. Two questions that arise are

(1) Why? i.e. what is the difference between the above functions that makes one integrable in closed form and not the other?

(2) How? i.e. can we always determine whether the antiderivative of a given elementary function is also an elementary function, and can we find it if it is?

As far as the second question is concerned, there are now algorithms that, given an elementary function, either compute an antiderivative or prove that no elementary antiderivative exists. These algorithms use a purely algebraic approach to the problem and have been described extensively in the literature ([3, 4, 7, 8, 10]), so we do not detail them in this note. We look however at the mathematical quantities that those algorithms compute with, and outline how they determine whether the integral is elementary.

Informally, we define the elementary functions to be the functions built from the rational functions by the successive adjunctions of a finite number of (nested) logarithms, exponentials, and roots of univariate polynomials. Since we can add a root of $t^2 + 1$, the usual trigonometric functions and their inverses are elementary functions. We will give a formal definition later in this paper.

---

*IBM Research Division, T.J.Watson Research Center, Yorktown Heights, NY 10598.

2. Rational Functions. We begin by reviewing the traditional partial fractions algorithm for integrating rational functions, since it provides the theoretical foundations for the other algorithms. Let $f \in \mathbf{C}(x)$ be our integrand, and write f as

$$f = P + \sum_{i=1}^{n} \sum_{j=1}^{e_i} \frac{a_{ij}}{(x - b_i)^j}$$

where $a_{ij}, b_i \in \mathbf{C}$ and the e_i's are positive integers. Computing $\int P$ poses no problem (it will for any other class of functions), and we have:

$$\int \frac{a_{ij}}{(x - b_i)^j} = \begin{cases} \dfrac{a_{ij}}{(1 - j)(x - b_i)^{j-1}}, & \text{if } j > 1 \\ a_{i1} \log(x - b_i), & \text{if } j = 1. \end{cases}$$

It is clear that computer integration algorithms do not perform such a factorization, and in fact, they work over an arbitrary field K of characteristic 0 not necessarily algebraically closed. However, we refer the reader to [4] for an algebraic description of the models used, and let our base field be $\mathbf{C}$ in this paper, since the mathematical quantities we are interested in appear more clearly. As we see from the above formulae, f has an elementary integral of the form

$$(*) \qquad \int f = v + c_1 \log(u_1) + \cdots + c_n \log(u_n)$$

where $v, u_i \in \mathbf{C}(x)$, and $c_i \in \mathbf{C}$. Over $\mathbf{C}$, $f(x)$ has a finite pole of order e_i at $x = b_i$ for $i = 1, \ldots, n$. Expanding f into a Laurent series at $x = b_i$, we get

$$f = \frac{a_{ie_i}}{(x - b_i)^{e_i}} + \cdots + \frac{a_{i2}}{(x - b_i)^2} + \frac{a_{i1}}{(x - b_i)} + \cdots$$

Integrating the term of order -1 gives $a_{i1} \log(x - b_i)$ so we get $c_i = a_{i1}$ and $u_1 = x - b_i$. We note that c_i is the *residue* of f at $x = b_i$. Integrating the terms of order less than -1 gives us the principal parts of the Laurent expansions of v at $x = b_i$:

$$v = \frac{a_{ie_i}}{(1 - e_i)(x - b_i)^{e_i-1}} + \cdots + \frac{-a_{i2}}{(x - b_i)} + \cdots$$

We view the polynomial part P of f as the principal part of the Laurent expansion of f at infinity. Then, $\int P$ is the principal part of the expansion of v at infinity. After proving that v has no other poles, we only have to interpolate for v. For rational functions this just means summing all its principal parts together, so

$$v = \int P + \sum_{i=1}^{n} \left(\frac{a_{ie_i}}{(1 - e_i)(x - b_i)^{e_i-1}} + \cdots + \frac{-a_{i2}}{(x - b_i)} \right).$$

Thus, we can view the integration of rational functions as a four-step process:

(i) Compute the principal parts of the Laurent series of f at its poles including at infinity,

(ii) Integrate these principal parts termwise, except the terms with exponents -1,

(iii) Interpolate, i.e. find a function $v \in \mathbf{C}(x)$ whose principal parts are the series found in step (ii), and with no other poles. Since differentiation and expansion into power series commute, $f - v'$ can have only simple poles,

(iv) Since the residue at a point $a \in \mathbf{C}$ of the logarithmic derivative of $g \in \mathbf{C}(x)$ is the order of g at $x = a$, find constants $c_i \in \mathbf{C}$ and functions $u_i \in \mathbf{C}(x)$ such that $residue_{x=a}(f - v') = c_1 order_{x=a}(u_1) + \cdots + c_n order_{x=a}(u_n)$ at any $a \in \mathbf{C}$. Then, the integral of $f dx$ is given by $(*)$.

In the case of rational functions, all those steps can be carried out (for step (iii) we sum the principal parts together, and for step (iv) we choose $u_a = x - a$, $c_a = residue_{x=a}(f - v')$), so rational functions are always integrable in closed form. However, these four steps can be applied to any subclass of elementary functions, and we describe them in the next sections. If any of those steps fail, then it is possible to prove that the integrand does not have an elementary antiderivative. It turns out that steps (iii) and (iv) can fail for purely algebraic functions, and steps (ii) and (iv) can fail for purely transcendental elementary functions, so steps (ii), (iii) and (iv) can fail for arbitrary mixed elementary functions.

3. Liouville's Principle. We now present the basic principle that allows the previous algorithm to apply to larger classes of functions. We have seen that the integral of a rational function can always be expressed as a rational function plus a linear combinations of logarithms of rational functions with constant coefficients. While this is not necessarily true for other classes of functions (e.g. $\int e^{x^2} dx$), Liouville's principle informally states that if a function f has an elementary antiderivative, then it has one of the form given by $(*)$.

A formal statement and proof of Liouville's theorem can be found in [7, 9], but the following informal version is sufficient to motivate the four-steps approach:

THEOREM (LIOUVILLE). *Let f be an elementary function of a variable x, and F be a field of characteristic zero such that $f, x \in F$ and $\dfrac{dg}{dx} \in F$ for any $g \in F$. If $\int f dx$ is an elementary function, then there exist $v \in F$, $c_1, \ldots, c_n \in \overline{K}$, and $u_1, \ldots, u_n \in \overline{K}F$ such that*

$$(\dagger) \qquad f dx = dv + \sum_{i=1}^{n} c_i \frac{du_i}{u_i}$$

where $K = \{c \in F$ such that $\dfrac{dc}{dx} = 0\}$ and $\overline{K}$ is an algebraic closure of K.

Thus algorithms can look for an integral in that specific form, and Liouville's theorem guarantees that if no integral is found in that form, then there is no elementary antiderivative. In the algorithm we used for rational functions, step (iii) finds the potential v of $(\dagger)$ by interpolating from the multiples poles of f, while step (iv) finds the c_i's and u_i's by looking at the residues of f. This separation is possible since dv can have only multiple poles (except at $x = \infty$) while du/u has only simple poles. Those basic facts will remain true for higher elementary functions once we have the adequate framework for poles and residues.

It should be noted that Liouville's principle does not state that any elementary integral must be of the form (†), but that there must be one of that form, which may not always be the most natural one. For example, with $F = \mathbf{C}(x)$, we have $\int \dfrac{dx}{1 + x^2} = \arctan(x)$, but it can also be expressed as $\dfrac{\sqrt{-1}}{2} \log(\dfrac{\sqrt{-1} + x}{\sqrt{-1} - x})$.

4. Algebraic Functions. We now describe how Risch ([8]) generalized the four-step algorithm to algebraic functions. Let $f \in \mathbf{C}(x, y)$ be our integrand where x is the integration variable and y is a function satisfying $F(x, y) = 0$ for a given irreducible polynomial $F \in \mathbf{C}[X, Y]$. Any algebraic function can be viewed in this way, for example, $\sqrt{x} \in \mathbf{C}(x, y)$ with $y^2 - x = 0$, while some algebraic functions may not be expressible in terms of radicals.

In order to define the notions of poles and series around a point, we first need a precise notion of a point. Since algebraic functions are n-valued almost everywhere where $n = \deg_Y(F(X, Y))$, elements of $\mathbf{C}$ do not define precise points. For example, it is undefined whether $1/(y - 1)$ has a pole or a zero at $x = 1$ where $y^2 - x = 0$. In order to eliminate this multiple-value problem, we consider f as a function on the *Riemann surface* of $F(X, Y)$ (see [1] for a definition). Intuitively, this surface can be seen as an n-sheet covering of the Riemann sphere (i.e. $\mathbf{C} \bigcup \{\infty\}$), with only finitely many points where two or more sheets intersect. At any point on the Riemann surface, f is either single-valued or has a finite pole, and there exists a local series expansion of elements of $\mathbf{C}(x, y)$. Those series, called *Puiseux expansions*, are similar to Laurent series with the exponents being of the form i/r for $i = a \ldots \infty$, and r the number of sheets intersecting at that point (the *ramification index* of the point). Those expansions can be computed using the algorithm in [11]. Since they are series in $x^{\pm 1/r}$ with complex coefficients, their principal parts can be integrated with respect to x which completes part (ii).

However, interpolating v from its principal parts is harder than for rational functions: those principal parts give us all the zeros and poles $P_1, \ldots, P_m$ of v on the Riemann surface and the orders ν_i of v at those points. The formal sum $\sum_{i=1}^{m} \nu_i P_i$ is called a *divisor* of $\mathbf{C}(x, y)$. Using the Bliss-Coates algorithm ([2, 6]), we compute a basis for the finite $\mathbf{C}$-vector space $L(D)$ of functions having order greater than or equal to ν_i at P_i and no poles at any other point. Computing the Puiseux expansions of a generic element of $L(D)$ at $P_1, \ldots, P_m$ and equating with the expansions of v yields a system of linear equations over $\mathbf{C}$. If it has no solution, then step (iii) fails and f has no elementary integral, otherwise we get v such that $f - v'$ has only simple poles.

¿From Liouville's theorem, if $h = f - v'$ has an elementary integral, it must have one of the form $c_1 \log(u_1) + \cdots c_n \log(u_n)$, where $c_i \in \mathbf{C}$ and $u_i \in \mathbf{C}(x, y)$. The residues of $h\,dx$ can be computed from its puiseux expansions at its poles. Since the residue of du/u at any point on the Riemann surface is the order of u at that point, we must have

$$residue_{P_i}(h\,dx) = c_1\, order_{P_i}(u_1) + \cdots + c_n\, order_{P_i}(u_n)$$

at each pole P_i of hdx. Let $(q_1, \ldots, q_r)$ be any basis for the **Z**-module generated by the residues of hdx and, at each pole P_i of hdx, write

$$residue_{P_i}(hdx) = a_{i,1}q_1 + \cdots + a_{i,r}q_r$$

with $a_{i,j} \in \mathbf{Z}$. We are then looking for functions $u_j \in \mathbf{C}(x, y)$ with orders exactly $a_{i,j}$ at the P_i's and 0 everywhere else. If there exists such a function u_j for a given j, then we say that the divisor $D_j = \sum_i a_{i,j}P_i$ is the divisor of u_j and that D_j is *principal*. Using the Bliss-Coates algorithm ([2, 6]), we can test whether the D_j's are principal. If we find $u_1, \ldots, u_r$ such that the divisor of u_j is D_j, then $w = hdx - q_1\dfrac{du_1}{u_1} - \cdots - q_r\dfrac{du_r}{u_r}$ is a differential of the first kind, so either $w = 0$, in which case

$$\int hdx = q_1 \log(u_1) + \cdots + q_r \log(u_r)$$

or $w \neq 0$ and it can be proven that hdx has no elementary antiderivative.

A new difficulty arises if some D_j is not principal. Suppose then that for each j, there exist a positive integer b_j and $u_j \in \mathbf{C}(x, y)$ such that b_jD_j is the divisor of u_j. The order of u_j at P_i is then $a_{i,j}b_j$ so we have

$$residue_{P_i}(hdx) = \frac{q_1}{b_1}order_{P_i}(u_1) + \cdots + \frac{q_r}{b_r}order_{P_r}(u_r)$$

so $w = hdx - \dfrac{q_1}{b_1}\dfrac{du_1}{u_1} - \cdots - \dfrac{q_r}{b_r}\dfrac{du_r}{u_r}$ is a differential of the first kind so, as above, either $\sum_{j=1}^{r} \dfrac{q_j}{b_j} \log(u_j)$ is an integral of hdx, or hdx has no elementary antiderivative.

Thus the problem is not to determine whether D is principal for a given divisor D, but to determine whether bD is principal for some positive integer b. Risch ([8]) completes his algorithm by describing how reducing the curve $F(X, Y) = 0$ to curves over finite fields determines a bound B such that if mD is not principal for $m = 1 \ldots B$, then mD is not principal for $m > B$, in which case we say that D has *infinite order*. We then test whether any of the D_j's has infinite order. If this is a case for some j, it can be proven that hdx has no elementary antiderivative, otherwise, we get a differential of the first kind as explained above.

Examples. 1. Consider

$$\int fdx = \int \frac{3x^8 + x - 1}{x^2\sqrt{x^8 + 1}}dx.$$

$f \in \mathbf{C}(x, y)$ with $F(x, y) = y^2 - x^8 - 1 = 0$.

(i) The integrand has potential poles above $x = 0$, $x = \infty$ and $x = \omega_i = e^{(2i+1)\pi/8}$ for $i = 0 \ldots 7$, and the principal parts of its Puiseux expansions at those points are:

$$
\begin{cases}
\quad -x^{-2} + x^{-1} + \cdots & \text{at} \quad (x = 0, y = 1) \\
\quad x^{-2} - x^{-1} + \cdots & \text{at} \quad (x = 0, y = -1) \\
\quad 3x^2 + 0x + 0 + \cdots & \text{at} \quad (x = \infty, \dfrac{y}{x^4} = 1) \\
\quad -3x^2 + 0x + 0 + \cdots & \text{at} \quad (x = \infty, \dfrac{y}{x^4} = -1) \\
\dfrac{(\omega_i - 4) \prod_{j \neq i} (\omega_j - \omega_i)^{-\frac{1}{2}}}{\omega_i{}^2} (x - \omega_i)^{-\frac{1}{2}} + \cdots & \text{at} \quad (x = \omega_i, y = 0)
\end{cases}
$$

(ii) Integrating them yields the expansions

$$
\begin{cases}
\quad\quad x^{-1} + \cdots & \text{at} \quad (x = 0, y = 1) \\
\quad\quad -x^{-1} + \cdots & \text{at} \quad (x = 0, y = -1) \\
x^3 + 0x^2 + 0x + \cdots & \text{at} \quad (x = \infty, \dfrac{y}{x^4} = 1) \\
-x^3 + 0x^2 + 0x + \cdots & \text{at} \quad (x = \infty, \dfrac{y}{x^4} = -1)
\end{cases}
$$

(iii) A basis over $\mathbf{C}$ for the vector space of functions which have order greater than or equal to -1 above $x = 0$, greater than or equal to -3 above $x = \infty$ and no other pole is

$$
\{v_1, v_2, v_3, v_4, v_5, v_6\} = \{1, x, x^2, x^3, \frac{1}{x}, \frac{y}{x}\}
$$

and the Puiseux expansions of a generic element $\sum_{j=1}^{6} c_j v_j$ at those points are

$$
\begin{cases}
\quad\quad (c_5 + c_6)x^{-1} + \cdots & \text{at} \quad (x = 0, y = 1) \\
\quad\quad (c_5 - c_6)x^{-1} + \cdots & \text{at} \quad (x = 0, y = -1) \\
(c_4 + c_6)x^3 + c_3 x^2 + c_2 x + \cdots & \text{at} \quad (x = \infty, \dfrac{y}{x^4} = 1) \\
(c_4 - c_6)x^3 + c_3 x^2 + c_2 x + \cdots & \text{at} \quad (x = \infty, \dfrac{y}{x^4} = -1)
\end{cases}
$$

Equating terms with the expansions found in step (ii), we get the following system of linear equations for the c_i's:

$$
\begin{cases}
c_5 + c_6 = 1 \\
c_5 - c_6 = -1 \\
c_4 + c_6 = 1 \\
c_4 - c_6 = -1 \\
c_3 = c_2 = 0 \\
\quad c_1 \text{ arbitrary}
\end{cases}
$$

which has $(c, 0, 0, 0, 0, 1)$ for solution for any $c \in \mathbf{C}$ (c is arbitrary since it is the coefficient of a constant). So we get $v = \dfrac{y}{x}$, and $h = f - v' = \dfrac{y}{x^9 + x}$ has only simple poles.

(iv) From the expansions of step (i), we find that $h\,dx$ has residues 1 at $P = (x = 0, y = 1)$, -1 at $Q = (x = 0, y = -1)$, and 0 everywhere else, so we look for a function $u \in \mathbf{C}(x, y)$ which has a simple pole at Q, a simple zero at P, and no other zero or pole, i.e. we test whether $D = P - Q$ is principal. No such function exists in $\mathbf{C}(x, y)$, so D is not principal, and iD is not principal for $i = 2, 3$. However, $u = \dfrac{x^4}{1 + y}$ has order 4 at P, -4 at Q, and 0 everywhere else, so $4D$ is principal, so $w = h\,dx - \dfrac{du}{4u}$ is a differential of the first kind. Since $w = 0$, we have

$$\int h = \frac{1}{4} \log(\frac{x^4}{1 + \sqrt{x^8 + 1}})$$

so

$$\int f\,dx = \int \frac{3x^8 + x - 1}{x^2 \sqrt{x^8 + 1}}\,dx = \frac{\sqrt{x^8 + 1}}{x} + \frac{1}{4} \log(\frac{x^4}{1 + \sqrt{x^8 + 1}}).$$

2. Consider

$$I_2 = \int g\,dx = \int \frac{x\,dx}{\sqrt{1 - x^3}}$$

$g \in \mathbf{C}(x, y)$ with $F(x, y) = y^2 + x^3 - 1 = 0$.

(i) The integrand has potential poles above $x = \omega_i = e^{2i\pi/3}$ for $i = 0, 1, 2$, and the principal parts of its Puiseux expansion at those points are:

$$\frac{\omega_i}{\prod_{j \neq i} (\omega_j - \omega_i)^{-\frac{1}{2}}} (x - \omega_i)^{\frac{1}{2}} + \cdots \qquad \text{at} \quad (x = \omega_i, y = 0).$$

(ii) and (iii) We see from those expansions that g has only simple poles, so we choose $v = 0$.

(iv) From the above expansions, $g\,dx$ has residue 0 everywhere, so if it has an elementary derivative, it must be of the form $c \log(u)$ where $u \in \mathbf{C}(x, y)$ has order 0 everywhere, so $u \in \mathbf{C}$, so $du = 0$, and since $w = g\,dx - c\dfrac{du}{u} = g\,dx$ is nonzero, $g\,dx$ has no elementary antiderivative.

5. Elementary Functions. Let f be an arbitrary elementary function. In order to generalize the four-step algorithm, we build an algebraic model in which f behaves like either a rational or algebraic function. For that purpose, we now need to formally define elementary functions.

A *differential field* is a field k with a given map $a \to a'$ from k into k, satisfying $(a + b)' = a' + b'$ and $(ab)' = a'b + ab'$. Such a map is called a *derivation* on k. An element $a \in k$ which satisfies $a' = 0$ is called a *constant*. The constants of k form a subfield of k.

A differential field K is a *differential extension* of k if $k \subseteq K$, and the derivation on K extends the one on k. Let K be a differential extension of k, and $\theta \in K$. We say that θ is an *elementary monomial* over k, if θ is transcendental over k, $k(\theta)$ and k have the same subfield of constants, and there exists $\eta \in k$ such that either

(i) $\theta' = \dfrac{\eta'}{\eta}$ in which case we say that θ is logarithmic over k, and write $\theta = \log(\eta)$, or

(ii) $\theta' = \eta'\theta$, in which case we say that θ is exponential over k, and write $\theta = \exp(\eta)$.

A differential extension K of k is an *elementary extension of k*, if there exist $\theta_1, \ldots, \theta_m \in K$ such that $K = k(\theta_1, \ldots, \theta_m)$ and for each $i = 1 \ldots m$, either

(i) θ_i is algebraic over $k(\theta_1, \ldots, \theta_{i-1})$, or

(ii) θ_i is an elementary monomial over $k(\theta_1, \ldots, \theta_{i-1})$.

A function f is called an elementary function in x over a field k if $\dfrac{da}{dx} = 0$ for any $a \in k$, and there is an elementary (w.r.t $' = d/dx$) extension K of $k(x)$ such that $f \in K$.

Elementary extensions are useful for modeling any function as a rational (or algebraic) function of one main variable over the other variables. Given an elementary integrand $f dx$, an integration algorithm constructs first a field k containing all the constants appearing in f, then the rational function field $k(x)$, and finally builds a tower $L = k(x)(\theta_1, \ldots, \theta_m)$ where the θ_i's are all the elementary monomials and algebraic functions needed to express f. The derivation used at every step is $' = \dfrac{d}{dx}$.

If $t = \theta_m$ is transcendental over $E = k(x)(\theta_1, \ldots, \theta_{m-1})$, then $f \in E(t)$ so f can be seen as a univariate rational function over E, the major difference with the complex case being that E is not constant with respect to $'$. The notion of Laurent series remains however well defined at points of $\overline{E}$ (an algebraic closure of E), so step (i) can be carried out. Integrating their principal parts can be done in a similar way than in the rational function case. However, since the coefficients are not constants, this may require recursively integrating elements of $\overline{E}$ (this process must terminate since $\overline{E}$ has a lower transcendence degree than $E(t)$ over k) or finding whether some linear first order differential equation with coefficients in $\overline{E}$ has a solution in $\overline{E}$. If the recursive integration fails (i.e. proves that there is no elementary antiderivative), then it can be proven that f has no elementary antiderivative. The interpolation step ((iii)) is exactly as for rational functions, i.e. an element of $\overline{E}(t)$ is the sum of its principal parts (up to an element of $\overline{E}$). Once we are reduced to an integrand h with at most simple poles, its residues can be defined in terms of its Laurent series, and step (iv) is similar to the rational function case, with the additional criterion that if one of the residues of $h dx$ is not a constant (w.r.t. d/dx), then $h dx$ has no elementary antiderivative. See [4, 7] for the detailed algorithms.

If $y = \theta_m$ is algebraic over $E = k(x)(\theta_1, \ldots, \theta_{m-1})$, then, by the primitive element theorem, we can ensure that $t = \theta_{m-1}$ is transcendental over $E = K(x)(\theta_1, \ldots, \theta_{m-2})$. Thus, $f \in E(t, y)$ so f can be seen as a univariate algebraic function over E, with $F(t, y) = 0$ where $F \in E[T, Y]$. We view $f \in L$ as an algebraic function over E in the same way than we viewed a transcendental elementary function as a rational function over E, and the algebraic function integration procedure described earlier can be generalized to this case. The notions of points of the Riemann surface defined by y, and of Puiseux expansions at those points remain well defined over $\overline{E}$ ([5]), so step (i) can be carried out. Integrating them may require integrating the coefficients recursively as above, so we may prove that f has no elementary antiderivative at this stage. Otherwise, step (iii) and (iv) are similar to the algebraic function case, with the difference, as above, that if one of the residues is not a constant, then the integrand has no elementary antiderivative.

6. Implementations in Computer Algebra. It should be noted that the algorithms presented here, while conceptually simple, are not very well suited for efficient implementations since they require a lot of superfluous computations in the algebraic closures of arbitrarily complicated fields. Although Davenport ([6]) has a partial implementation in the computer algebra system REDUCE of an algorithm essentially similar to the one we describe here for algebraic functions, there are no reported implementations of those techniques in the general case.

However, many so-called "rational algorithms" have appeared in the last decade, which are more efficient and easier to implement. While based on Liouville's theorem and on a similar analysis of poles and residues, those algorithms compute only in a minimal algebraic extension necessary to express the integral if it exists. The transcendental and rational function cases of those algorithms are now implemented in most computer algebra systems, and in addition, the complete algebraic case, and part of the general case are now implemented in the Scratchpad computer algebra system. We refer the reader to [3, 4, 10] for descriptions of those algorithms.

7. Acknowledgements. This paper is based on a talk presented at the workshop on Computer Assisted Proofs in Analysis in Cincinnati from March 22 to March 25 1989. I would like to thank the organizers for their invitation to present these results there.

REFERENCES

[1] AHLFORS, L. V.,, *Complex Analysis*, International series in pure and applied mathematics, McGraw-Hill, New York, 1966.
[2] BLISS, G. A., *Algebraic Functions*, Dover Publications, New York, 1966.
[3] BRONSTEIN, M., *Integration of Elementary Functions*, Ph.D. thesis, Dpt. of Mathematics, Univ. of California, Berkeley, 1987. To appear in the Journal of Symbolic Computation.
[4] BRONSTEIN, M., *Symbolic Integration: towards Practical Algorithms*, in *Proceedings of CADE '88*, Academic Press, 1989. In Press.
[5] CHEVALLEY, C., *Algebraic Functions of One Variable*, Math. Surveys Number VI, American Mathematical Society, New York, 1951.
[6] DAVENPORT, J. H., *On the Integration of Algebraic Functions*, Lecture Notes in Computer Science No. 102, Springer-verlag, New York, 1981.
[7] RISCH, R., *The Problem of Integration in Finite Terms*, Transactions American Mathematical Society, 139 (1969), pp. 167–189.
[8] RISCH, R., *The Solution of the Problem of Integration in Finite Terms*, Bulletin American Mathematical Society, 76 (1970), pp. 605–608.

[9]   ROSENLICHT, M., *Integration in Finite Terms*, American Mathematical Monthly, 79 (1972), pp. 963–972.

[10] TRAGER, B., *Integration of Algebraic Functions*, Ph.D. thesis, Dpt. of EECS, Massachusetts Institute of Technology, 1984.

[11] WALKER, R. J., *Algebraic Curves*, Springer-Verlag, New York, 1978.

A COMPUTER-ASSISTED APPROACH TO SMALL-DIVISORS PROBLEMS ARISING IN HAMILTONIAN MECHANICS

ALESSANDRA CELLETTI† AND LUIGI CHIERCHIA‡

One of the most powerful and versatile tools in the study of invariant surfaces for conservative dynamical systems relies upon KAM theory ([14], [1], [16], [17], [7], [19], [12], [20]). However, because of the apparently stringent quantitative requirements, such theory has been (and often still is) considered not too well suited for concrete applications. Nevertheless, in [3], [5], [6], [21] and especially in [4], [18], [9], [2], it has been shown how refinements and implementations of KAM theory may yield quantitative rigorous results that are in *good* agreement with the numerical expectations.

The most satisfying among the above quoted results have been obtained with the aid of computers used to perform long *perturbative* calculations whose numerical errors were controlled by means of the so called *interval arithmetic* (see Appendix B for a short review of the basic concepts involved in interval arithmetic).

Here we want to discuss, with special emphasys on the role of computers, some new results in this area. In particular we will see that, eventhough mechanical computations allow to establish rigorous results, there are intrinsic difficulties in proving computer-assisted theorems whose proofs are based on algorithms involving a lot of divisions by small quantities.

In [4] we provided a rigorous technique for constructing analytic invariant homotopically non trivial tori and circles for certain classes of Hamiltonian systems (with any number of degrees of freedom) and of area preserving twist maps. On these surfaces, called KAM surfaces, the time evolution is linear with rationally independent frequencies. Applying our technique to the so-called standard map ([11]),

(1)
$$F \ : \ (x,y) \in \mathbf{T} \times \mathbf{R} \longrightarrow (x',y') = (x+y+\varepsilon \ sin \ x, \ y+\varepsilon \ sin \ x), \ , \qquad (\mathbf{T} \equiv \mathbf{R}/2\pi\mathbf{Z}) \ ,$$

and to a forced pendulum ([10]) with Hamiltonian (in standard symplectic coordinates)

$$(2) \quad H(x,t,y;\varepsilon) \equiv \frac{y^2}{2} + \varepsilon \ [cos \ x \ + \ cos(x-t)] \ , \qquad (x,t,y) \in \mathbf{T}^2 \times \mathbf{R} \ ,$$

we established in [4], with the aid of computer-assisted estimations, the existence of the "golden-mean" KAM tori (namely the surfaces with rotation numbers $\omega = (\sqrt{5} - 1)\pi$ and $\omega = \frac{\sqrt{5}-1}{2}$, respectively) for $|\varepsilon| \leq \rho_0$, with ρ_0 given, respectively,

---

†Forschungsinstitut für Mathematik, ETH-Zentrum, CH-8092 Zürich. Supported by Istituto Nazionale di Alta Matematica.

‡Dipartimento di Matematica, II Universita' di Roma, 00173 Roma, Italia.

by 0.65 and 0.015. We recall that, for such surfaces, the (numerically) expected "breakdown thresholds" (i.e. the numerical value $\varepsilon_c > 0$ immediately above which such surfaces cease to exist) are, respectively, 0.971 and 0.027. We also showed that such surfaces are actually *analytic in ε in the full disk* $\mathcal{D}_{\rho_0} \equiv \{\varepsilon \in \mathbf{C} : |\varepsilon| \leq \rho_0\}$, and provided explicit polynomial approximations in such regions.

A similar result, concerning a rigorous lower bound of ε_c for the standard map, has been announced by De la Llave and Rana ([9], see also [18]). However, their method, eventhough is not completely unrelated to ours, seems not to yield much information about ε-analyticity properties ([8]). For a more general introduction on this subject and for further references see [4].

In the rest of this note we shall present the results for the standard map (1) and we refer to the Appendix C for the results on the forced pendulum (2).

As in [4], we relate the existence of ω-KAM tori for (1) to (smooth) solutions of the equation

$$(3) \qquad D^2 u - \varepsilon \sin(\theta + u) = 0, \qquad 1 + u_\theta \neq 0,$$

where $u(\theta) = u(\theta; \varepsilon)$ is periodic in θ, $Du(\theta) \equiv u(\theta + \omega/2) - u(\theta - \omega/2)$ and u is normalized by fixing the θ-average, $\langle u \rangle$, to be zero. It is easy to see that u solves (3) if and only if the curve $\theta \to \Gamma_\omega(\theta) = (x, y) = (\theta + u(\theta), \omega + u(\theta) - u(\theta - \omega))$ (which, since $1 + u_\theta \neq 0$, is a graph) is F-invariant with $F^n(x, y) = \Gamma_\omega(\theta + n\omega)$, $n \in \mathbf{Z}$.

To solve equation (3), we construct a polynomial approximate solution of (3), which is used as initial guess of the Newton-KAM algorithm of [4]. The initial guess is obtained as follows: Consider the ε-power expansion of u near $\varepsilon = 0$, $u = \sum_{k \geq 1} u_k(\theta) \varepsilon^k$; we take as approximate solution of (3) the N_0-truncation, $v_{N_0} \equiv \sum_{k=1}^{N_0} u_k(\theta) \varepsilon^k$. The functions $u_1(\theta), ..., u_{N_0}(\theta)$ can be constructed explicitly, via the recursive procedure presented in Appendix A.

The efficiency of the KAM algorithm depends of course on the choice of the initial approximation v_{N_0}; namely, one should take N_0 as large as possible since $v_{N_0} \xrightarrow{N_0 \to \infty} u$. In [4] we constructed explicitly v_{N_0} up to $N_0 = 38$ for the standard map (and up to $N_0 = 24$ for the forced pendulum). The formulae used in [4] for the computation of v_{N_0}, though being very general, present serious combinatorial problems if N_0 is large (see Appendix A) and require a large amount of computer time.

However in [4] we were left with the doubt that using faster machines or more efficient formulae for the computation of v_{N_0}, one might obtain rigorous results much closer to the numerical threshold.

In this note we implement a new procedure that we learned in [13] for the recursive computation of the functions $u_1(\theta), ..., u_{N_0}(\theta)$. Such formulae allow to construct a higher order approximation v_{N_0} with a reasonable amount of computer time. For example, the construction of v_{38} requires now ~ 2 seconds of CPU time (on a VAX 8650), despite the 6 minutes of the computer program used in [4].

However, in [4] we did not realize that even if we have at hand much powerful formulae one cannot control the computer errors for more than the order $N_0 = 40$. To illustrate this new hindrance, let us write the Fourier representation of the function $v_{N_0}(\theta; \varepsilon)$ as

$$v_{N_0}(\theta; \varepsilon) \equiv \sum_{k=1}^{N_0} u_k(\theta)\varepsilon^k = \sum_{k=1}^{N_0} \sum_{j \in \mathfrak{M}_k} \hat{u}_{k,j} \; sin(j\theta)\varepsilon^k$$

for some real coefficients $\hat{u}_{k,j}$ and for a suitable set of integers $\mathfrak{M}_k$. Since the number of Fourier coefficients $\hat{u}_{k,j}$ at the order $N_0 = 40$ is 420, the explicit computation of v_{N_0} requires the use of a computer. However since the computer works with a finite precision, one has to control the round-off and propagation errors introduced by the machine implementing the interval arithmetic procedure which is shortly described in Appendix B.

Roughly speaking to take care of the computer errors, we first reduce every computation to *elementary operations* (i.e. sum, subtraction, multiplication and division). Around each result of an elementary operation we construct an interval which certainly contains the real result. Subsequent operations are performed between intervals rather than between real numbers. Let us call $(\hat{u}_{k,j}^{Down}, \hat{u}_{k,j}^{Up})$ the interval around the coefficient $\hat{u}_{k,j}$. During the computation of v_{N_0} the relative amplitude $|\frac{\hat{u}_{k,j}^{Up} - \hat{u}_{k,j}^{Down}}{\hat{u}_{k,j}^{Down}}|$ of the Fourier coefficients corresponding to the small divisors (namely the approximants $j = 1, 3, 5, 8, ...$ of the golden ratio) becomes remarkably large. The effect of the propagation of such enlargement produces (around the order $N_0 = 40$) a relative amplitude of the order of unity, making impossible a careful computation of v_{N_0} with $N_0 > 40$.

Starting with v_{40} as initial approximant we are able to state the following result. (Notice that the program without interval arithmetic runs in ~ 2 seconds of CPU time on a VAX 8650, while the program with interval arithmetic requires about 15 seconds of CPU time).

THEOREM (on the standard map). *Let* $\omega = (\sqrt{5} - 1)\,\pi$ *and let* $\xi = 0.1 \cdot 2^{-8}$ ($\sim$ $3.906 \cdot 10^{-4}$), $\rho_1 = 0.66$. *Then equation (3) has a unique real-analytic (both in* θ *and* ε*) solution* u *with vanishing mean-value on* $\mathbf{T}$. *For such a function one has* $|u|_{\xi,\rho_1} < 0.2586$, $0.3641 < |u_\theta(\pi; 0.66)| < |u_\theta|_{\xi,\rho_1} < 0.3807$. *Furthermore, if* $N_0 = 40$ *and* $v \equiv v_{N_0} \equiv \sum_{k=1}^{N_0} u_k(\theta)\, \varepsilon^k$, $\langle u_k \rangle = 0$, *one has*

$$|u - v|_{\xi,\rho_1} < 1.415 \cdot 10^{-5}, \quad |u_\theta - v_\theta|_{\xi,\rho_1} < 8.468 \cdot 10^{-4}.$$

Let us stress again that while the computational limits of the work in [4] were dictated by the (relatively) large numbers of operations involved in the computation of v_{38}, in the theorem above the limit came in because of the blowing up of the interval arithmetic: The small-divisors appearing in the $\varepsilon - \theta$ expansion of the (golden-mean) invariant curve produce a (constant) growth of the "controlling-intervals" (produced by the interval arithmetic as described in appendix B performed over double-precision operations) leading to a relative round-off error of size one for terms of the series corresponding to orders (in ε) larger than ~ 40.

However we do believe that this hindrance is of pure technical nature and we certainly believe that using more sophisticated computations (like arbitrary precision evaluation) our theoretical method would yield much better (possibly "optimal") results.

Our belief is based on the following facts. We had the Vax 8650 of the ETH-Zürich evaluate (*without controlling round-off errors*) the first 340 coefficients in the ε-expansion $u(\theta, \varepsilon) = \sum_{k \geq 1} u_k(\theta)\varepsilon^k$ (such a computation took about $1^h 13^m$ of CPU time). Now, *assuming* that the automatic round-off errors do not propagate in a catastrophyc way (e.g. assuming that all the quantities computed are exact up to an error of order 10^{-9}), then our computer-assisted KAM algorithm (cfr. [4]) would yield the following

PSEUDO-THEOREM (on the standard map). *Let* $\omega = (\sqrt{5} - 1)\,\pi$ *and let* $\xi = 0.014 \cdot 2^{-9}$ ($\sim 2.734 \cdot 10^{-5}$), $\rho_1 = 0.875$. *Then equation* (3) *has a unique real-analytic (both in θ and ε) solution u with vanishing mean-value on* $\mathbb{T}$. *For such a function one has* $|u|_{\xi,\rho_1} < 0.4185$, $0.8824 < |u_\theta(\pi; 0.875)| \leq |u_\theta|_{\xi,\rho_1} < 0.8933$. *Furthermore, if $N_0 = 340$ and $v \equiv v_{N_0} \equiv \sum_{k=1}^{N_0} u_k(\theta)\,\varepsilon^k$*, $\langle u_k \rangle = 0$, *one has*

$$|u - v|_{\xi,\rho_1} < 1.3731 \cdot 10^{-10}, \qquad |u_\theta - v_\theta|_{\xi,\rho_1} < 3.9003 \cdot 10^{-8}.$$

Notice that the pseudo-result is in agreement of the 90% with the numerical expectation.

From the point cf view of computer-assisted techniques, it is certainly a challenging and interesting problem to try to implement further the above ideas in an *efficient way* and to actually prove the above "pseudo-theorem". At this regard it seems likely that the use of parallel-computations might be of significant help.

Appendix A: Series expansion for the solution and estimate of the error term.

As mentioned before the existence of a KAM torus with rotation number ω for the standard map (1) is related to solutions of the equation

(A.1) $\quad D^2 u - \varepsilon \, sin(\theta + u) = 0$, $\quad 1 + u_\theta \neq 0$, $\quad Du(\theta) \equiv u(\theta + \frac{\omega}{2}) - u(\theta - \frac{\omega}{2})$.

To find a solution of (A.1) we expand u in power series around $\varepsilon = 0$ as $u(\theta; \varepsilon) \equiv \sum_{k \geq 1} u_k(\theta)\varepsilon^k$. As approximate solution we take the N_0-truncation $v^{(0)} \equiv v_{N_0} \equiv \sum_{k=1}^{N_0} u_k(\theta)\varepsilon^k$. Then $v^{(0)}$ will satisfy (A.1) up to a small error term $e^{(0)} \equiv e_{N_0}$:

(A.2) $\qquad\qquad D^2 v^{(0)} - \varepsilon \, sin(\theta + v^{(0)}) \equiv e^{(0)}$, where $e^{(0)} \simeq O(\varepsilon^{N_0+1})$. To solve equation $(A.1)$ we apply the Newton iteration procedure presented in [4]: under suitable hypotheses, starting with $v^{(0)}$ one can construct a new approximation $v^{(1)}$ satisfying $D^2 v^{(1)} - \varepsilon \sin(\theta + v^{(1)}) \equiv e^{(1)}$, with an error term $|e^{(1)}| \sim O(|e^{(0)}|^2)$. The iterative application of the Newton step provides a sequence of approximants $\{v^{(j)}\}$ (with relative error bounds $e^{(j)}$), provided the condition $1 + v_\theta^{(j)} \neq 0$ holds for every j.

To control the sequences $\{v^{(j)}\}$, $\{e^{(j)}\}$ we apply a KAM algorithm, which given upper bounds $V^{(j)}$, $E^{(j)}$ on the norms of $v^{(j)}$ and $e^{(j)}$ provides upper bounds on the norms of $v^{(j+1)}$ and $e^{(j+1)}$. If the algorithm converges, namely if the conditions

$$|(1 + v_\theta^{(j)})^{-1}| < \infty \quad \forall j \geq 0 \quad and \quad \lim_{j \to \infty} E^{(j)} = 0$$

are satisfied, then a solution of $(A.1)$ is obtained as a uniform limit of the $v^{(j)}$'s.

To implement such scheme we need to know as input data the norms of $v^{(0)} \equiv v_{N_0}$, $v_\theta^{(0)}$ and $e^{(0)}$. The size of the initial approximation $v^{(0)} \equiv v_{N_0} \equiv \sum_{k=1}^{N_0} u_k(\theta) \varepsilon^k$ (and hence of $v_\theta^{(0)}$) can be obtained by a recursive construction of the functions $u_k(\theta)$.

In [4] the u_k's, which are odd trygonometric polynomials of degree k, were computed according to the formula (obtained inserting $\sum u_k \varepsilon^k$ in $(A.1)$)

$$D^2 u_{k+1} = \sum_{h \in \mathcal{H}_k} (\partial_\theta^{|h|} \sin\theta) \prod_{i=1}^{k} \frac{u_i^{h_i}}{h_i!} \ , \qquad \mathcal{H}_k \equiv \{h \in \mathbf{N}^k : h_1 + 2h_2 + \ldots + k h_k = k\} \ .$$

Such a formula is very general (the sinus could be replaced by any real analytic periodic function) but, in actual computations, it presents serious combinatorial problems if k is large.

A much faster way which we learned in [13] is the following. Let $\sum_{k \geq 0} c_k(\theta) \varepsilon^k$ denote the ε-power series expansion of $exp(i(\theta + u))$. By differentiation w.r.t. ε one obtains the relations

$$(A.3) \qquad c_k = \frac{i}{k} \sum_{n=1}^{k} n \, u_n \, c_{k-n} \ , \qquad k \geq 1 \ (c_0 = e^{i\theta}) \ .$$

One can compute recursively the c_k's and u_k's (and hence $v_{N_0} \equiv \sum_{k=1}^{N_0} u_k \varepsilon^k$) since $(A.1)$ implies that

$$(A.4) \qquad D^2 u_{k+1} = \frac{1}{2i} (c_k - \overline{c_k}) \ ,$$

where, as usual, $\overline{c_k}$ denotes the complex analytic function $\theta \to \overline{c_k(\overline{\theta})}$. Notice that in these formulae there is no combinatorics involved so that the computing time will all be spent in the Fourier analysis of $(A.3)$, $(A.4)$.

Our next task is the estimate of the error term relative to the initial approximant $v^{(0)} \equiv v_{N_0}$

$$|e_{N_0}|_{\xi,\rho} \equiv \sup_{|Im\theta|\leq\xi, \ |\varepsilon|\leq\rho} |e_{N_0}(\theta;\varepsilon)| \ ,$$

where $e_{N_0} \equiv e^{(0)}$ is defined by the l.h.s. of $(A.2)$.

If $a(z)$ is a z-power series $\sum_{k\geq 0} a_k z^k$, we set $P_{z,N_0}(a) \equiv \sum_{k\geq N_0} a_k z^k$ and $P'_{z,N_0} = id. - P_{z,N_0}$. Then the following simple lemma (which substitutes Lemma 7 of [4]) holds.

LEMMA. *For any positive ξ, ρ and any $N_0 \in \mathbf{Z}_+$, one has*

$(A.5)$ $\ |P_{\varepsilon,N_0}(sin(\theta + v_{N_0}))|_{\xi,\rho} \leq [(P_{\rho,N_0}(cosh \ V_{N_0}))^2 + (P_{\rho,N_0}(sinh \ V_{N_0}))^2]^{1/2} \ ,$

where $V_{N_0} \equiv \sum_{k=1}^{N_0} U_k \rho^k$ *and* $U_k = U_k(\xi)$ *is an upper bound on* $|u_k|_\xi \equiv \sup_{|Im\theta|\leq\xi} |u_k(\theta)|$.

Proof.

$$\begin{aligned} |P_{\varepsilon,N_0}(sin(\theta + v_{N_0}))| &= |sin\theta \ P_{\varepsilon,N_0}(cos \ v_{N_0}) \ + \ cos\theta \ P_{\varepsilon,N_0}(sin \ v_{N_0})| \\ &\leq [|P_{\varepsilon,N_0}(cos \ v_{N_0})|^2 \ + \ |P_{\varepsilon,N_0}(sin \ v_{N_0})|^2]^{1/2} \ . \end{aligned}$$

Now, expanding in power series one obtains

$$|P_{\varepsilon,N_0}(cos \ v_{N_0})|_{\xi,\rho} \leq P_{\rho,N_0}(cosh \ V_{N_0}), \quad |P_{\varepsilon,N_0}(sin \ v_{N_0})|_{\xi,\rho} \leq P_{\rho,N_0}(sinh \ V_{N_0}) \,.\square$$

The point of this Lemma is that the r.h.s. of $(A.4)$ can be exactly calculated in terms of $U_1, ..., U_{N_0}$. In fact, since P_{N_0} acts linearly, it is enough to look at $P_{\rho,N_0}(exp(\pm V_{N_0})) = exp(\pm V_{N_0}) - P'_{\rho,N_0}(exp(\pm V_{N_0-1}))$. But

$$P'_{\rho,N_0-1}(exp(\pm V_{N_0})) = \sum_{k=0}^{N_0-1} a_k^{\pm} \rho^k \ , \quad a_0^{\pm} \equiv 1, \quad a_k^{\pm} \equiv \pm\frac{1}{k} \sum_{n=1}^{k} nU_n a_{k-n}^{\pm} \ ,$$

where such a formula is proved as before by setting $exp(\pm V_{N_0}) = \sum_{k\geq 0} a_k^{\pm} \rho^k$ and differentiating w.r.t. ρ.

Appendix B: Interval arithmetic.

In floating point notation real numbers are represented by the computer with a sign-exponent-fraction representation. Since the number of digits in the fraction and exponent is fixed, the result of any operation is usually an approximation of the real result. To take care of the round-off and propagation errors introduced by the machine we perform the so-called interval arithmetic technique ([15]).

The first step consists in the reduction of any expression of our programs into a sequence of *elementary* operations, i.e. sum, subtraction, multiplication and division (eventually using a truncated Taylor expansion for the functions sin, exp, etc.).

In double precision a real number is represented by 64 bits and the result of an elementary operation is guaranteed up to 1/2 of the last significant bit of the mantissa ([22]).

The idea of the interval arithmetic is to produce an interval around each result of an elementary operation and to perform subsequent computations on intervals. For example, along the line of [15] let us add $[a_1, b_1]$ to $[a_2, b_2]$; the result of such operation is the interval $[a_3, b_3] \equiv [a_1 + b_1, a_2 + b_2]$ such that if $x \in [a_1, b_1]$ and $y \in [a_2, b_2]$ then $x + y \in [a_3, b_3]$. However, since the endpoints a_3, b_3 are the result of an elementary operation one has to link the subroutines for operations on intervals to a procedure which constructs strict lower and upper bounds, say $\tilde{a}_3, \tilde{b}_3$, on a_3, b_3 respectively. The new extrema of the final interval $[\tilde{a}_3, \tilde{b}_3]$ are obtained decreasing or increasing by one unit the last significant bit of a_3 and b_3.

Appendix C: Results on the forced pendulum.

In this appendix we describe the results of the application of KAM theorem to the forced pendulum system described by the Hamiltonian

$$(C.1) \quad H(x, t, y; \varepsilon) \equiv \frac{y^2}{2} + \varepsilon \left[\cos x + \cos(x - t) \right], \qquad (x, t, y) \in \mathbf{T}^2 \times \mathbf{R}.$$

The equation to be solved to find the ω-KAM torus for $(C.1)$ is given by

$$(C.2) \quad D^2 u - \varepsilon \left[\sin(\theta + u) + \sin(\theta + u - t) \right] = 0, \quad 1 + u_\theta \neq 0, \quad (\langle u \rangle = 0),$$

where $u(\theta, t) = u(\theta, t; \varepsilon)$ is periodic in (θ, t), $Du \equiv \omega u_\theta + u_t$, $\langle \cdot \rangle \equiv (\theta, t)$-average. A solution of $(C.2)$ yields an H-invariant torus $\Gamma_\omega(\theta, t) \equiv (x, t, y) = (\theta + u, t, \omega + Du)$ and the H-flow at time s is given by $\mathcal{F}_H^s(x, t, y) = \Gamma_\omega(\theta + \omega s, t + s)$.

The formulae presented in the Appendix A for the standard map concerning the recursive construction of the approximant v_{N_0} and the estimate of the error term e_{N_0} transpose easily to the Hamiltonian case. For the forced pendulum the maximal order we are able to reach is $N_0 = 40$ (requiring ~ 5 minutes of CPU time for the program without interval arithmetic and ~ 55 minutes for the program with interval arithmetic).

THEOREM (on the forced pendulum). *Let* $\omega = (\sqrt{5} - 1)/2$ *and let* $\xi = 0.08 \cdot 2^{-10}$ ($\sim 7.812 \cdot 10^{-5}$), $\rho_1 = 0.018$. *Then equation* $(C.2)$ *admits a unique real-analytic solution* u *with vanishing mean-value on* $\mathbf{T}^2$. *For such solution one has* $|u|_{\xi, \rho_1} < 0.2325$, $0.3161 < |u_\theta(\pi, 0; 0.018)| < |u_\theta|_{\xi, \rho_1} < 0.3370$. *Furthermore, if* $N_0 = 40$ *and* $v \equiv v_{N_0} \equiv \sum_{k=1}^{N_0} u_k(\theta, t) \varepsilon^k$, $\langle u_k \rangle = 0$, *one has*

$$|u - v|_{\xi, \rho_1} < 1.672 \cdot 10^{-5}, \quad |u_\theta - v_\theta|_{\xi, \rho_1} < 9.208 \cdot 10^{-4}.$$

Analogously to the standard map the opening of the intervals around the Fourier coefficients of v_{N_0} does not allow to construct rigorously a higher order approximation. However, let us call again *pseudo-theorem* the result obtained constructing a higher order approximant *without* interval arithmetic. Starting with v_{60} (which requires ~ 54 minutes of CPU time for the non-rigorous program) we obtain the following

PSEUDO-THEOREM (on the forced pendulum). *Let* $\omega = (\sqrt{5} - 1)/2$ *and let* $\xi = 0.06 \cdot 2^{-10}$ $(\sim 5.859 \cdot 10^{-5})$, $\rho_1 = 0.0197$. *Then equation* $(C.2)$ *admits a unique real-analytic solution* u *with vanishing mean-value on* $\mathbf{T}^2$. *For such solution one has* $|u|_{\xi,\rho_1} < 0.2641$, $0.3869 < |u_\theta(\pi,0;0.0197)| < |u_\theta|_{\xi,\rho_1} < 0.4059$. *Furthermore, if* $N_0 = 60$ *and* $v \equiv v_{N_0} \equiv \sum_{k=1}^{N_0} u_k(\theta,t)\, \varepsilon^k$, $\langle u_k \rangle = 0$, *one has*

$$|u - v|_{\xi,\rho_1} < 4.2192 \cdot 10^{-6}, \quad |u_\theta - v_\theta|_{\xi,\rho_1} < 3.1563 \cdot 10^{-4}.$$

Notice that in this case the result is in agreement of the 72% with the numerical expectation.

REFERENCES

[1] ARNOLD V.I., *Proof of a Theorem by A.N. Kolmogorov on the invariance of quasi-periodic motions under small perturbations of the Hamiltonian*, Russ. Math. Surveys **18**, 9 (1963).
[2] CELLETTI A., *Analysis of resonances in the spin-orbit problem in Celestial Mechanics*, Ph.D. Thesis, ETH Zürich (1989).
[3] CELLETTI A., CHIERCHIA L., *Rigorous estimates for a computer-assisted KAM theory*, J. Math. Phys. **28**, 2078 (1987).
[4] CELLETTI A., CHIERCHIA L., *Construction of analytic KAM surfaces and effective stability bounds*, Commun. Math. Phys. **118**, 119 (1988).
[5] CELLETTI A., FALCOLINI C., PORZIO A., *Rigorous numerical stability estimates for the existence of KAM tori in a forced pendulum*, Ann. Inst. Henri Poincare' **47**, 85 (1987).
[6] CELLETTI A., GIORGILLI A., *On the numerical optimization of KAM estimates by classical perturbation theory*, J. Appl. Mathem. and Phys. (ZAMP) **39**, 743 (1988).
[7] CHIERCHIA L., GALLAVOTTI G., *Smooth prime integrals for quasi-integrable Hamiltonian systems*, Nuovo Cimento **67 B**, 277 (1982).
[8] DE LA LLAVE R., private communication.
[9] DE LA LLAVE R., RANA D., *Proof of accurate bounds in small denominators problems*, preprint (1986).
[10] ESCANDE D.F., DOVEIL F., *Renormalization method for computing the threshold of the large-scale stochastic instability in two degrees of freedom Hamiltonian systems*, J. Stat. Physics **26**, 257 (1981).
[11] GREENE J.M., *A method for determining a stochastic transition*, J. of Math. Phys. **20**, 1183 (1979).
[12] HERMAN M., *Sur le courbes invariantes par le difféomorphismes de l'anneau*, Astérisque **2**, 144 (1986).
[13] HERMAN M., *Recent results and some open questions on Siegel's linearization theorems of germs of complex analytic diffeomorphisms of* $\mathbf{C}^n$ *near a fixed point*, preprint (1987).
[14] KOLMOGOROV A.N., *On the conservation of conditionally periodic motions under small perturbation of the Hamiltonian*, Dokl. Akad. Nauk. SSR **98**, 469 (1954).
[15] LANFORD III O.E., *Computer assisted proofs in analysis*, Physics A **124**, 465 (1984).
[16] MOSER J., *On invariant curves of area-preserving mappings of an annulus*, Nach. Akad. Wiss. Göttingen, Math. Phys. Kl. II **1**, 1 (1962).
[17] MOSER J., *Minimal solutions of variational problems on a torus*, Ann. Inst. Henri Poincare' **3**, 229 (1986).

[18] RANA D., *Proof of accurate upper and lower bounds to stability domains in small denominators problems*, Ph.D. Thesis, Princeton (1987).

[19] RÜSSMANN H., *On the existence of invariant curves of twist mappings of an annulus*, Springer Lecture Notes in Math. **1007**, 677 (1983).

[20] SALAMON D., ZEHNDER E., *KAM theory in configuration space*, Comment. Math. Helvetici **64**, 84 (1989).

[21] WAYNE C.E., *The KAM theory of systems with short range interactions I*, Comm. Math. Phys. **96**, 311 (1984).

[22] (no author listed), *Vax Architecture handbook*, Digital Equipment Corporation (1981).

ON A COMPUTER ALGEBRA AIDED PROOF IN BIFURCATION THEORY

CARMEN CHICONE†AND MARC JACOBS†

Abstract. Some bifurcation theory for the zeros of an analytic function depending on a vector of parameters is presented. The emphasis is on problems for which the Weierstrass Preparation Theorem is not applicable and where the techniques require the analysis of ideals in rings of convergent power series. Several examples are given to illustrate the methods of this analysis. In particular, the utility of the Gröbner basis algorithms which are currently implemented in several computer algebra packages is explained. Several problems for future research which are natural candidates for computer assisted proofs are also given.

Key words. period function, center, bifurcation, quadratic system, Hamiltonian system, linearization

AMS(MOS) subject classifications. 58F14, 58F22, 58F30, 34C15

1. Introduction. There are a number of ways that computer algebra can assist in the analysis of bifurcation problems. Here we are going to describe a class of bifurcation problems where computer algebra is not only useful to handle the complexities of the computations involved but, more importantly, where it can be used to aid in the proofs of bifurcation theorems. Two of our main goals are to demonstrate how the analysis of ideals in rings of convergent power series arise naturally in bifurcation theory and to describe how some portions of this analysis can be carried out with the currently available computer implemented algorithms; especially, Buchberger's algorithm [3] for the computation of Gröbner bases.

Usually we begin with a differential equation model of a physical phenomenon where we wish to observe the behavior of a certain observable quantity which can be represented by the zeros of a function $\xi \mapsto F(\xi)$, where ξ is the coordinate for some subspace of the state space of the dynamical system defined by the differential equation. In realistic situations the model equations will depend on a number of parameters given by a vector $\lambda = (\lambda_1, \lambda_2, \dots, \lambda_N)$ of real numbers. The problem is to consider, for the changing values of λ, the number and position of the solutions of the equation $F(\xi, \lambda) = 0$. Since F may be nonlinear, it is often extremely difficult to obtain the solutions for the equation in the full domain of the function F. However, as a step in this direction, there may be some values of the parameter λ where the number of zeros can be determined. If λ_* is such a point, then we may ask for the number of zeros of F when λ is very close to λ_*, i.e., we may ask for the number of *local zeros* of $F(\xi, \lambda) = 0$ near the point (ξ_*, λ_*). More precisely, we say F has a *bifurcation to K-local zeros at* (ξ_*, λ_*) if for each $\epsilon > 0$ and $\delta > 0$ there is a point λ such that $|\lambda - \lambda_*| < \delta$, and $F(\xi, \lambda) = 0$ has K solutions $\xi_1, \xi_2, \dots, \xi_K$ satisfying $|\xi_k - \xi_*| < \epsilon$ for $k = 1, 2, \dots, K$. If, in addition, there is some choice of ϵ and δ such that for each λ with $|\lambda - \lambda_*| < \delta$, $F(\xi, \lambda) = 0$ has at most K local zeros, then

†Department of Mathematics, The University of Missouri, Columbia, MO 65211. This research was sponsored by the Air Force Office of Scientific Research under grant AF-AFOSR-89-0078.

we say F has K *local zeros at* (ξ_*, λ_*), cf. [1]. When F is analytic in both ξ and λ, we can always translate the point ξ_* to the origin of the coordinate system and consider the power series development of F to have the form

$$F(\xi, \lambda) = f_0(\lambda) + f_1(\lambda)\xi + f_2(\lambda)\xi^2 + \cdots.$$

It is the number of zeros of this power series which lie near $\xi = 0$ which we wish to determine.

Two fundamental examples which lead to a bifurcation problem of this type are the bifurcation of limit cycles from a center or a weak focus and the bifurcation of critical points of the period function from a weak center. For these examples, we begin with an autonomous differential equation in the plane depending on a vector λ of parameters

$$\dot{x} = f(x, y, \lambda), \quad \dot{y} = g(x, y, \lambda)$$

where we have the following hypotheses: (1) f and g are both analytic, (2) $f(0, 0, \lambda)$ and $g(0, 0, \lambda)$ vanish identically, (3) the linear part of the vector field (f, g) is $(-y, x)$ when $\lambda = 0$, and (4) the origin of the phase plane is, for the limit cycle case, either a center or a focus and, for the case of the critical points of the period function, a center for all values of the parameter λ. We consider a transverse section Σ to the flow of the vector field through the origin of the plane which, for simplicity, we will always assume is an interval $[0, \omega)$ of the positive x-axis and we let ξ denote the distance coordinate along Σ.

For the limit cycle bifurcation problem we define $\xi \mapsto h(\xi, \lambda)$ to be the Poincaré return map on Σ and $d(\xi, \lambda) := h(\xi, \lambda) - \xi$, the associated succession function. Under our assumptions d is analytic and can be developed into a power series of form

$$d(\xi, \lambda) = d_1(\lambda)\xi + d_2(\lambda)\xi^2 + \cdots.$$

The zeros of d correspond to periodic trajectories of the differential equation.

For the case of the bifurcation of critical points of the period function we consider instead of the return map the derivative of the period function $\xi \mapsto P_\xi(\xi, \lambda)$. Here P assigns to the periodic trajectory through $(\xi, 0)$ its minimum period. The series representation of $P(\xi, \lambda)$ has the form

$$P(\xi, \lambda) = 2\pi + p_2(\lambda)\xi^2 + p_3(\lambda)\xi^3 + \cdots.$$

Thus, in this case, the bifurcation function is

$$P_\xi(\xi, \lambda) = 2p_2(\lambda)\xi + 3p_3(\lambda)\xi^2 + \cdots,$$

and the zeros of $P_\xi(\xi, \lambda)$ are the critical points of the period function.

In both of the examples just described the bifurcation function can not be obtained in closed form except in special cases. However, the coefficients of the power series expansions can be obtained, in principle, to any desired order.' Thus, in each case, our problem is to determine the number of local zeros from knowledge of a finite number of explicitly computed Taylor coefficients and from the qualitative analysis of the differential equation.

In §2 we give some of the basic bifurcation theory useful in the analysis of problems of the type just described. In §3 we will show how this bifurcation theory can be applied to the problem of bifurcation of critical points of the period function of the quadratic systems which have centers. While, in §4 we will show how the same methods can be used to study the bifurcation of critical points of the period function for Hamiltonian systems of the form

$$\ddot{u} + g(u, \lambda) = 0$$

where g is a polynomial in u and λ is again a vector of parameters. We will also mention in this section some problems for further research which offer promising applications for computer algebra assisted proofs.

2. Bifurcation Theory. The basic tool for the analysis of the local zeros of a power series is provided by the following theorem.

WEIERSTRASS PREPARATION THEOREM. *Let* $(\xi, \lambda) \mapsto F(\xi, \lambda)$ *be an analytic function for* $(\xi, \lambda) \in \mathbf{R} \times \mathbf{R}^N$ *and suppose* F *has the series representation*

$$F(\xi, \lambda) = f_0(\lambda) + f_1(\lambda)(\xi - \xi_*) + f_2(\lambda)(\xi - \xi_*)^2 + \cdots$$

near the point (ξ_*, λ_*). *If* $f_0(\lambda_*) = f_1(\lambda_*) = \cdots = f_{K-1}(\lambda_*) = 0$ *but* $f_K(\lambda_*) \neq 0$, *then*

$$F(\xi, \lambda) = \left[(\xi - \xi_*)^K + a_{K-1}(\lambda)(\xi - \xi_*)^{K-1} + \cdots + a_0(\lambda) \right] u(\xi, \lambda)$$

with $a_i(\lambda)$ *and* $u(\xi, \lambda)$ *analytic,* $a_i(\lambda_*) = 0$ *for* $i = 0, 1, \ldots, K-1$ *and* $u(\xi_*, \lambda_*) \neq 0$.

In other words, under the hypotheses of the theorem, F has at most K local zeros at (ξ_*, λ_*). If additional conditions are satisfied by the coefficients $a_i(\lambda)$ so that the polynomial

$$(\xi - \xi_*)^K + a_{K-1}(\lambda)(\xi - \xi_*)^{K-1} + \cdots + a_0(\lambda)$$

has K real roots near ξ_* for some choice of λ in each interval centered at λ_*, then F has K local zeros at (ξ_*, λ_*).

In order to illustrate a typical situation and the application of the Preparation Theorem we consider an abstract example. Suppose we start with a differential equation model for which we have defined our analytic bifurcation function $\xi \mapsto F(\xi, \lambda_1, \lambda_2)$ and that we are able to compute :

$$F(\xi, \lambda_1, \lambda_2) = (\lambda_2 - \lambda_1^2)\xi + \cdots.$$

By the Preparation Theorem the only local zero when $\lambda_2 - \lambda_1^2 \neq 0$ is $\xi = 0$. To investigate the number of local zeros near points on the curve $\lambda_2 - \lambda_1^2 = 0$ we compute another term of the power series. Suppose we find

$$F(\xi, \lambda_1, \lambda_2) = (\lambda_2 - \lambda_1^2)\xi + \lambda_2 \xi^2 + \cdots.$$

Now, if $\lambda_2 - \lambda_1^2 = 0$ but $\lambda_2 \neq 0$ there is at most one other local zero aside from $\xi = 0$. The only remaining point from which we do not know the local bifurcations is the origin of the parameter space. The bifurcation analysis at this point of the parameter space can be quite subtle. It is often the case in the examples that, as a consequence of the definition of F, we have $F(\xi, 0, 0) \equiv 0$. This implies that all the Taylor coefficients of the bifurcation function vanish at the origin of the parameter space. In particular, at this point of the parameter space the Preparation Theorem is not applicable. If the bifurcation problem were generic so that perturbations have to be considered in the direction of an arbitrary smooth vector field, any number of local zeros can be obtained. But, since the bifurcation problem arises from a specific vector field with the dependence on the parameters given from the outset, the number of local zeros is restricted. It is perhaps tempting to conjecture that, since the bifurcation function degenerates to the zero function at just one point, that there remains at most one nontrivial local zero. However, this may not be the case. In fact, suppose we compute further and find

$$F(\xi, \lambda_1, \lambda_2) = (\lambda_2 - \lambda_1^2)\xi + \lambda_2\xi^2 - \lambda_1\xi^3 + \cdots .$$

Consider the polynomial

$$F_3(\xi, \lambda_1, \lambda_2) = (\lambda_2 - \lambda_1^2)\xi + \lambda_2\xi^2 - \lambda_1\xi^3$$

and its evaluation for $t \geq 0$ along the path given by

$$\lambda_1(t) = -t, \quad \lambda_2(t) = 0.$$

We have

$$F_3(\xi, -t, 0) = t\xi(-t + \xi^2)$$

and it follows that there are *two* local zeros; they are given by $\xi = \pm\sqrt{t}$. Now, under the assumption that $F(\xi, 0, 0) \equiv 0$, we conclude that for each $k = 1, 2, \ldots$, the Taylor coefficient of order k can be represented in the form

$$f_k(\lambda_1, \lambda_2) = a_k(\lambda_1, \lambda_2)\lambda_1 + b_k(\lambda_1, \lambda_2)\lambda_2$$

where a_k and b_k are analytic. Formally, we then have

$$F(\xi, \lambda_1, \lambda_2) = (\lambda_2 - \lambda_1^2)\xi + \lambda_2\xi^2(1 + b_4\xi^2 + \cdots) - \lambda_1\xi^3(1 - a_4\xi + \cdots).$$

Thus, for small ξ, the analytic function F appears to behave like the polynomial

$$F_3(\xi, \lambda_1, \lambda_2) = (\lambda_2 - \lambda_1^2)\xi + \lambda_2\xi^2 - \lambda_1\xi^3$$

and there should be at most two nontrivial local zeros. These remarks would be a proof that there are at most two local zeros for the bifurcation problem if the convergence of the representation of F in its rearranged form were established. Fortunately, this can be done. In fact, we have the following theorem, cf. [4].

INFINITE ORDER BIFURCATION THEOREM. *If*

$$F(\xi, \lambda) = f_0(\lambda) + f_1(\lambda)(\xi - \xi_*) + f_2(\lambda)(\xi - \xi_*)^2 + \cdots$$

is analytic and the ideal I generated by the set of all the Taylor coefficients $f_i(\lambda)$, for $i = 0, 1, \ldots,$ in the ring $\mathbf{R}\{\lambda_1, \ldots, \lambda_N\}_{\lambda_*}$ of convergent power series expanded at λ_* is generated by the finite set $\{f_0(\lambda), f_1(\lambda), \ldots, f_K(\lambda)\}$, then F has at most K local zeros at (ξ_*, λ_*).

In the example, it is clear that the ideal in $\mathbf{R}\{\lambda_1, \lambda_2\}_0$ of all the Taylor coefficients is generated by the polynomials $\lambda_2 - \lambda_1^2$, λ_2, and λ_1 corresponding to the first three Taylor coefficients. However, in practice, the first few Taylor coefficients may well be nontrivial polynomials (or power series) where it will not be at all obvious whether or not they generate the ideal of all Taylor coefficients. There is another aspect of this which is illustrated by the example. Note that the *variety* of all the Taylor coefficients was apparent from the assumption $F(\xi, 0, 0) \equiv 0$ after the computation of the second coefficient, since the variety of the ideal $(\lambda_2 - \lambda_1^2, \lambda_2)$ is already just the origin of the parameter space. Thus, the variety of the sequence of ideals $(f_0), (f_0, f_1), (f_0, f_1, f_2), \ldots$ may stabilize before the sequence of ideals itself stabilizes. Of course, in a ring of convergent power series, we are assured, by an extension of the Hilbert Basis Theorem cf. [2, p. 345], that a tower of ideals as above will stabilize at a finite height. Computer algebra is usually essential in the computation of the Taylor coefficients of the bifurcation function F in a realistic problem. However, it is in the application of the Infinite Order Bifurcation Theorem where we find computer algebra assisted proofs of bifurcation theorems are feasible. In the next section we describe an example where the bifurcation analysis can be completed with this aid.

3. Critical Points of the Period Function. In a classic paper, N.N. Bautin [1] considered the bifurcation of local limit cycles from a weak focus or center of a quadratic system as described in the introduction. Here one begins with the quadratic differential equation in the Bautin normal form

$$\dot{x} = \lambda_1 x - y - \lambda_3 x^2 + (2\lambda_2 + \lambda_5)xy + \lambda_6 y^2$$
$$\dot{y} = x + \lambda_1 y + \lambda_2 x^2 + (2\lambda_3 + \lambda_4)xy - \lambda_2 y^2.$$

This normal form is chosen so that the conditions for the stationary point at the origin to be a center are expressed as reasonably simple polynomial relations in the coefficients. In fact, the origin will be a center, i.e., every trajectory of the differential equation starting near the origin is periodic, provided $\lambda_1 = 0$ and one of the following conditions holds: (1) $\lambda_4 = 0$ and $\lambda_5 = 0$, (2) $\lambda_3 = \lambda_6$, (3) $\lambda_5 = 0, \lambda_4 + 5\lambda_3 - 5\lambda_6 = 0$ and $\lambda_3\lambda_6 - 2\lambda_6^2 - \lambda_2^2 = 0$, or (4) $\lambda_2 = 0$, and $\lambda_5 = 0$. When $\lambda_1 = 0$ the origin is called a weak focus. Recall that for Bautin the succession function is defined on a segment of the positive x−axis and he considers the series representation

$$d(\xi, \lambda) = d_1(\lambda)\xi + d_2(\lambda)\xi^2 + d_3(\lambda)\xi^3 + \cdots.$$

Using the Preparation Theorem we have at most one limit cycle bifurcating from a weak focus of order one ($d_1 = 0$, $d_3(\lambda) \neq 0$), at most two bifurcate from a weak focus of order two ($d_1 = d_3 = 0$, $d_5 \neq 0$), and at most three from a weak center of order three ($d_1 = d_3 = d_5 = 0$, $d_7 \neq 0$). When $d_1 = d_3 = d_5 = d_7 = 0$ all the Taylor coefficients vanish and we have a center. In particular, the conditions listed above are satisfied. This fact can be shown by finding an integral for the differential equation in each such case. The theorem of Bautin determines the local bifurcations from one of these centers; it is the prototype theorem for the infinite order bifurcation problems where the Preparation Theorem is not applicable.

BAUTIN'S THEOREM. *The first nonvanishing coefficient d_k of the series for the succession function is the coefficient of an odd order term. The first four odd order coefficients are*

$$d_1 = e^{2\pi\lambda_1} - 1,$$

$$d_3 = -\frac{\pi}{4}\lambda_5(\lambda_3 - \lambda_6) \ (\text{mod } (\lambda_1)),$$

$$d_5 = \frac{\pi}{24}\lambda_2\lambda_4(\lambda_3 - \lambda_6)(\lambda_4 + 5\lambda_3 - 5\lambda_6) \ (\text{mod } (\lambda_1, d_3)),$$

$$d_7 = -\frac{25\pi}{32}\lambda_2\lambda_4(\lambda_3 - \lambda_6)^2(\lambda_3\lambda_6 - 2\lambda_6^2 - \lambda_2^2) \ (\text{mod } (\lambda_1, d_3, d_5)),$$

where $(\lambda_1, d_3, \ldots, d_k)$ denotes the ideal in the polynomial ring $R[\lambda_1, \lambda_2, \ldots, \lambda_6]$. If $k > 7$, then d_k is in the ideal $(\lambda_1, d_3, d_5, d_7)$. In particular, at most three local limit cycles bifurcate from a center of a quadratic system.

We now describe the problem of bifurcation of critical points of the period function from a quadratic isochrone. This was previously discussed by us [4]; here we outline the most important features of the analysis. We will be especially interested in those aspects of this analysis which lead to applications of computer algebra.

For the bifurcation problem as described in the introduction, we will consider the bifurcation of critical points for the period function of a quadratic system. Recall that we insist that the differential equation have a center at the origin of the phase space for all values of the parameters in the problem. For quadratics, the centers are determined from the Bautin normal form. In [4] we show that no critical points of the period function bifurcate from a weak center, i.e., a center where the second order period coefficient vanishes, when the center occurs at a parameter value corresponding to the first three cases for a center in the Bautin normal form. For the fourth case, where $\lambda_2 = \lambda_5 = 0$, a linear change of coordinates transforms the differential equation to the Loud system [7]:

$$\dot{x} = -y + Bxy, \quad \dot{y} = x + Dx^2 + Fy^2.$$

This system has a center at the origin of the phase plane for all values of the parameters and the positive x–axis is a section for the flow of the differential equation for all values of the parameters.

In order to apply the theory we have developed we must obtain the series expansion of the period function. It is computationally efficient to first consider the dehomogenized Loud system

$$\dot{x} = -y + xy, \quad \dot{y} = x + Dx^2 + Fy^2.$$

In the case when $B \neq 0$ this system is obtained from the Loud system by a simple rescaling of the coordinates. The period coefficients of the Loud system can be obtained subsequently by homogenization. To compute the period coefficients we express the dehomogenized Loud system in polar coordinates

$$\dot{r} = r^2 \alpha(\theta, \lambda), \quad \dot{\theta} = 1 + r\beta(\theta, \lambda)$$

to obtain the integral representation of the period function

$$P(\xi, \lambda) = \int_0^{2\pi} \frac{d\theta}{1 + r(\theta, \xi, \lambda)\beta(\theta, \lambda)}$$

where

$$\frac{dr}{d\theta} = \frac{r^2 \alpha(\theta, \lambda)}{1 + r\beta(\theta, \lambda)}, \quad r(0, \xi, \lambda) = \xi.$$

Using the differential equation for $r(\theta, \xi, \lambda)$ we expand the solution with initial condition $r(0, \xi, \lambda) = \xi$ in powers of ξ. This series is substituted into the integrand of the integral representation of P and the resulting series is integrated term by term to obtain

$$P(\xi, D, F) = 2\pi + p_2(D, F)\xi^2 + p_3(D, F)\xi^3 + \cdots$$

where

$$p_2(D, F) = \frac{\pi}{12}(10D^2 + 10DF - D + 4F^2 - 5F + 1),$$

$$p_4(D, F) = \frac{\pi}{1152}(1540D^4 + 4040D^3F + 1180D^3 + 4692D^2F^2 + 1992D^2F + 2768DF^3$$
$$+ 453D^2 + 228DF^2 + 318DF - 2D + 784F^4 - 616F^3 - 63F^2 - 154F + 49),$$

$$p_6(D, F) = \frac{\pi}{1244160}(4142600D^6 + 17971800D^5F + 6780900D^5 + 34474440D^4F^2$$
$$+ 22992060D^4F + 4531170D^4 + 37257320D^3F^3 + 28795260D^3F^2$$
$$+ 10577130D^3F + 1491415D^3 + 24997584D^2F^4 + 14770932D^2F^3$$
$$+ 7686378D^2F^2 + 2238981D^2F + 339501D^2 + 10527072DF^5 + 367584DF^4$$
$$+ 1400478DF^3 + 598629DF^2 + 228900DF - 663D + 2302784F^6$$
$$- 1830576F^5 - 213972F^4 - 126313F^3 - 53493F^2 - 114411F + 35981).$$

Since we are interested in the critical points of the period function we must work with the bifurcation function $P_\xi(\xi, B, D, F)$. From symmetry considerations it is not difficult to see that the first nonzero period coefficient is the coefficient of an even power of ξ and, after the computation of the period coefficients, we see that if $B = 0$ then the origin is a weak center only if both D and F vanish. Of course, in this case, the system is linear and the center is isochronous. When $B \neq 0$ we can apply the Preparation Theorem to obtain the following results: If $p_2(D, F) = 0$ and $p_4(D, F) \neq 0$ at most one critical point of the period function bifurcates from the origin. If $p_2(D, F) = 0$, $p_4(D, F) = 0$ and $p_6(D, F) \neq 0$ then (D, F) is one of the three points

$$(-\frac{3}{2}, \frac{5}{2}), \, (\frac{-11 + \sqrt{105}}{20}, \frac{15 - \sqrt{105}}{20}), \, (\frac{-11 - \sqrt{105}}{20}, \frac{15 + \sqrt{105}}{20})$$

and, at these points, at most two critical points of the period function bifurcate from the origin. It turns out that in all cases the maximum number of critical points can be obtained by an appropriate bifurcation. If $p_2(D, F) = 0, p_4(D, F) = 0$ and $p_6(D, F) = 0$, then (D, F) is one of the four points

$$(0, 1), \, (0, \frac{1}{4}), \, (-\frac{1}{2}, \frac{1}{2}), \, (-\frac{1}{2}, 2),$$

and, at these points, $p_k(D, F) = 0$ for all $k = 2, 3, 4, \ldots$ [7]. In other words, these four points correspond to the nonlinear isochrones, the full set being those points (B, D, F) where the ratios $(D/B, F/B)$ give one of the four listed points. It is at these points and at the origin of the parameter space that the Preparation Theorem does not apply. However, we have a theorem analogous to Bautin's Theorem.

QUADRATIC ISOCHRONE BIFURCATION THEOREM [4]. *The kth Taylor coefficient of $P_\xi(\xi, B, D, F)$ is a homogeneous polynomial of degree k in $\mathbb{R}[B, D, F]$ and this polynomial is in the ideal*

$$(p_2(B, D, F), p_4(B, D, F), p_6(B, D, F)).$$

In particular, at most two critical points of the period function bifurcate from an isochrone of a quadratic system.

Here, it turns out that the maximum number of critical points which can bifurcate from a nonlinear isochrone is only one while the maximum number which can bifurcate from the linear isochrone is two. This difference is accounted for by a stronger version of the theorem which states that the ideal of all the Taylor coefficients in the ring of convergent power series at a given nonlinear isochrone is actually generated by the polynomials p_2 and p_4; this is proved in [4].

While we will not give a formal proof of this theorem here we will give an outline of the proof which contains the important ideas. We begin with the dehomogenized Loud system. Our strategy is to find an ideal membership theorem for the polynomial ideal

$$I := (p_2(D, F), p_4(D, F), p_6(D, F))$$

and to show that each of the Taylor coefficients meets this ideal membership test. Then, we will use this result to obtain the same conclusion for the homogenized ideal.

The basic ideal membership theorem for the dehomogenized ideal is the content of the following theorem.

IDEAL MEMBERSHIP THEOREM. *If p is a polynomial in $\mathbf{R}[D, F]$, then $p \in I$ if and only if (1) p vanishes at each isochrone and (2)*

$$\frac{d}{dt}p\left(t - \frac{1}{2}, \frac{1}{2} - t\right)\bigg|_{t=0} = 0.$$

The proof of this result uses some interesting algebra and, in particular, the theorem of M. Noether [9, p. 163]. For this, recall that the maximal ideal $M_{a,b}$ at a point $(a, b) \in \mathbf{R}[x, y]$ is the ideal generated by $\{x - a, y - b\}$. Noether's theorem states: *If I is a polynomial ideal whose zero set contains only a finite number of zeros (a_i, b_i), then if, for each i, σ_i is the smallest exponent such that the maximal ideal at the ith zero (denoted by M_i for short) satisfies the condition*

$$M_i^{\sigma} \subset (I, M_i^{\sigma+1}),$$

then

$$I = \bigcap_i (I, M_i^{\sigma}).$$

In the present case, the values of these exponents can be readily calculated with a computer algebra system which includes the standard Gröbner basis procedures [3]. In fact, this is just a problem in the repeated testing of monomials for membership in a polynomial ideal, and such procedures are provided in many of the standard computer algebra packages such as MACSYMA, MAPLE, and REDUCE. Now, the first condition of the Ideal Membership Theorem states that the polynomial p vanishes on the variety of I. Ordinarily, the variety will contain both real and complex zeros of the polynomial generators. However, it turns out that in the present case we can remain in the field of real numbers because the union of the four isochrones is exactly this variety. The derivative condition is derived from Noether's theorem. The key facts are that at the isochrone $(-1/2, 1/2)$ the three polynomials generating I have second order contact, while at the remaining isochrones they meet pairwise transversally. In particular, at the isochrone $(-1/2, 1/2)$, the Noether exponent has value two, while at the other isochrones the value is one. The derivative condition of the ideal membership theorem derives from the second order contact at the isochrone $(-1/2, 1/2)$ and this condition requires the verification of the identity

$$\frac{d}{d\lambda}p_k(\lambda)\bigg|_{\lambda=0} = 0$$

for the coefficients of the period function of the system

$$\dot{x} = -y + xy, \qquad \dot{y} = x - \frac{1}{2}x^2 + \frac{1}{2}y^2 + \lambda(x^2 - y^2).$$

Since all the trajectories starting at $(\xi, 0)$ are periodic, for ξ in the domain of the period function, we have $y(P(\xi, \lambda), \xi, \lambda) \equiv 0$. After differentiation of this identity with respect to λ we compute

$$\dot{y}(P(\xi, 0), \xi, 0)P_\lambda(\xi, 0) + y_\lambda(P(\xi, 0), \xi, 0) \equiv 0.$$

But, $P(\xi, 0) \equiv 2\pi$ because when $\lambda = 0$ the system is an isochrone. Thus, $p'_k(0) = 0$ exactly when $y_\lambda(2\pi, \xi, 0) \equiv 0$. Using the variational equations for the system one can show

$$y_\lambda(2\pi, \xi, 0) = \int_{-\infty}^{\infty} \frac{f(\xi, t)}{g(\xi, t)} \, dt$$

where

$$f(\xi, t) := \xi^2 (1 - (1 - \xi)^2 t^2)(1 - (6 - 6\xi + \xi^2)t^2 + (1 - \xi)^2 t^4)$$

and

$$g(\xi, t) := (1 + t^2)^2 (1 + (1 - \xi)^2 t^2)^2.$$

A computer check using the residue calculus shows the integral vanishes identically. This establishes that the ideal of all the Taylor coefficients in the case of the dehomogenized Loud system are in the ideal I.

Now define the ideal

$$J := (p_2(B, D, F), \ p_4(B, D, F), \ p_6(B, D, F))$$

for the homogenized system which corresponds to I. In order to establish the ideal membership result for this case we first consider the homogenization of the ideals (I, M_i^g). These are

$$Q_1 := (D, F - B), \quad Q_2 := (D, 4F - B)$$

$$Q_3 := (2D + B, F - 2B), \quad Q_4 := (D + F, (2D + B)^2).$$

It is not difficult to show using some standard theorems in ideal theory [8, pp. 64-65] and the results for the dehomogenized ideal that, with

$$J_1 := \bigcap_{i=1}^{4} Q_i,$$

we have for each $k = 2, 3, 4, \ldots,$ $p_k \in J_1$. However, this is not sufficient to establish the desired result since, unfortunately, $J_1 \neq J$. At this point the crucial steps using computer algebra are made. If we define a new ideal

$$J_0 := (J, B^4),$$

then, using a Gröbner basis computation, we can show by checking inclusions of the generators that

$$J = J_0 \cap J_1.$$

Finally, by another Gröbner basis computation we can check that every homogeneous monomial of degree seven belongs to J_0. This implies that the homogeneous polynomials p_k for all $k = 2, 3, 4, \ldots,$ belong to J_0. Then, since all the Taylor coefficients belong to both J_0 and J_1, it follows that they belong to J. This gives the desired result.

4. Polynomial Potential Functions. In this section we examine the bifurcation of critical periods for conservative second order scalar differential equations of the form

$$\ddot{u} + g(u, \lambda) = 0$$

when $\lambda \mapsto g(u, \lambda)$, $\lambda \in \mathbf{R}^N$ is linear, and $u \mapsto g(u, \lambda)$ is a polynomial. In fact, if we denote the potential energy by $V(u, \lambda)$, then

$$V(u, \lambda) := \int_0^u g(s, \lambda)\, ds$$

and the total energy is given by the Hamiltonian

$$H(u, v, \lambda) := \frac{1}{2}v^2 + V(u, \lambda), \text{ where } v = \dot{u}.$$

We will assume that the polynomial potential function has the form

$$V_N(u, \lambda) = \frac{1}{2}u^2 + \sum_{i=1}^{N-2} \lambda_{i+2} u^{i+2}.$$

For such potential functions we have the fundamental

POLYNOMIAL POTENTIAL ISOCHRONE THEOREM [4]. *The second order scalar differential equation $\ddot{u}+g(u, \lambda) = 0$ with $\lambda := (\lambda_3, \lambda_4, \dots, \lambda_N)$, and potential energy*

$$V_N(u, \lambda) = \frac{1}{2}u^2 + \sum_{i=1}^{N-2} \lambda_{i+2} u^{i+2}$$

has an isochronous center at the origin if and only if $\lambda = 0$.

In order to use our bifurcation results we need to obtain a series expansion for the period function in a neighborhood of $\xi = 0$. One can show [4] that for any potential function $V(u)$ which is analytic on $\mathbf{R}$ with $V'(0) = 0$ and $V''(0) = 1$ the function h defined by

$$X = h(u) := sgn(u)\sqrt{2V(u)}$$

when $V(u) \geq 0$, is analytic on each connected component of $\{u | V(u) \geq 0\}$ which contains the origin. Moreover, the inverse function $X \mapsto h^{-1}(X)$ is defined and analytic on the connected component of $\{\, u \,|\, h'(u) > 0\,\}$ which contains $u = 0$, and its MacLaurin series has the form

$$h^{-1}(X) = X + \sum_{k=2}^{\infty} d_k X^k.$$

Once the coefficients d_k in this expansion are known, the Maclaurin series for the period function $X \mapsto P(X)$ is determined by

$$P(X) = 2\pi + \sum_{k=1}^{\infty} p_{2k} X^{2k}$$

where

$$p_{2k} = 2\pi \frac{1 \cdot 3 \cdot 5 \cdots (2k+1)}{2 \cdot 4 \cdot 6 \cdots 2k} d_{2k+1}, \ k \geq 1.$$

Returning now to the case of a polynomial potential function of the form

$$V_N(u, \lambda) = \frac{1}{2} u^2 + \sum_{i=1}^{N-2} \lambda_{i+2} u^{i+2},$$

we define a function ϕ by the relation

$$\phi(u) = 2 \sum_{i=1}^{N-2} \lambda_{i+2} u^i.$$

Next we need the power series for the analytic solution $u = u(X)$ of the equation

$$X = h(u) = u\sqrt{1 + \phi(u)}$$

for $|X|$ sufficiently small. Most computer algebra systems have a *series reversion procedure* available which can be used to obtain the power series for $u(X)$. For our purposes we can formulate this result as follows.

SERIES REVERSION LEMMA. *Let $f : \mathbf{R} \to \mathbf{R}$ be analytic at 0 with $f(0) = 1$, and let $p \in \mathbf{R}[t]$ have the form*

$$p(t) = p_1 t + p_2 t^2 + \cdots + p_n t^n.$$

Then the function $t \mapsto z := tf(p(t))$, $t \in \mathbf{R}$, and its inverse $z \mapsto t := F(z)$, are analytic at 0. Moreover, the power series expansion for $F(z)$ has the form

$$F(z) = z + F_2 z^2 + F_3 z^3 + \cdots$$

where $F_k \in \mathbf{R}[p_1, p_2, \ldots, p_n]$.

An algorithm for calculating the coefficients F_k is given in [5,6]. One first obtains the coefficients v_k in the expansion

$$tf(p(t)) = t + v_2 t^2 + v_3 t^3 + \cdots,$$

and then the coefficients F_k can be calculated from the

Lagrange–Henrici Series Reversion Algorithm:
Input: n_{max} and v_k, for $k = 2, 3, \ldots, n_{max}$, and $u_0 = 1$.
Output: F_k, for $k = 2, 3, \ldots, n_{max}$.
for $n = 2$ **until** n_{max} **step** 1 **do**
begin
    for $m = 1$ **until** $n - 1$ **step** 1 **do**

$$u_m := \frac{1}{m} \sum_{k=1}^{m} [(1-n)k - m] u_{m-k} v_{k+1};$$

$$F_n := \tfrac{1}{n} u_{n-1};$$
$$\textbf{end;}$$

If we define $f(t) := \sqrt{1+t}$, then we can apply the Lagrange-Henrici Series Reversion Algorithm to calculate the coefficients in the power series expansion of the solution

$$u = u(X) = X + d_2 X^2 + d_3 X^3 + d_4 X^4 + \cdots$$

of

$$X = h(u) = u\sqrt{1 + \phi(u)}.$$

As we pointed out earlier, the period function for this problem has the form

$$P(X, \lambda) = 2\pi + p_2(\lambda) X^2 + p_4(\lambda) X^4 + \cdots,$$

and

$$\mathfrak{m} := (p_2, p_4, p_6, \dots) = (d_3, d_5, d_7, \dots).$$

To simplify the notation we define

$$q_m(\lambda_3, \lambda_4, \dots, \lambda_N) := d_{2m+1}(\lambda_3, \lambda_4, \dots, \lambda_N)$$

for $m \geq 1$. While the ideal $(d_2, d_3, d_4, \dots)$ is very easy to analyze, since it is radical in $\mathbf{R}[\lambda_3, \lambda_4, \dots, \lambda_N]$, the ideal $\mathfrak{m} = (d_3, d_5, d_7, \dots)$ which we must study is generally substantially more difficult to understand. The next lemma will assist us in detecting the structure of the ideal $\mathfrak{m}$.

POTENTIAL REVERSION COEFFICIENT LEMMA [4]. *The following relations are valid:*

$$\frac{\partial q_m}{\partial \lambda_{i+2}} = -\frac{1}{(2m)!} \frac{\partial^{2m}}{\partial u^{2m}} (1 + \phi(u))^{-\frac{2m+3}{2}} u^i \bigg|_{u=0},$$

and in general, if $\lambda = (\lambda_3, \lambda_4, \dots, \lambda_N)$, $\mathbf{k} = (k_3, k_4, k_5, \dots, k_N)$, *and* $|\mathbf{k}| = \sum_{i \geq 3} k_i$, *then*

$$D_\lambda^{\mathbf{k}} q_m := \frac{\partial^{|\mathbf{k}|} q_m}{\partial \lambda_3^{k_3} \partial \lambda_4^{k_4} \cdots \partial \lambda_N^{k_N}}$$

$$= \frac{\mu_{m,|\mathbf{k}|}}{(2m+1)!} \frac{\partial^{2m}}{\partial u^{2m}} (1 + \phi(u))^{-\frac{2m+2|\mathbf{k}|+1}{2}} u^{k_3 + 2k_4 + 3k_5 + \cdots + (N-2)k_N} \bigg|_{u=0},$$

where

$$\mu_{m,|\mathbf{k}|} = (-1)^{|\mathbf{k}|} (2m+1)(2m+3) \cdots (2m + 2|\mathbf{k}| - 1).$$

Moreover,

$$\frac{\partial q_m}{\partial \lambda_{i+2}} \bigg|_{\lambda=0} = -\delta_{i,2m},$$

and for $|\mathbf{k}| \geq 2$ *we have*

$$D_\lambda^{|\mathbf{k}|} q_m \bigg|_{\lambda=0} = \frac{\mu_{m,|\mathbf{k}|}}{2m+1} \delta_{k_3 + 2k_4 + 3k_5 + \cdots + (N-2)k_N, 2m}.$$

Using the Potential Reversion Coefficient Lemma one can prove the

EVEN POTENTIAL THEOREM [4]. *Assume that the polynomial potential function $V(u)$ is even, i.e., let*

$$V(u) = \frac{1}{2}u^2 + \sum_{i=1}^{n-1} \lambda_{2i+2} u^{2i+2}.$$

Then $n-1$ is the smallest integer k such that $\mathfrak{m}_k = (q_1, q_2, \ldots, q_k) = \mathfrak{m}$ over the polynomial ring $\mathbf{R}[\lambda_4, \lambda_6, \ldots, \lambda_{2n}]$. The differential equation $\ddot{u} + g(u, \lambda) = 0$ corresponding to the potential function V has at most $n-2$ local critical periods which bifurcate from the origin. Moreover, there are perturbations with exactly k local critical periods for each $k \leq n-2$.

Application of the Infinite Order Bifurcation Theorem requires the calculation of $K = K(N)$, the smallest positive integer k such that

$$\mathfrak{m}_k := (q_1, q_2, \ldots, q_k) = \mathfrak{m} := (q_1, q_2, q_3, \ldots).$$

For a general polynomial potential function of the form

$$V_N(u, \lambda) = \frac{1}{2}u^2 + \sum_{i=1}^{N-2} \lambda_{i+2} u^{i+2}$$

it is easy to prove with the aid of the Potential Reversion Coefficient Lemma that $K(N)$ must be at least as large as the number of unknowns, i.e.,

$$K(N) \geq N - 2.$$

Unfortunately when $N \geq 6$, this inequality is strict. Obtaining a precise formula for $K(N)$ remains a difficult unsolved problem. However, some rather extensive computer algebra computations suggest the following conjecture:

If $N = deg\, V_N \geq 3$, then the complex variety of the ideal $\mathfrak{m}_{N-2}$ is $\{0\}$, and

$$K(N) = N - 2 + \left[\frac{N-4}{2}\right],$$

where

$$[x] = \begin{cases} \text{largest integer } \leq x \text{ if } x > 0 \\ 0 \text{ otherwise.} \end{cases}$$

For the cases $3 \leq N \leq 6$ the following theorem gives a full account of the possibilities.

THEOREM [4]. *The differential equation $\ddot{u} + g(u, \lambda) = 0$ corresponding to the potential V_6 can have at most four critical periods bifurcating from the origin. There are perturbations which will produce k critical periods for $k \leq 3$. For the potential V_5 at most two critical periods bifurcate from the origin, and there are* perturbations with k critical periods for $k \leq 2$. Finally, for the potential V_4 at most one critical period bifurcates from the origin, and there are perturbations with one critical period.

While we are not yet able to prove the general conjecture, we do have the tools to analyze the ideal membership problem for any particular N. We illustrate the procedure for calculating $K(N)$ for the cases $N = 5$ and $N = 6$ which are mentioned in the preceding Theorem. Consider the four parameter potential function

$$V_6(u) := \frac{1}{2}u^2 + \lambda_3 u^3 + \lambda_4 u^4 + \lambda_5 u^5 + \lambda_6 u^6.$$

Since $K(6) \geq 4$, we use series reversion to calculate d_3, d_5, d_7, and d_9, or equivalently, q_1, q_2, q_3, and q_4 to obtain

$$q_1(\lambda_3, \lambda_4, \lambda_5, \lambda_6) = \frac{5}{2}\lambda_3^2 - \lambda_4,$$

$$q_2(\lambda_3, \lambda_4, \lambda_5, \lambda_6) = \frac{231}{8}\lambda_3^4 - \frac{63}{2}\lambda_3^2\lambda_4 + 7\lambda_3\lambda_5 + \frac{7}{2}\lambda_4^2 - \lambda_6,$$

$$q_3(\lambda_3, \lambda_4, \lambda_5, \lambda_6) = \frac{9}{2}\lambda_5^2 - 99\lambda_3\lambda_4\lambda_5 + \frac{429}{2}\lambda_3^3\lambda_5 - \frac{6435}{8}\lambda_3^4\lambda_4 + \frac{1287}{4}\lambda_3^2\lambda_4^2 + \frac{7293}{16}\lambda_3^6$$
$$- \frac{33}{2}\lambda_4^3 - \frac{99}{2}\lambda_3^2\lambda_6 + 9\lambda_4\lambda_6$$

$$q_4(\lambda_3, \lambda_4, \lambda_5, \lambda_6) = \frac{1062347}{128}\lambda_3^8 - \frac{323323}{16}\lambda_3^6\lambda_4 + \frac{46189}{8}\lambda_3^5\lambda_5 + \frac{230945}{16}\lambda_3^4\lambda_4^2$$
$$- \frac{12155}{8}\lambda_3^4\lambda_6 - \frac{12155}{2}\lambda_3^3\lambda_4\lambda_5 - \frac{12155}{4}\lambda_3^2\lambda_4^3$$
$$+ \frac{2145}{2}\lambda_3^2\lambda_4\lambda_6 + \frac{2145}{4}\lambda_3^2\lambda_5^2 + \frac{2145}{2}\lambda_3\lambda_4^2\lambda_5 - 143\lambda_3\lambda_5\lambda_6$$
$$+ \frac{715}{8}\lambda_4^4 - \frac{143}{2}\lambda_4^2\lambda_6 - \frac{143}{2}\lambda_4\lambda_5^2 + \frac{11}{2}\lambda_6^2.$$

In order to prove that $K(6) = 5$ we will show that $q_j \in \mathfrak{m}_4 := (q_1, q_2, q_3, q_4)$ if $j \geq 6$, and that $q_5 \notin \mathfrak{m}_4$. Before proving this we first observe (for any $N \geq 3$) that the Potential Reversion Coefficient Lemma implies that the polynomials $q_j(\lambda_3, \ldots, \lambda_N)$ are weighted homogeneous polynomials [2] of degree $2j$ with weighting pattern $(1, 2, 3, \ldots, N - 2)$. To assist in the analysis of the ideal $\mathfrak{m} := (q_1, q_2, q_3, \ldots)$ we introduce a weight preserving polyomorphism

$$\lambda_3 = L_1, \quad \lambda_4 = \frac{5}{2}L_1^2 - L_2,$$

$$\lambda_5 = L_3, \quad \lambda_6 = -28L_1^4 + 14L_1^2 L_2 + 7L_1 L_3 + \frac{7}{2}L_2^2 - L_4$$

with inverse

$$L_1 = \lambda_3, \quad L_2 = q_1(\lambda_3, \lambda_4, \lambda_5, \lambda_6), \quad L_3 = \lambda_5, \quad L_4 = q_2(\lambda_3, \lambda_4, \lambda_5, \lambda_6).$$

This transformation maps the weighted homogeneous polynomials $q_j(\lambda_3, \lambda_4, \lambda_5, \lambda_4)$ of degree $2j$ into weighted homogeneous polynomials in (L_1, L_2, L_3, L_4) of degree

$2j$ with weight pattern $(1,2,3,4)$. Using the same notation for the polynomials q_j in the new coordinates we have

$$q_1(L_1, L_2, L_3, L_4) = L_2,$$

$$q_2(L_1, L_2, L_3, L_4) = L_4,$$

$$q_3(L_1, L_2, L_3, L_4) = 36L_1L_2L_3 + 954L_1^6 - 222L_1^3L_3 + \frac{9}{2}L_3^2 - 621L_1^4L_2 - \frac{45}{2}L_1^2L_2^2$$
$$- 15L_2^3 + 27L_1^2L_4 + 9L_2L_4$$

$$q_4(L_1, L_2, L_3, L_4) = 66L_1L_3L_4 + 341L_1L_2^2L_3 + 561L_1^2L_2L_4 - 5214L_1^3L_2L_3 - 11704L_1^8$$
$$+ 4136L_1^5L_3 - 374L_1^2L_3^2 + 25146L_1^6L_2 - \frac{21021}{2}L_1^4L_2^2$$
$$- \frac{1639}{2}L_1^2L_2^3 + \frac{143}{2}L_2L_3^2 - \frac{187}{2}L_2^4 - 407L_1^4L_4 + 33L_2^2L_4 + \frac{11}{2}L_4^2.$$

It is now easy to verify that the complex variety of the ideal $\mathfrak{m}_4$ is $\{0\}$. Thus by the Hilbert Nullstellensatz [9] there is a smallest positive integer ν such that every monomial $L_1^{i_1}L_2^{i_2}L_3^{i_3}L_4^{i_4}$ with $i_1 + 2i_2 + 3i_3 + 4i_4 \geq \nu$ is in $\mathfrak{m}_4$. This observation will guide our proof that $K(6) = 5$. Each of the polynomials q_j in the new indeterminants L_1, L_2, L_3, L_4 is a linear combination of monomials $v = L_1^{i_1}L_2^{i_2}L_3^{i_3}L_4^{i_4}$ with $i_1 + 2i_2 + 3i_3 + 4i_4 = 2j$. We will show that every such monomial v is in $\mathfrak{m}_4$ if $j \geq 6$, and this will be sufficient to prove that $q_j \in \mathfrak{m}_4$ for $j \geq 6$. We define $q_{j0}(L_1, L_3) := q_j(L_1, 0, L_3, 0)$ for $j = 1, 2, 3, \ldots$, and let $\mathfrak{m}_{40}$ be the ideal

$$\mathfrak{m}_{40} := (q_{30}, q_{40}) = (954L_1^6 - 222L_1^3L_3 + \frac{9}{2}L_3^2, 4136L_1^5L_3 - 11704L_1^8 - 374L_1^2L_3^2).$$

over $\mathbf{R}[L_1, L_3]$. Since L_2 and L_4 are in $\mathfrak{m}_4$, it follows that for $j \geq 5$ we have $q_j \in \mathfrak{m}_4$ over $\mathbf{R}[L_1, L_2, L_3, L_4]$ if and only if $q_{j0} \in \mathfrak{m}_{40}$ over $\mathbf{R}[L_1, L_3]$. This observation considerably reduces the amount of work necessary to analyze the ideal $\mathfrak{m}_4$. The following identities can be verified

$$1420L_1^3L_3^2 - 957L_3^3 = (\frac{7049}{4}L_1^3 - \frac{638}{3}L_3)q_{30}(L_1, L_3) + \frac{25281}{176}L_1q_{40}(L_1, L_3),$$

$$319L_1^5L_3 - 72L_1^2L_3^2 = \frac{133}{48}L_1^2q_{30}(L_1, L_3) + \frac{159}{704}q_{40}(L_1, L_3),$$

$$L_1^2L_3^3 = (\frac{5417}{15324}L_1^2L_3 - \frac{42427}{15324}L_1^5)q_{30}(L_1, L_3) + (\frac{355}{224752}L_3 - \frac{4611}{20432}L_1^3)q_{40}(L_1, L_3),$$

$$L_3^4 = (\frac{2}{9}L_3^2 - \frac{60541}{45972}L_1^3L_3 - \frac{47215}{11493}L_1^6)q_{30}(L_1, L_3) - (\frac{99623}{674256}L_1L_3 + \frac{18815}{56188}L_1^4)q_{40}(L_1, L_3),$$

$$L_1^{11} = (\frac{1037}{245184}L_1^2L_3 - \frac{1643}{61296}L_1^5)q_{30}(L_1, L_3) + (\frac{183}{3596032}L_3 - \frac{2041}{899008}L_1^3)q_{40}(L_1, L_3).$$

From these relationships one can also deduce that $L_1^8L_3$ and $L_1^5L_3^2$ are both in $\mathfrak{m}_{40}$. Therefore every monomial $v_0 = L_1^{i_1}L_3^{i_3}$ which has weighted degree $i_1 + 3i_3 = 11$ belongs to the ideal $\mathfrak{m}_{40}$. Since every monomial v_0 of weighted degree $2j \geq 12$ is either a multiple of a monomial of weighted degree 11 or a multiple of L_3^4 which has weighted degree 12, it follows that each such monomial $v_0 \in \mathfrak{m}_{40}$. Now the q_{j0}

are weighted homogeneous polynomials of degree $2j$ with weight pattern $(1,3)$ in (L_1, L_3). Consequently $q_{j0} \in \mathfrak{m}_{40}$ if $j \geq 6$. Next we show that $q_{50} \notin \mathfrak{m}_{40}$, and this will complete the proof that $K(6) = 5$. We calculate a Gröbner basis [3], $\mathfrak{G}$, for $\mathfrak{m}_{40}$ using the lexicographic ordering $L_3 \prec L_1$,

$$\mathfrak{G} = \left\{ 636L_1^6 - 148L_1^3L_3 + 3L_3^2, 319L_1^5L_3 - 72L_1^2L_3^2, 957L_3^3 - 1420L_1^3L_3^2, -L_1^2L_3^3, -L_3^4 \right\}$$

In the original coordinates $(\lambda_3, \lambda_4, \lambda_5, \lambda_6)$ the polynomial q_5 is

$$
\begin{aligned}
q_5(\lambda_3, \lambda_4, \lambda_5, \lambda_6) = &\frac{2414425}{16}\lambda_3^7\lambda_5 - \frac{65189475}{128}\lambda_3^8\lambda_4 - \frac{676039}{16}\lambda_3^6\lambda_6 \\
&+ \frac{16900975}{32}\lambda_3^6\lambda_4^2 - \frac{3380195}{16}\lambda_3^4\lambda_4^3 + \frac{440895}{16}\lambda_3^4\lambda_5^2 \\
&+ \frac{42010995}{256}\lambda_3^{10} - \frac{4199}{8}\lambda_4^5 + \frac{440895}{8}\lambda_3^4\lambda_4\lambda_6 \\
&- \frac{2028117}{8}\lambda_3^5\lambda_4\lambda_5 + \frac{440895}{16}\lambda_3^2\lambda_4^4 - \frac{20995}{2}\lambda_3^3\lambda_5\lambda_6 \\
&+ \frac{3315}{4}\lambda_3^2\lambda_6^2 + \frac{440895}{4}\lambda_3^3\lambda_4^2\lambda_5 + \frac{1105}{2}\lambda_3\lambda_5^3 + \frac{1105}{2}\lambda_4^3\lambda_6 \\
&- \frac{62985}{4}\lambda_3^2\lambda_4^2\lambda_6 - \frac{62985}{4}\lambda_3^2\lambda_4\lambda_5^2 - \frac{195}{2}\lambda_5^2\lambda_6 \\
&- \frac{20995}{2}\lambda_3\lambda_4^3\lambda_5 + 3315\lambda_3\lambda_4\lambda_5\lambda_6 + \frac{3315}{4}\lambda_4^2\lambda_5^2 - \frac{195}{2}\lambda_4\lambda_6^2.
\end{aligned}
$$

If we transform this polynomial using the weight preserving polyomorphism constructed above, and then put $L_2 \equiv 0 \equiv L_4$, then we see that

$$q_{50}(L_1, L_3) = 213772L_1^{10} - 130L_1L_3^3 + 9295L_1^4L_3^2 - 85228L_1^7L_3.$$

Then the normal form algorithm in [3] can be used to show that q_{50} is not reduced to 0 relative to the Gröbner basis $\mathfrak{G}$. Hence $q_{50} \notin \mathfrak{m}_{40}$.

Next we show that $K(5) = 3$. In this case the potential function has the form

$$V_5(u, \lambda_3, \lambda_4, \lambda_5) = \frac{1}{2}u^2 + \lambda_3 u^3 + \lambda_4 u^4 + \lambda_5 u^5.$$

Since $K(5) \geq 3$, we again use series reversion to calculate d_3, d_5, and d_7 or equivalently q_1, q_2, and q_3 and obtain

$$q_1(\lambda_3, \lambda_4, \lambda_5) = \frac{5}{2}\lambda_3^2 - \lambda_4,$$

$$q_2(\lambda_3, \lambda_4, \lambda_5) = \frac{231}{8}\lambda_3^4 - \frac{63}{2}\lambda_3^2\lambda_4 + 7\lambda_3\lambda_5 + \frac{7}{2}\lambda_4^2,$$

$$q_3(\lambda_3, \lambda_4, \lambda_5) = \frac{9}{2}\lambda_5^2 - 99\lambda_3\lambda_4\lambda_5 + \frac{429}{2}\lambda_3^3\lambda_5 - \frac{6435}{8}\lambda_3^4\lambda_4 + \frac{1287}{4}\lambda_3^2\lambda_4^2 + \frac{7293}{16}\lambda_3^6 - \frac{33}{2}\lambda_4^3.$$

We will show that $q_j \in \mathfrak{m}_3$ for $j = 1, 2, 3, \ldots$. Since $K(5) \geq 3$, this will show $K(5) = 3$. Again we introduce a weight preserving polyomorphism

$$\lambda_3 = L_1, \quad \lambda_4 = \frac{5}{2}L_1^2 - L_2, \quad \lambda_5 = L_3$$

with inverse

$$L_1 = \lambda_3, \quad L_2 = \frac{5}{2}\lambda_3^2 - \lambda_4, \quad L_3 = \lambda_5.$$

This transformation maps the weighted homogeneous polynomials $q_j(\lambda_3, \lambda_4, \lambda_5)$ of degree $2j$ into weighted homogeneous polynomials in (L_1, L_2, L_3) of degree $2j$ with weight pattern $(1, 2, 3)$. Using the same notation for the polynomials q_j in the new coordinates we have

$$q_1(L_1, L_2, L_3) = L_2,$$

$$q_2(L_1, L_2, L_3) = 7L_1L_3 - 28L_1^4 + 14L_2L_1^2 + \frac{7}{2}L_2^2,$$

$$q_3(L_1, L_2, L_3) = \frac{9}{2}L_3^2 - 33L_1^3L_3 + 99L_2L_1L_3 + 198L_1^6 - 495L_2L_1^4 + 198L_2^2L_1^2 + \frac{33}{2}L_2^3.$$

As in the case $N = 6$ we consider only the polynomials q_{j0}, and we show that

$$q_{j0} \in \mathfrak{m}_{30} := (q_{20}, q_{30}) = (7L_1L_3 - 28L_1^4, \frac{9}{2}L_3^2 - 33L_1^3L_3 + 198L_1^6), \quad j = 2, 3, 4, \ldots$$

over $\mathsf{R}[L_1, L_3]$. The following identities can be verified

$$3L_3^2 + 11L_1^3L_3 = \frac{33}{7}L_1^2 q_{20}(L_1, L_3) + \frac{2}{3}q_{30}(L_1, L_3),$$

$$L_1L_3^2 = (\frac{132}{161}L_1^3 + \frac{11}{161}L_3)q_{20}(L_1, L_3) + \frac{8}{69}L_1 q_{30}(L_1, L_3),$$

$$L_3^3 = (\frac{638}{483}L_1^2L_3 - \frac{484}{161}L_1^5)q_{20}(L_1, L_3) + (\frac{2}{9}L_3 - \frac{88}{207}L_1^3)q_{30}(L_1, L_3).$$

It follows easily from these observations that L_1^7 and $L_1^4L_3$ are also in $\mathfrak{m}_{30}$. Consequently every monomial $v_0 = L_1^{i_1}L_3^{i_3}$ with weighted degree $i_1 + 3i_3 = 7$ belongs to $\mathfrak{m}_{30}$. Since $L_3^3 \in \mathfrak{m}_{30}$ it follows that every monomial in the two variables L_1, L_3 of weighted degree $2j \geq 8$ also belongs to $\mathfrak{m}_{30}$, and we can conclude that $K(5) = 3$.

The case $N = 6$ gives the polynomial potential of smallest degree where there is a "jump" in $K(N)$, i.e. where $K(N) > N - 2$. It is perhaps worth noting that this "jump" cannot be removed by passing to the larger ring of convergent power series as was done in the remark following the Quadratic Isochrone Bifurcation Theorem. It is also easy to prove that $K(3) = 1$ and that $K(4) = 2$ so that the conjectured formula for $K(N)$ is valid for $3 \leq N \leq 6$. Moreover, Gröbner basis computer calculations prove that the conjectured formula for $K(N)$, viz.,

$$K_C(N) := N - 2 + \left[\frac{N - 4}{2}\right],$$

is a lower bound for $K(N)$ and that $q_{K_C(N)+1} \in \mathfrak{m}_{K_C(N)}$ for $7 \leq N \leq 11$, but we have not shown that $K(N)$ is actually equal to $K_C(N)$ for these N (note that our use of the greatest integer function requires $[x] = 0$ when $x \leq 0$). Also one can show that the complex variety of $\mathfrak{m}_k$ is not $\{0\}$ when $k < N - 2$ so that $K(N) \geq N - 2$ for all $N \geq 3$. Part of our conjecture is that for the general polynomial potential V_N

of degree N (of the form we have been considering) the complex variety of $\mathfrak{m}_{N-2}$ is $\{0\}$. This conjecture has been proved for $3 \leq N \leq 11$, but as yet we are lacking a proof for the general case. It is also worth noting that from the Potential Reversion Coefficient Lemma one can prove that there are weight preserving polyomorphisms $L \longleftrightarrow \lambda$ such that the indeterminants of odd weight are unchanged, i.e., $L_1 = \lambda_3$, $L_3 = \lambda_5$, etc. and such that the indeterminants of even weight $L_2, L_4, \ldots$ are all in the ideal $\mathfrak{m}_{N-2}$. If such a transformation is made, as in the cases $N = 5$ and $N = 6$ above, then the ideals $\mathfrak{m}_{N-2}$ and $\mathfrak{m}$ over $\mathbf{R}[\lambda_3, \lambda_4, \ldots, \lambda_{N-2}]$ can be understood by studying the ideal $\mathfrak{m}_{N-2,0}$ over $\mathbf{R}[L_1, L_3, \ldots, L_{2\nu+1}]$ and the polynomials $q_{j0}(L_1, L_3, \ldots, L_{2\nu+1})$ where $2\nu + 1$ is the largest odd integer not exceeding $N - 2$.

REFERENCES

[1]   N. N. BAUTIN, *On the number of limit cycles which appear with the variation of coefficients from an equilibrium position of focus or center type*, Amer. Math. Soc. Transl., 100 (1954), pp. 1–19.

[2]   E. BRIESKORN AND H. KNÖRRER, *Plane Algebraic Curves*, J. Stillwell translator, Birkhäuser Verlag, Boston, 1986.

[3]   B. BUCHBERGER, *Gröbner bases: An algorithmic method in polynomial ideal theory*, in *Multidimensional Systems Theory*, N. K. Bose, Ed., D. Reidel, Boston, 1985.

[4]   C. CHICONE AND MARC JACOBS, *Bifurcation of critical periods for plane vector fields*, Trans. Amer. Math. Soc., 312 (1989), pp. 433–486.

[5]   P. HENRICI, *Applied and Computational Complex Analysis*, Wiley-Interscience, New York, 1974.

[6]   D. KNUTH, *The Art of Computer Programming*, Addison–Wesley, Reading, 1981.

[7]   W. S. LOUD, *Behavior of the period of solutions of certain plane autonomous systems near centers*, Contributions to Differential Equations, 3 (1964), pp. 21–36.

[8]   B. L. VAN DER WAERDEN, *Algebra*, F. Ungar, New York, 1950.

[9]   B. L. VAN DER WAERDEN, *Algebra*, F. Ungar, New York, 1970.

MACSYMA PROGRAM TO IMPLEMENT AVERAGING USING ELLIPTIC FUNCTIONS

VINCENT T. COPPOLA AND RICHARD H. RAND*

Abstract. The method of averaging is applied to the system:

$$x'' + \alpha(\tau)\, x + \beta(\tau)\, x^3 + \epsilon\, g(x, x', \tau) = 0$$

where $\tau = \epsilon t$ is slow time, and where $\epsilon \ll 1$. This involves the laborous manipulation of Jacobian elliptic functions, a process which is most easily and accurately accomplished using computer algebra. We present the listing of a MACSYMA program which implements the method to $0(\epsilon)$, as well as the results of a run for which $g(x, x', \tau)$ is taken as a general cubic polynomial in x and x'.

Introduction. In this paper we treat a class of problems which involve perturbing off of the Jacobian elliptic function solution of the strongly nonlinear oscillator

$$(1) \qquad\qquad x'' + \alpha\, x + \beta\, x^3 = 0$$

In particular, we investigate the following nonautonomous perturbation of eq. (1):

$$(2) \qquad\qquad x'' + \alpha(\tau)\, x + \beta(\tau)\, x^3 + \epsilon\, g(x, x', \tau) = 0$$

where τ represents "slow time",

$$(3) \qquad\qquad \tau = \epsilon\, t, \quad \epsilon \ll 1$$

and where primes represent differentiation with respect to t. The functions $\alpha(\tau)$, $\beta(\tau)$, and $g(x, x', \tau)$ must be specified for a particular problem but are otherwise arbitrary.

We shall use the method of averaging implemented on the computer algebra system MACSYMA to obtain approximate equations governing the solutions to eq. (2).

Although the method of averaging has been treated in numerous references (e.g. [13-15, 17-19, 21-23]), most treatments deal almost exclusively with perturbations off of the simple harmonic oscillator. A few authors have considered perturbations

---

*Department of Theoretical and Applied Mechanics, Cornell University, Ithaca, NY 14853

off of nonlinear systems using elliptic functions. Kuzmak [16] looks for periodic solutions in eq. (1) using a multiple scale method, where α and β are slowly varying parameters. Garcia-Margallo and Bejarano [9] find limit cycles in a generalized van der Pol oscillator using generalized harmonic balance. Davis [8] investigates second order ordinary differential equations using elliptic functions. Cap [2] applies the method of averaging to perturbations of the mathematical pendulum. Pocobelli [20] studies the slowly varying pendulum. Greenspan and Holmes [12] and Guckenheimer and Holmes [13] apply the Melnikov method to perturbations of eq. (1) where $\alpha < 0$. Nayfeh [18], Kevorkian & Cole [15] and Sanders & Verhulst [22] also treat such problems.

Perturbations of eq. (1) with $\alpha = 0$ (a purely nonlinear oscillator) have appeared in the literature. Chirikov [3] studies resonance overlap under multiple harmonic excitations. Garcia-Margallo and Bejarano [10] employ generalized harmonic balance in order to approximate limit cycles. Yuste and Bejarano [24] use first order averaging as a means to find transitory behavior as the motion is attracted to a limit cycle.

In most of these references the authors have obtained general expressions for approximate equations of motion in terms of integrals. Although the problem is thus reduced to the evaluation of these integrals, little use has been made of these treatments because the evaluations require complicated algebraic manipulations of Jacobian elliptic functions. By using the computer algebra system MACSYMA, we have been able to efficiently evaluate the associated integrals.

Variation of Parameters. In preparation for the method of averaging, we use the method of variation of parameters to express the influence of the slowly varying and order ϵ terms in eq. (2) on the solution of the unperturbed eq. (1). In contrast to the method of averaging, the computations presented in this section are exact. However, the results are intractable and unenlightening. The method of averaging (introduced in the next section) replaces the results obtained in this section by more useful equations, which are, however, approximate (valid in the small ϵ limit.)

The unperturbed solution (the general solution to eq. (1) for τ fixed and $\epsilon = 0$) is expressed in terms of Jacobian elliptic functions. A brief summary of elliptic functions is given in Appendix A, which also gives the notation we use for elliptic functions and elliptic integrals. We are interested in systems (2) in which the unperturbed system (1) is oscillatory. It can be shown [4] that the unperturbed solution for such systems can be expressed as:

(3.1) $$x = \mu\, r\, cn(u, k)$$

(3.2) $$x' = -\mu\, r\, a\, sn(u, k)\, dn(u, k)$$

where

$$k^2 = \frac{\beta\, r^2}{2\, a^2}, \quad a^2 = \beta\, r^2 + \alpha, \quad u = a\, t + u_o$$

$$r \geq 0,\ a \geq 0,\ \mu = \pm 1$$

In eqs. (3), r and u_o are arbitrary constants of integration. As usual in the method of variation of parameters, we look for a solution to eq. (2) in the form of eqs. (3), where the two arbitrary constants r and u_o are allowed to vary in time. This results in first order differential equations on $r(t)$ and $u_o(t)$. Unfortunately the resulting expression for $du_o(t)/dt$ is not periodic, and is thus unsuitable for averaging. This may be remedied by replacing u_o by φ, the angle variable of action-angle variables [11], where

(4)
$$u = 4\, K(k)\, \varphi$$

The proportionality factor $4K(k)$ in eq. (4) takes into account the dependence of frequency on amplitude in eq. (1), thus eliminating "phase shear". The resulting equations on r and φ then become:

(5.1) $$\frac{dr}{dt} = -\epsilon\, g\, \frac{sn\ dn}{\sqrt{\alpha + \beta\, r^2}} - \epsilon\frac{d\alpha}{d\tau}\frac{r\,(1 - cn^2)}{2\,(\alpha + \beta r^2)} - \epsilon\frac{d\beta}{d\tau}\frac{r^3\,(1 - cn^4)}{4\,(\alpha + \beta\, r^2)}$$

(5.2) $$\frac{d\varphi}{dt} = \frac{\sqrt{\alpha + \beta\, r^2}}{4\, K}$$

$$+ \epsilon\, g\ \frac{2\,\alpha\,(\alpha + \beta\, r^2)\,(Z\ sn\ dn + cn) + \alpha\,\beta\, r^2\, cn^3 + \beta^2\, r^4\, cn}{4\, K\, r\,(\alpha + \beta\, r^2)^{3/2}\,(2\,\alpha + \beta\, r^2)}$$

$$+ \epsilon\ \frac{d\alpha}{d\tau}\ \frac{\alpha\, Z\, cn^2 - (2\,\alpha + \beta\, r^2)\, Z - \alpha\, cn\ sn\ dn}{4\, K\,(\alpha + \beta\, r^2)\,(2\,\alpha + \beta\, r^2)}$$

$$+ \epsilon\ \frac{d\beta}{d\tau}\ \frac{[\alpha\,\beta\, r^2(Z\, cn^4 + Z - cn\ sn\ dn\,(2 + cn^2)) + 2\,\alpha^2 Z - \beta^2 r^4 cn\ sn\ dn]}{8\, K\,\beta\,(\alpha + \beta\, r^2)\,(2\alpha + \beta\, r^2)}$$

where $cn = cn(4\, K\, \varphi, k)$, $sn = sn(4\, K\, \varphi, k)$, $dn = dn\,(4\, K\, \varphi, k)$, and $Z = Z(4\, K\, \varphi, k)$. Eqs. (5) are periodic in φ and are thus in the proper form for averaging.

The Method of Averaging. In order to explain the method of averaging, we write eqs. (5) in the abbreviated form:

$$(6.1) \qquad\qquad r' = \epsilon\, F_1(r, \varphi)$$

$$(6.2) \qquad\qquad \varphi' = \Omega(r) + \epsilon\, F_2(r, \varphi)$$

The method is based on positing a near-identity transformation from (r, φ) to $(\bar{r}, \bar{\varphi})$:

$$(7.1) \qquad\qquad r = \bar{r} + \epsilon w_1(\bar{r}, \varphi) + \epsilon^2\, v_1(\bar{r}, \varphi) + O(\epsilon^3)$$

$$(7.2) \qquad\qquad \varphi = \bar{\varphi} + \epsilon\, w_2(\bar{r}, \varphi) + O(\epsilon^2)$$

where the generating functions w_1, v_1, and w_2 are to be chosen so as to simplify the resulting equations. Substituting eqs. (7) into eqs. (6) and co llecting terms gives equations of the form:

$$(8.1) \quad \bar{r}' = \epsilon\left[F_1(\bar{r}, \varphi) - \Omega(\bar{r})\frac{\partial w_1}{\partial \varphi} \right] + \epsilon^2 \left[H_1(\bar{r}, \varphi) - \Omega(\bar{r})\frac{\partial v_1}{\partial \varphi} \right] + O(\epsilon^3)$$

$$(8.2) \quad \bar{\varphi}' = \Omega(\bar{r}) + \epsilon\left[F_2(\bar{r}, \varphi) + \frac{d\Omega}{dr}(\bar{r}) - \Omega(\bar{r})\frac{\partial w_2}{\partial \varphi} \right] + O(\epsilon^2)$$

The transformation functions w_1, w_2, and v_1 are chosen so that eqs. (8) are in averaged form, i.e.,

$$(9.1) \qquad\qquad \bar{r}' = \epsilon\, \bar{F}_1(\bar{r}) + \epsilon^2\, \bar{H}_1(\bar{r}) + O(\epsilon^3)$$

$$(9.2) \qquad\qquad \bar{\varphi}' = \Omega(\bar{r}) + \epsilon\, \bar{F}_2(\bar{r}) + O(\epsilon^2)$$

where $\bar{F}_1, \bar{F}_2$, and $\bar{H}_1$ are the means of F_1, F_2, and H_1 taken over one period in the periodic variable φ.

Computer Algebra. We have written a computer algebra (MACSYMA) program that implements the averaging procedure (6)-(9) for eqs. (5). The user first inputs expressions for α, β and g, which may contain symbolic parameters. The computer then generates F_1, F_2, and Ω from eqs. (5) (where eqs. (3) are substituted for x and x' in g). Using an elliptic function integration subroutine that we developed, the program finds $\bar{F}_1, \bar{F}_2, w_1$, and w_2 (which completes the first order averaging).

The listing for this first order averaging program is given in Appendix B, and a sample run is given in Appendix C.

This program was then extended to include second order averaging of the r equation. We found it essential to proceed in several steps in order to prevent excessive intermediate expression swell. First, H_1 is computed and its terms are divided up among several pre-identified categories. These categories group together terms whose means are computed in like manners. The mean of H_1 is then computed category by category. After this second order averaging is completed, the program outputs the averaged system and the first order transformation.

The second order averaging program consists of 460 lines of code. Typical runs on a Symbolic 3670 computer take from one to six hours. For $\alpha = \beta = 1$ and g consisting of three ter ms ($x', x\,x'^2$, x'^3), there are 497 second order terms to be averaged. For more information on the program, see [4 − 7] which contains many applications. Electronic copies of both programs are available from the authors via BITNET. Our current BITNET addresses are:

$DUGY@CRNLVAX5$ (for VTC), $RHRY@CRNLVAX5$ (for RHR)

Results.

In order to illustrate the use of the program, we present the results of applying it to eq. (2) when $g(x, x', \tau)$ takes the general form:

$$(10) \quad g(x,x',t) = a_{00}(\tau) + a_{10}(\tau)\,x + a_{01}(\tau)\,x' + a_{20}(\tau)\,x^2 + a_{11}(\tau)\,x\,x'$$
$$+ a_{02}(\tau)\,x'^2 + a_{30}(\tau)\,x^3 + a_{21}(\tau)\,x^2\,x' + a_{12}(\tau)\,x\,x'^2$$
$$+ a_{03}(\tau)\,x'^3$$

The averaged eqs. (9) become, to lowest order in ϵ (dropping the bars on r and φ for convenience):

$$(11.1) \quad r' = \epsilon\,\frac{d\alpha}{d\tau}\,\frac{1}{\beta\,r}\,(\frac{E}{K} - 1) - \epsilon\,\frac{d\beta}{d\tau}\,\frac{1}{6\,\beta^2\,r}\,\left[\beta\,r^2 + 4\,\alpha\,(\frac{E}{K} - 1)\right]$$

$$+ \epsilon\, a_{01}(\tau)\, \frac{1}{3\,\beta\, r}\, \left[2\,\alpha\, (\frac{E}{K} - 1) - \beta\, r^2\right]$$

$$- \epsilon\, a_{11}(\tau)\, \frac{\sqrt{2}\, \pi\, H(k^2 - 1)}{8\, K\, r}\, \left[\frac{\alpha + \beta\, r^2}{\beta}\right]^{3/2}$$

$$+ \epsilon\, a_{21}(\tau)\, \left[\frac{8\,\alpha^2}{15\,\beta^2\, r}\, (1 - \frac{E}{K}) + \frac{2\,\alpha\, r}{15\,\beta}\, (5 - 6\,\frac{E}{K}) + \frac{r^3}{5}\, (1 - 2\,\frac{E}{K})\right]$$

$$+ \epsilon\, a_{03}(\tau)\, \left[\frac{8\,\alpha^3}{35\,\beta^2\, r}\, (\frac{E}{K} - 1) + \frac{2}{35}\,\frac{\alpha^2}{\beta}\, r\, (16\,\frac{E}{K} - 15) + \frac{\alpha\, r^3}{35}\, (16\,\frac{E}{K} - 23) - \frac{\beta}{7}\, r^5\right]$$

$$(11.2) \qquad \varphi' = \frac{\sqrt{\alpha + \beta\, r^2}}{4\, K} + O(\epsilon)$$

where $H(k^2 - 1)$ is the Heaviside step function with argument $k^2 - 1$, i.e., $H(k^2 - 1) = 1$ when the system point has $k^2 > 1$ and $H(k^2 - 1) = 0$ otherwise. In (11.1), K and E respectively represent complete elliptic integral s of the first and second kinds.

The generality of the foregoing example may obviate the need for the pro gram in many cases. Consider, for example, the system

$$(12) \qquad x'' + x + x^3 + 0.035\, x' - 0.6\, x^2 x' + 0.1\, x'^3 = 0$$

This system may be cast in the form (2) by choosing $\epsilon = 0.1$ and

$$(13) \qquad \alpha(\tau) = 1,\ \beta(\tau) = 1,\ g(x, x', \tau) = x'\, (0.35 - 6\, x^2 + x'^2)$$

Then we find from eq. (11.1)

$$(14)\ r' = \epsilon\, \left[\frac{769}{210}\, \frac{1}{r}\, (\frac{E}{K} - 1) + \frac{r}{420}\, (2400\,\frac{E}{K} - 2089) + \frac{r^3}{7}\, (20\,\frac{E}{K} - 13) - \frac{r^5}{7}\right]$$

Numerical evaluation of the right hand side of eq. (14) shows that it has two zeros at about $r = r_1 \simeq 1.12675$ and $r = r_2 \simeq 0.83984$, and thus that the original system (12) is predicted to have two limit cycles of approxim ate amplitudes r_1 and r_2. This result is in agreement with numerical integration of eq. (12).

It is interesting to note that the usual approach to averaging (based on perturbations off of the simple harmonic oscillator) fails to give correct predictions for this examp le. In order to use such an approach, we first write the example (12) in the form:

$$(15) \qquad x'' + x + \epsilon (\lambda x^3 + \delta x' + \rho x^2 x' + \eta x'^3) = 0$$

such that

$$(16) \qquad \epsilon\lambda = 1, \ \epsilon\delta = 0.035, \ \epsilon\rho = -06, \ \epsilon\eta = 0.1$$

Eq. (15) may be averaged using previously published computer algebra p rograms in [21]. This involves assuming a solution of the form

$$(17) \qquad x = r(t) \cos [t + \theta(t)]$$

and results in the following slow flow for $r(t)$, valid to $0(\epsilon^2)$:

$$(18) \qquad r' = -\epsilon \frac{r}{8} \left[4\delta + (\rho + 3\eta)r^2 \right] + \epsilon^2 \frac{r^5}{32} \left[\lambda (\rho - 3\eta) \right] + O(\epsilon^3)$$

If we keep only $0(\epsilon)$ terms in (18), and use (16), we obtain

$$(19) \qquad r' = -\frac{r}{8} (0.14 - 0.3 \, r^2)$$

which results in the incorrect prediction of only one limit cycle at $r \simeq 0.68$. If, on the other hand, we keep both $0(\epsilon)$ and $0(\epsilon^2)$ terms in (18), we obtain

$$(20) \qquad r' = -\frac{r}{.8} (0.14 - 0.3 \, r^2 + 0.225 \, r^4)$$

the right hand side of which has no real roots. Thus both $0(\epsilon)$ and $0(\epsilon^2)$ approximations based on trigonometric averaging fail to give qualitatively correct limit cycle results for example (12).

Appendix A: Jacobian Elliptic Functions

Jacobian elliptic functions involve a collection of identities which are similar to those for trigonometric functions but are more complicated algebraically. The use of computer algebra makes manipulation of these identities easier, permitting investigations to proceed on problems which were previously avoided because of the large quantities of algebra involved. All formulas and conventions concerning Jacobian elliptic functions in this paper are taken from Byrd and Friedman's Handbook of Elliptic Integrals for Engineers and Physicists [1].

For the reader's convenience, we offer a brief comparison of elliptic functions with the more familiar trigonometric functions. Corresponding to sin(u) and cos(u) are three fundamental elliptic functions sn(u,k), cn(u,k), and dn(u,k). Each of the elliptic functions depends on the modulus k as well as the argument u. These reduce to sin(u), cos(u), and 1 respectively, when $\overset{,}{k} = 0$. The sn and sin functions share common properties as do cn and cos. These are summarized in Table A. The dn function has no trigonometric counterpart. Note that the elliptic functions sn and cn may be thought of as generalizations of sin and cos where their period depends on the modulus k.

The argument u is identified as the incomplete elliptic integral of the first kind which is usually denoted $F(\theta, k)$. This identification shows that u also depends on k. The value of k normally ranges from 0 to 1, but we allow $k^2 \epsilon[-\infty, \infty]$. For the interpretation of the elliptic functions on this range, see [4] or [1].

Table A. Properties of Jacobi elliptic functions

Function f

Property	sn(u,k)	sin(u)	cn(u,k)	cos(u)	dn(u,k)
Max. value	1	1	1	1	1
Min. value	-1	-1	-1	-1	$(1-k^2)^{1/2}$
Period	4K(k)	2π	4 K(k)	2π	2 K(k)
Odd/Even	Odd	Odd	Even	Even	Even
df/du	cn dn	cos	-sn dn	-sin	$-k^2$ sn cn
$f\|k=0$	sin	sin	cos	cos	1
$f\|k=1$	tanh	—	sech	—	sech

K(k) = complete elliptic integral of the first kind

$$K(0) = \pi/2, \quad K(1) = +\infty$$

The elliptic functions also satisfy the following identities which correspond to $sin^2 + cos^2 = 1$:

$(A1)$
$$sn^2 + cn^2 = 1$$

$(A2)$
$$k^2 sn^2 + dn^2 = 1$$

$(A3)$
$$1 - k^2 + k^2 cn^2 = dn^2$$

In addition to the sn, cn, and dn functions, there are three other frequently encountered elliptic functions. First, there is the amplitude function am(u,k) $(= \theta)$ which is the inverse of $F(\theta, k)$ $(= u)$. This function maps the elliptic argument u onto a trignometric argument θ so that the period 4 K(k) in u equals the period 2π in θ.

Second, there is $E(\theta, k)$, the incomplete elliptic integral of the second kind. It is often written in abbreviated notation as E(u) since θ depends on u (via the am function) and the dependence on k is understood. Both E(u) and u are not periodic in u. The complete elliptic integral of the second kind is denoted E(k).

Finally, there is the Jacobi Zeta function $Z(\theta, k)$, a linear combination of E(u) and u:

$(A4)$
$$Z(\theta, k) = E(\theta, k) - \frac{E(k)}{K(k)} F(\theta, k)$$

This function is periodic in u with period 2 K(k) and has zero mean. It is often written in abbreviated notation as Z(u) in the same manner as E(u).

Appendix B: MACSYMA Program Listing

```
/* Averaging using Elliptic Functions */

/* x'' + alpha(tau) x + beta(tau) x^3 + e g(x,x',tau) = 0 */

/* where tau = eps t and g is polynomial in x and x' */

/* Variable names and their meanings */
/* */
/* X = variable in differential equation */
/* Y = dx/dt */
/* T = time */
/* TAU = eps*t */
/* G = g(x,x',tau) = g(x,y,tau) is a perturbation */
/* AL = alpha(tau) */
/* BE = beta(tau) */
/* RR = amplitude */
/* AA = instantaneous time frequency */
/* U0 = phase angle constant */
/* K = modulus of the elliptic functions */
/* U = argument of elliptic functions */
/* CN(U,K) = a Jacobi elliptic function */
/* CN'(U,K) = - sn(u,k) dn(u,k) = derivative of cn(u,k) w.r.t.  argument u */
/* ZZ = jacobi zeta function */
/* KC = complete elliptic integral of the first kind */
/* EC = complete elliptic integral of the second kind */
/* PHI = angle variable [where u = 4 kc phi] */
/* MU = +1 or -1 depending on whether the system point is within left(-1) */
/*     or right(+1) separtrix loop when k^2 > 1 */
/* WITHINSEP = H(k^2-1) where H is the Heaviside step function */
/*     = a flag telling whether an orbit is within the double */
/*     homoclinic loop separtrix (+1) or not (0) */

/* The 2nd order differential equation is written as 2 1st order equations:  */
/* */
/* x' = y */
/* y' = - eps g(x,y,tau) - alpha(tau) x - beta(tau) x^3 */

/* For alpha and beta fixed */
/* [and proper interpretation of elliptic variables */
/* for k between -infinity and infinity] */
/* the differential equation is solved exactly by:  */
/* */
/* x = mu rr cn(u,k) y = mu rr aa cn'(u,k) u = aa t + u0 */
/* aa^2 = al + be rr^2 k^2 = be rr^2/2/aa^2 */
/* */
/* where initial conditions determine the values of rr and u0 */

/* But, we must use slow flow equation for phi [rather than u0 or u] */
```

```
/* where initial conditions determine the values of rr and u0 */

/* But, we must use slow flow equation for phi [rather than u0 or u] */

/* The Variational equations to be averaged are:  */
/* */
/* diff(rr,t) = eps F[1] + O(eps^2) */
/* diff(phi,t) = aa/4/kc + eps F[2] + O(eps^2) */

/* Averaging uses a near-identity transformation as follows:  */
/* */
/* rr=rbar+eps*W[1](rbar,phi) */
/* phi=phibar+eps*W[2](rbar,phi) */

/* Symbols used in the computation */

/* XX = cn function */
/* YY = cn' function */
/* ZZ = Jacobian zeta function */
/* SS = arcsin(k*sn(u,k)) = arcsin(k*sqrt(1-xx^2)) */
/* S0 = integral of SS with respect to u */
/* S2 = integral of SS*XX^2 with respect to u */
/* UU = argument */
/* TH = LN(THETA(U)/THETA(0)) */

/* Symbols used on output */
/* */
/* CNF = cn elliptic function */
/* CNP = cn' elliptic function = - sn dn */
/* SNF = sn elliptic function */
/* ZETA = jacobi zeta function */
/* THETA = an elliptic theta function */
/* S0 = see above */
/* S2 = see above */

AVERAGE():=BLOCK([X,Y,XX,YY,F,FI,W,DPHIW,H,V,FBAR,W1MEAN,W1INT,W1C,K,AL,BE,
    RR,XX,YY,ZZ,SS,UU,TH,TAU,KC,EC,EPS,RBAR,KBAR,PHI,CNF,SNF,CNP,
    THETA,ZETA,ARCSIN],

/* Input problem */

PRINT("AVERAGING OF X'' + ALPHA(TAU) X + BETA(TAU) X^3 + EPS G(X,X',TAU) = 0")
PRINT(" WHERE TAU = EPS T AND G IS POLYNOMIAL IN X AND X'"),
PRINT(" "),ALVAL:READ("ENTER ALPHA(TAU):"),
PRINT(" "),BEVAL:READ("ENTER BETA(TAU): "),
PRINT(" "),PRINT("ENTER G(X,X',TAU) USING Y=X':"),
G:READ(),
PRINT(" "),
PRINT("UNPERTURBED SOLUTION IS: X = RR CN(U,K) AND X' = Y = RR AA CN'(U,K)"),
PRINT("WHERE RR = AMPLITUDE AND U = 4 KC PHI = PHASE"),
PRINT(" "),PRINT("AVERAGING WILL USE A NEAR-IDENTITY TRANSFORMATION FROM"),
PRINT("(RR,PHI) TO (RBAR,PHIBAR) AS FOLLOWS:"),
```

```
PRINT(" "),
PRINT("RR=RBAR+EPS*W[1](RBAR,PHI)"),
PRINT("PHI=PHIBAR+EPS*W[2](RBAR,PHI)"),

/* Kill variables used by the routine */

KILL(K,AL,BE,RR,XX,YY,ZZ,SS,UU,TH,TAU,KC,EC,EPS,RBAR,KBAR,PHI,
    CNF,SNF,CNP,THETA,ZETA,ARCSIN,WITHINSEP,FF),

/* Set AL and BE dependency and check for numeric k value */

DEPENDS([AL,BE],TAU),
IF ALVAL = 0 THEN K:SQRT(1/2),
IF BEVAL = 0 THEN K:0,

/* Set WITHINSEP flag to zero if no double homoclinic loops exist */
/* in unperturbed system */

IF SCALARP(ALVAL) AND ALVAL>=0 THEN WITHINSEP:0,

/* Create REDUC routine that reduces an expression involving YY (cn') */
/* into its even and odd components and apply the identity */
/* (cn')^2 = (1-cn^2)*(1-k^2+k^2*cn^2) */

REDUC(EXPR):=BLOCK([EVEN,ODD,VAL],
    EVEN:EXPAND((EXPR+EV(EXPR,YY=-YY))/2),
    ODD:EXPAND((EXPR-EVEN)/YY),
    ODD:YY*EXPAND(EV(ODD,YY=SQRT((1-XX^2)*(1-K^2+K^2*XX^2)))),
    EVEN:EXPAND(EV(EVEN,YY=SQRT((1-XX^2)*(1-K^2+K^2*XX^2)))),
    VAL:EVEN+ODD
),

/* Substitute x = r cn and y = r a cn' into g */

G:EV(G,X=RR*XX,Y=RR*SQRT(AL+BE*RR^2)*YY),

/* Compute F[1] and F[2] */

F[1]:-REDUC(YY*G)/SQRT(BE*RR^2+AL),
IF DIFF(ALVAL,TAU)#0 THEN
    F[1]:F[1]+DIFF(AL,TAU,1)*RR*(XX-1)*(XX+1)/(2*(BE*RR^2+AL)),
IF DIFF(BEVAL,TAU)#0 THEN
    F[1]:F[1]+DIFF(BE,TAU,1)*RR^3*(XX-1)*(XX+1)*(XX^2+1)/(4*(BE*RR^2+AL)),

F[2]:-REDUC(G*(2*AL*BE*RR^2*YY*ZZ+2*AL^2*YY*ZZ-AL*BE*RR^2*XX^3
    -BE^2*RR^4*XX-2*AL*BE*RR^2*XX-2*AL^2*XX))
    /(4*RR*(BE*RR^2+AL)^(3/2)*(BE*RR^2+2*AL)*KC),
IF DIFF(ALVAL,TAU)#0 THEN
    F[2]:F[2]+DIFF(AL,TAU,1)*(AL*XX^2*ZZ-BE*RR^2*ZZ-2*AL*ZZ+AL*XX*YY)
    /(4*(BE*RR^2+AL)*(BE*RR^2+2*AL)*KC),
IF DIFF(BEVAL,TAU)#0 THEN
    F[2]:F[2]+DIFF(BE,TAU,1)*(AL*BE*RR^2*XX^4*ZZ+AL*BE*RR^2*ZZ+2*AL^2*ZZ
    +AL*BE*RR^2*XX^3*YY+BE^2*RR^4*XX*YY+2*AL*BE*RR^2*XX*YY)
    /(8*BE*(BE*RR^2+AL)*(BE*RR^2+2*AL)*KC),
```

```
/* If k=0 then simplify above by setting be=0 and zz=0 */

IF SCALARP(K) AND K = 0 THEN (
    F[1]:EV(F[1],BE=0),
    F[2]:EV(F[2],ZZ=0,BE=0)
),

/* Integrate F[ii] w.r.t.  phi (GENINT) and find mean FBAR[ii] */

FOR II:1 THRU 2 DO (
    FI[II]:GENINT(F[II],K),
    FF[II]:EV(FI[II],XX=0,YY=0,ZZ=0,TH=0),
    FBAR[II]:RATCOEF(FF[II],UU)+WITHINSEP*RATCOEF(FF[II],SS)*%PI/2/KC
),

/* Find transformation or not */

KILL(W,W1MEAN,W1INT),PRINT(" "),
Q1:READ("DO YOU WISH TO FIND THE TRANSFORMATION? (Y/N)"),

/* If q1<>n then find transformation by computing w[1] and w[2] */

IF Q1#N THEN (

    /* Compute W[1] and integrate w.r.t.  phi (GENINT). Find mean of W[1] */

    W[1]:1/SQRT(AL+BE*RR^2)*( EV(FI[1],UU=0)
    -WITHINSEP*RATCOEF(FF[1],SS)*%PI/2/KC*UU) ),
    W1INT:GENINT(W[1],K),
    W1MEAN:RATCOEF(W1INT,UU,1)+WITHINSEP*RATCOEF(W1INT,SS)*%PI/2/KC,

    /* When withinsep=1 then ss-%pi/2/kc*uu has zero mean.*/
    /* Make mean of W[1]=0 by adding a constant */

    W[1]:W[1]-W1MEAN,

    /* Find diff(w[1],phi) */

    DPHIW[1]:4*KC/SQRT(AL+BE*RR^2)*(F[1]-FBAR[1]),

    /* Find W1C=diff(aa/4/kc,rr) and simplify if possible */

    W1C:SQRT(BE*RR^2+AL)*(BE*RR^2*KC+2*AL*KC-2*AL*EC)
    /(4*RR*(BE*RR^2+2*AL)*KC^2),
    IF ALVAL=0 THEN (W1C:EV(W1C,AL=0),W1C:EV(W1C,ABS(RR)=RR)),
    IF BEVAL=0 THEN W1C:0,

    /* Find diff(w[2],phi) */

    DPHIW[2]:4*KC/SQRT(AL+BE*RR^2)*(F[2]-W[1]*W1C-FBAR[2]),

    /* Find W[2] */

    W[2]:1/SQRT(AL+BE*RR^2)*( EV(FI[2],UU=0)
     -WITHINSEP*RATCOEF(FF[2],SS)*%PI/2/KC*UU
     -W1C*(W1INT-W1MEAN*UU)

    /* W[2] has not been set to have zero mean */
),
```

```
/* Create a list of substitution rules for output */

IF SCALARP(K) AND K = 0 THEN
    PF:[XX=COS(2*%PI*PHI),YY=-SIN(2*%PI*PHI),UU=2*%PI*PHI,ABS(RR)=RBAR,
    RR=RBAR,KC=%PI/2,AL=ALVAL,BE=BEVAL]
ELSE
PF:[XX=CNF(4*KC*PHI),YY=CNP(4*KC*PHI),ZZ=ZETA(4*KC*PHI),
    TH=LOG(THETA(4*KC*PHI)/THETA(0)),SS=ARCSIN(KBAR*SNF(4*KC*PHI)),
    UU=4*KC*PHI,ABS(RR)=RBAR,RR=RBAR,AL=ALVAL,BE=BEVAL],
IF NOT SCALARP(K) THEN PFK:[K=SQRT(BEVAL*RBAR^2/2/(BEVAL*RBAR^2+ALVAL))]
    ELSE PFK:[],

/* Change results to output form */

FOR II:1 THRU 2 DO FBAR[II]:RATSIMP(EV(FBAR[II],PF,PFK,DIFF)),

/* Save averaged system into Rflow and Phiflow */

RFLOW:EPS*FACTOR(FBAR[1]),
PFLOW:EV(1/4/KC*SQRT(AL+BE*RR^2),PF)+EPS*FACTOR(FBAR[2]),

/* Print avg. eqns, kbar^2, kc, ec (VAL contains this list of output) */

DERIVABBREV:TRUE,
VAL:[DIFF(RBAR(T),T)=RFLOW,DIFF(PHIBAR(T),T)=PFLOW,
    'KBAR^2=FACTOR(BEVAL*RBAR^2/2/(ALVAL+BEVAL*RBAR^2))],
IF K=0 THEN VALCOMP:[KC=%PI/2,EC=%PI/2] ELSE VALCOMP:[KC=KC(KBAR),EC=EC(KBAR)]
VAL:APPEND(VAL,VALCOMP),

PRINT("THE AVERAGED EQUATIONS ARE"),PRINT(" "),PRINT(VAL),PRINT(" "),

/* If q1<>n then simplify transformation */

IF Q1#N THEN (

    /* Change to output form */

    PRINT("SIMPLIFYING TRANSFORMATION"),
    FOR II:1 THRU 2 DO W[II]:EXPAND(EV(W[II],PF,PFK,DIFF)),

    /* Save transformations into Rtrans and Phitrans (TRANSF contains both) */

    RTRANS:RR=RBAR+EPS*MAP(FACTOR,W[1]),
    PTRANS:PHI=PHIBAR+EPS*MAP(FACTOR,W[2]),
    TRANSF:[CTRANS,PTRANS],

    /* Input to see transformation */

    Q2:READ("DO YOU WISH TO SEE THE TRANSFORMATION? (Y/N)"),
    IF Q2#N THEN (

    /* Print transformation */

    PRINT("THE TRANSFORMATION IS :"),PRINT(" "),PRINT(TRANSF)
    )
)
)$
```

```
/* F1,F2,W1,W2 integrator */

/* Routine to integrate integrands of the form :  */

/* (a) xx^m (b) xx^m yy (c) zz xx^m (d) zz xx^m yy */
/* (e) ss xx^m (f) ss xx^m yy */

/* Symbols */

/* XX = cn function */
/* YY = cn' function (derivative of cn w.r.t.  argument) */
/* ZZ = Jacobian zeta function */
/* SS = arcsin(k*sn(u,k)) = arcsin(k*sqrt(1-xx^2)) */
/* SO = integral of SS with respect to u */
/* S2 = integral of SS*XX^2 with respect to u */
/* UU = argument and hence 1st elliptic integral */
/* TH = ln(theta(u)/theta(0)) */
/* KC,EC = complete elliptic integrls of 1st,2nd kinds */
/* K = modulus */

/* V contains the expression to be integrated.  */
/* Expressions are integrated w.r.t u.  */
/* For integrations w.r.t.  phi, multiply by 1/4/kc */

GENINT(V,K):=BLOCK([TEMP,STERMS,XTERMS,ZTERMS,XYT,XT,SYT,ST,ZYT,ZT,
    VALX,VALZ,VALS,VAL],
TEMP:EXPAND(V),

/* V is assumed to be in REDUC form, ie., V is linear in YY */

/* Separate V into categories: */
/* */
/* XT contains terms in V of the form (a) */
/* XYT contains terms in V of the form (b) */
/* ZT contains terms in V of the form (c) */
/* ZYT contains terms in V of the form (d) */
/* ST contains terms in V of the form (e) */
/* SYT contains terms in V of the form (f) */

STERMS:EXPAND(DIFF(TEMP,SS)),
ZTERMS:EXPAND(DIFF(TEMP,ZZ)),
XTERMS:EXPAND(TEMP-SS*STERMS-ZZ*ZTERMS),
XYT:EXPAND(DIFF(XTERMS,YY)),
XT:EXPAND(XTERMS-YY*XYT),
SYT:EXPAND(DIFF(STERMS,YY)),
ST:EXPAND((STERMS-YY*SYT)),
ZYT:EXPAND(DIFF(ZTERMS,YY)),
ZT:EXPAND(ZTERMS-YY*ZYT),

/* Create XYINT function to integrate form (b) */

XYINT(VV):=BLOCK(VV:EXPAND(VV),EXPAND(INTEGRATE(VV,XX))),

/* Integrate forms (a) [using CNINT routine] and (b) */
```

```
VALX:CNINT(XT,K)+XYINT(XYT),

/* Integrate form (d) by integration by parts */

ARG:XYINT(ZYT),
VALZ:ZZ*ARG-CNINT(ARG*(1-K^2-EC/KC+K^2*XX^2),K)

Integrate form (f) by integration by parts */

ARG:XYINT(SYT),
VALS:SS*ARG-CNINT(ARG*K*XX,K),

/* Create a general Integration By Parts routine for forms (c) and (e) */

INTBYPARTS(VV,K,TYPE):=BLOCK([ARG,UUT,ZZT,SST,YYT,XXT,VALP],

    /* VV just contains the XX terms of forms (c) and (e) */
    /* TYPE indicates either form (c) or (e) */

    /* Find DERIV, the derivative of TYPE w.r.t.  u */

    IF TYPE = ZZ THEN DERIV:1-K^2-EC/KC+K^2*XX^2,
    IF TYPE = SS THEN DERIV:K*XX,

    /* Set ARG = integral of VV w.r.t.  u [using CNINT routine] */

    ARG:CNINT(VV,K),

    /* Separate ARG into categories:  */
    /* */
    /* UUT contains UU terms in ARG */
    /* ZZT contains ZZ terms in ARG */
    /* SST contains SS terms in ARG */
    /* YYT contains terms in ARG of the form (b) */
    /* XXT contains terms in ARG of the form (a) */

    UUT:DIFF(ARG,UU),
    ZZT:DIFF(ARG,ZZ),
    SST:DIFF(ARG,SS),
    YYT:DIFF(ARG,YY),
    XXT:EXPAND(ARG-UU*UUT-ZZ*ZZT-SS*SST-YY*YYT),

    /* Perform integration by parts */

    IF TYPE = ZZ THEN
    VALP:EXPAND(UUT*(UU*ZZ-TH)+ZZT*(ZZ^2/2)+EV(SST*S0*DERIV,XX=SQRT(S2/S0))
    +XYINT(YYT*DERIV)+CNINT(XXT*DERIV,K)),
    IF TYPE = SS THEN
    VALP:EXPAND(UUT*(UU*SS-S0)+ZZT*(ZZ*SS-K^2*S2-(1-K^2-EC/KC)*S0)
    +SST*(SS^2/2)+XYINT(YYT*DERIV)+CNINT(XXT*DERIV,K)),
    VALP:EXPAND(TYPE*ARG-VALP)
),

/* Integrate forms (c) and (e) using INTBYPARTS */

VALZ:VALZ+INTBYPARTS(ZT,K,ZZ),
VALS:VALS+INTBYPARTS(ST,K,SS),
```

```
/* Add together */

VAL:EXPAND(VALX+VALZ+VALS)
)$

/* CN function integrator */

/* Routine finds the integral of g(xx) where g is polynomial in XX */
/* and XX stands for the cn function */

/* Symbols */

/* XX = cn function */
/* YY = cn' function (derivative of cn w.r.t.  argument) */
/* ZZ = Jacobian zeta function */
/* SS = arcsin(k*sn(u,k)) = arcsin(k*sqrt(1-xx^2)) */
/* UU = argument and hence 1st elliptic integral */
/* KC,EC = complete elliptic integrls of 1st, 2nd kinds */
/* K = modulus */

CNINT(V,K):=BLOCK([TEMP,HI,IC,VAL],

/* Find highest power of cn in V and kill integration function IC */

TEMP:EXPAND(V),
HI:HIPOW(TEMP,XX),
KILL(IC),

/* IC[II] = integration function array that defines the integral of xx^ii */
/* It is defined recursively */

/* If k=0, then cn=cos so set the IC to use cosine routine */

IF SCALARP(K) AND EV(K) = 0 THEN (

    IC[0]:UU,
    IC[1]:-YY,
    IC[II]:=RATSIMP((II-1)/II*IC[II-2]-1/II*XX^(II-1)*YY)

) ELSE (

    IC[0]:UU,
    IC[1]:SS/K,
    IC[2]:1/K^2*(ZZ+(EC/KC-(1-K^2))*UU),
    IC[3]:1/2/K^3*((2*K^2-1)*SS-K*YY),
    IC[II]:=RATSIMP((II-2)*(2*K^2-1)*IC[II-2]+(II-3)*(1-K^2)*IC[II-4]
    -XX^(II-3)*YY)/K^2/(II-1)
),

/* Set VALue of the integral to zero */

VAL:0,

/* For each xx^ii expression found in V, substitute its integral IC[ii] */

FOR II:0 THRU HI DO VAL:VAL+RATCOEF(TEMP,XX,II)*IC[II],
VAL:EXPAND(VAL)
)$
```

Appendix C: Sample Run of MACSYMA Program

```
AVERAGE()$
AVERAGING OF X'' + ALPHA(TAU) X + BETA(TAU) X^3 + EPS G(X,X',TAU) = 0
    WHERE TAU = EPS T AND G IS POLYNOMIAL IN X AND X'

ENTER ALPHA(TAU):
A(TAU);

ENTER BETA(TAU):
1;

ENTER G(X,X',TAU) USING Y=X':
DEL*Y;

UNPERTURBED SOLUTION IS: X = RR CN(U,K) AND X' = Y = RR AA CN'(U,K)
WHERE RR = AMPLITUDE AND U = 4 KC PHI = PHASE

AVERAGING WILL USE A NEAR-IDENTITY TRANSFORMATION FROM
(RR,PHI) TO (RBAR,PHIBAR) AS FOLLOWS:

RR=RBAR+EPS*W[1](RBAR,PHI)
PHI=PHIBAR+EPS*W[2](RBAR,PHI)

DO YOU WISH TO FIND THE TRANSFORMATION? (Y/N)
N;

THE AVERAGED EQUATIONS ARE

 D                                                                2
[-- (RBAR(T)) = - EPS (2 KC A(TAU) DEL - 2 EC A(TAU) DEL + KC RBAR DEL
 DT

         D                      D
+ 3 KC (---- (A(TAU))) - 3 EC (---- (A(TAU))))/(3 KC RBAR),
        DTAU                   DTAU

                               2                     2
 D              SQRT(A(TAU) + RBAR )        2       RBAR
-- (PHIBAR(T)) = --------------------,  KBAR = ------------------,
 DT                     4 KC                                  2
                                                   2 (A(TAU) + RBAR )

KC = KC(KBAR), EC = EC(KBAR)]

[SYMBOLICS 3670 time = 211 seconds]
```

89

REFERENCES

[1] BYRD, P. AND FRIEDMAN, M., *Handbook of Elliptic Integrals for Engineer and Scientists*, Springer-Verlag, Berlin, 1954.

[2] CAP, F. F., *Averaging Methods for the Solution of Non-linear Differential Equations with Periodic Non-harmonic Solutions*, International Journal of Non-Linear Mechanics, vol. 9, 1973, pp.441–450.

[3] CHIRIKOV, B. V., *A Universal Instability of Many Dimensional Oscillator Systems*, Physics Reports, vol. 52, 1979, pp.263–379.

[4] COPPOLA, V.T., *Averaging of Strongly Nonlinear Oscillators Using Elliptic Functions*, Ph.D. dissertation, Cornell University, 1989.

[5] COPPOLA, V.T. AND RAND, R.H., *Symbolic Computation and Perturbation Methods Using Elliptic Functions*, Transactions of the Sixth Army Conference on Applied Mathematics and Computing (1989), pp. 639–676.

[6] COPPOLA, V.T. AND RAND, R.H., *Averaging Using Elliptic Functions: Approximation of Limit Cycles*, Acta Mechanica (1989) (to appear).

[7] COPPOLA, V.T. AND RAND, R.H., *Chaos in a System with Periodically Disappearing Separatrix*, in preparation.

[8] DAVIS, H.T., *Introduction to Nonlinear Differential and Integral Equations*, Dover, NY, 1962.

[9] GARCIA-MARGALLO, J. AND BEJARANO J., *Stability of Limit Cycles and Bifurcations of Generalized van der Pol Oscillators*, preprint.

[10] GARCIA-MARGALLO, J. AND BEJARANO, J., *A Generalization of the Method of Harmonic Balance*, Journal of Sound and Vibration, vol. 116(3), 1987, pp. 591–595.

[11] GOLDSTEIN, H., *Classical Mechanics*, second edition, Addison-Wesley, Reading, Mass., 1980.

[12] GREENSPAN, B. AND HOLMES, P., *Repeated Resonance and Homoclinic Bifurcation in a Periodically Forced Family of Oscillators*, SIAM Journal of Mathematical Analysis, vol. 15, no.1, 1983, pp 69–97.

[13] GUCKENHEIMER, J. AND HOLMES, P., *Nonlinear Oscillations, Dynamical Systems, and Bifurcations of Vector Fields*, Applied Mathematical Sciences, vol. 42, Springer-Verlag, NY, 1983.

[14] HAGEDORN, P., *Non-Linear Oscillations*, Oxford University Press, NY, 1982.

[15] KERVOKIAN, J. AND COLE, J.D., *Perturbation Methods in Applied Mathematics*, Applied Mathematical Sciences, vol. 34, Springer-Verlag, NY, 1981.

[16] KUZMAK, G.E., *Asymptotic Solutions of Nonlinear Second Order Differential Equations with Variable Coefficients*, P.M.M. (English translation), vol. 23, no. 3, 1959, pp. 515–526.

[17] MINORSKY, N., *Nonlinear Oscillations*, Van Nostrand, NY, 1962.

[18] NAYFEH, A., *Perturbation Methods*, Wiley-Interscience, NY, 1973.

[19] NAYFEH, A. AND MOOK, D.T., *Nonlinear Oscillations*, John Wiley and Sons, NY, 1979.

[20] POCOBELLI, G., *Electron Motion in a Slowly Varying Wave*, Phys. Fluids, vol. 24(12), 1981, pp. 2173–2176.

[21] RAND, R.H. AND ARMBRUSTER, D., *Perturbation Methods, Bifurcation Theory and Computer Algebra*, Applied Mathematical Sciences, vol. 65, Springer-Verlag, NY, 1987.

[22] SANDERS, J.A. AND VERHULST, F., *Averaging Methods in Nonlinear Dynamical Systems*, Applied Mathematical Sciences, vol. 59, Springer-Verlag, NY, 1985.

[23] STOKER, J.J., *Nonlinear Vibrations*, Wiley, NY, 1950.

[24] YUSTE, S. AND BEJARANO, J., *Construction of Approximate Analytical Solutions to a New Class of Nonlinear Oscillator Equations*, Journal of Sound and Vibration, vol. 110(2), 1986, pp. 347–350.

VALIDATED ANTI-DERIVATIVES

GEORGE F. CORLISS*

Abstract. We present an overview of two approaches to validated, one dimensional indefinite integration. The first approach is to find an inclusion of the integrand, then integrate this inclusion to obtain an inclusion of the indefinite integral. Inclusions for the integrand may be obtained from Taylor polynomials, Tschebyscheff polynomials, or other approximating forms which have a known error term. The second approach finds an inclusion of the indefinite integral directly as a linear combination of function evaluations plus an interval-valued error term. The second approach can be applied to any quadrature formula such as Gaussian or Newton-Cotes quadrature with a known error expression. In either approach, composite formulae improve the accuracy of the inclusion.

The result of the validated indefinite integration is an algorithm which may be represented as a character string, as a subroutine in a high level programming language such as Pascal-SC or Fortran, or as a collection of data. An example is given showing the application of validated indefinite integration to constructing a validated inclusion of the error function, erf (x).

Key words. Indefinite integration, validation, differentiation arithmetic, Taylor polynomials, Gaussian quadrature, error function, interval analysis

AMS(MOS) subject classifications. 65D30

1. Indefinite integration. Consider the problem of finding an anti-derivative

$$(1.1) \qquad g(x) = \int_a^x f(t)\, dt, \text{ for } a \le x \le b.$$

If the problem is easy, one uses the rules of calculus to write down a formula for g. If the problem is somewhat more complicated, one uses a symbolic processor like MACSYMA [14], Maple [2], REDUCE [6], or Scratchpad [17] to derive a formula for g. However, if the problem is non-elementary, or if its solution is too complex, one resorts to a numerical approach. For example, a high quality suite of numerical quadrature routines like QUADPACK [12] accepts a value of x and returns an approximation for $g(x)$. However, the numerical approach is not suitable for symbolic computation because it does not provide a formula for g.

This paper examines two methods for obtaining approximate, validated formulae for g. The formulae are *approximate* in the sense that there may be no closed form for g, so an approximation for g is the best that can be expected. The formulae are *validated* in the sense that an interval-valued function $G(x) := [\underline{G}(x), \overline{G}(x)]$ is given which satisfies

$$\underline{G}(x) \le g(x) \le \overline{G}(x), \text{ for every } x \in [a, b].$$

Furthermore, we seek formulae which are *accurate* in the sense that $\text{wid}(G(x)) := \overline{G}(x) - \underline{G}(x)$ is as small as possible.

---

*Department of Mathematics, Statistics and Computer Science, Marquette University, Milwaukee, WI 53233. This work was supported in part by IBM Deutschland, GmbH, and in part by the National Science Foundation under Grant No. CCC–8802429. The Government has certain rights to this material.

The first approach is discussed in Section 2. We compute a Taylor polynomial enclosure $F(t)$ of the integrand $f(t)$, and then integrate F. The result is a formula for G as a polynomial with interval coefficients.

The second approach is discussed in Section 3 where we enclose g directly, without first enclosing the integrand f. This is illustrated using Gaussian quadrature because of its accuracy, but any other numerical quadrature rule with a known error expression could also be used.

Both of the Taylor polynomial and the Gaussian quadrature approaches are semi-numeric. Like a symbolic processor, they produce an algorithm (a formula) which is executed to evaluate g. The difference is that the algorithms are constructed by numerical tools (Taylor polynomials or Gaussian quadrature).

Validated indefinite integration extends the scope of a symbolic processor. Current symbolic processors are very clever at solving problem (1.1) when there is a closed form expression for g. They find an expression which is mathematically correct (not approximate) which can be used for further manipulation. Caviness [1] includes a survey of the history and applications of the Risch integration algorithm [15, 16]. Cherry [3] discusses the integration of classes of transcendental elementary functions in terms of elementary functions and error functions. However, the approaches described in this paper are intended for integrands for which such symbolic approaches fail and a symbolic processor faces two undesirable alternatives:

Alternative I. Admit failure - the correct answer cannot be returned.

Alternative II. Return a numerical approximation, perhaps with a warning.

We provide a third alternative:

Alternative III. Return an interval-valued formula which is validated to contain the mathematically correct result.

A validated inclusion of $g(x)$ by $G(x)$ allows processing to continue with an approximation (not the mathematically correct result) which still supports a guarantee.

This paper summarizes results appearing in [4].

2. Taylor polynomial with error term.

THEOREM 1. *Let $f^{(p)}$ be a continuous function on $X = [a, b]$, and let*

$$\underline{L}_{p+1}(x - y) \leq f^{(p)}(x) - f^{(p)}(y) \leq \overline{L}_{p+1}(x - y), \text{ for all } x, y \in [a, b],$$

where $L_{p+1} := [\underline{L}_{p+1}, \overline{L}_{p+1}]$. Then

$$g(x) \in G(x) := f(a)(x - a) + f'(a)(x - a)^2/2! + f''(a)(x - a)^3/3! +$$
(2.1)
$$\cdots + f^{(p)}(a)(x - a)^{p+1}/(p + 1)! + L_{p+1}(x - a)^{p+2}/(p + 2)!$$

The proof appears in [4].

The formula for G in equation (2.1) is a polynomial in $(x - a)$ of degree $p + 2$. The coefficients of $(x - a)$, $(x - a)^2$, ..., $(x - a)^{p+1}$ are all real numbers, and the coefficient of $(x-a)^{p+2}$ is an interval. Viewed another way, $G(x) = [\underline{G}(x), \overline{G}(x)]$, where

$$\underline{G}(x) := f(a)(x-a) + f'(a)(x-a)^2/2! + f''(a)(x-a)^3/3! +$$
$$\cdots + f^{(p)}(a)(x-a)^{p+1}/(p+1)! + \underline{L_{p+1}}(x-a)^{p+2}/(p+2)!$$
$$\overline{G}(x) := f(a)(x-a) + f'(a)(x-a)^2/2! + f''(a)(x-a)^3/3! +$$
$$(2.2) \qquad \cdots + f^{(p)}(a)(x-a)^{p+1}/(p+1)! + \overline{L_{p+1}}(x-a)^{p+2}/(p+2)!$$

are real-valued polynomials in $x-a$ of degree $p+2$ which satisfy

$$\underline{G}(x) \le g(x) \le \overline{G}(x), \text{ for all } x \in X.$$

For example, let $g(x) = \int_0^x e^{-t^2}\, dt$. Then for $x \in [0,2]$,

$$(2.3) \qquad g(x) \in G(x) = x - x^3/3 + x^5/10 + [-1.5449, 1.3819]x^7.$$

In order to achieve the inclusion of g in a practical implementation, all the operations in equation (2.1) are performed as interval operations [10, 11]. The derivatives $f(a), f'(a), \ldots, f^{(p)}(a)$ which appear in equation (2.1) are computed using differentiation arithmetic [13]. Differentiation arithmetic uses recurrence relations derived from the expression for f to compute the Taylor coefficients $f(a), f'(a), f''(a)/2!$, $\ldots, f^{(p)}(a)/p!$ accurately and efficiently. The recurrence relations are computed using interval operations in order to capture the Taylor coefficients in tight interval inclusions. The calculations in equation (2.1) are easily arranged to take advantage of the accurate scalar product provided by Pascal-SC for micro-computers [8] or the IBM product ACRITH [7] for IBM 370 class mainframe computers.

The output from the pre-processor consists of an algorithm for the evaluation of G as expressed in equation (2.1). This output may take several different forms, depending on its desired use.

The pre-processing can be designed to return $p+2$ interval-valued coefficients for the polynomial G or to store the coefficients in an internal work area accessible to an evaluation routine. The client program can use these coefficients to evaluate G, or to manipulate G in any manner it desires.

Alternatively, the pre-processing can be designed to return or to print a character string containing the textual form of the expression for G. This output is in the style of a symbolic processor, except that a numerical algorithm constructed the expression. For our example, the text of equation (2.3) is returned.

Finally, the pre-processing can be designed to write a subroutine which can be called at run-time to evaluate G. Either Pascal-SC or Fortran is an attractive language for the output routine because these two languages provide access to libraries for interval arithmetic.

3. Gaussian quadrature. Our second approach computes an inclusion of g directly without first computing an inclusion of f. We illustrate this approach using the n-point, one panel Gaussian quadrature formula

$$(3.1) \qquad g(x) = \int_a^x f(t)\, dt = \frac{x-a}{2} \sum_{i=1}^n w_i f(t_i) + C_n \cdot \xi(x) \cdot (x-a)^{2n+1},$$

where w_i are the Gaussian weights on $[-1, 1]$, t_i are the Gaussian nodes on $[-1, 1]$, and $C_n = \frac{1}{(2n+1)!} \cdot \left[\frac{(n!)^2}{(2n)!}\right]^2$ is the Gaussian error coefficient. The function $\xi(x) := f^{(2n)}(\xi)$.

THEOREM 2. *Let $f^{(2n-1)}$ satisfy a Lipschitz condition on $X = [a, b]$:*

$$f^{(2n-1)}(x) - f^{(2n-1)}(y) \in L_{2n}(x - y), \quad \text{where } L_{2n} := [\underline{L}_{2n}, \overline{L}_{2n}].$$

Then

$$(3.2) \quad g(x) \in G(x) := \frac{x-a}{2} \sum_{i=1}^{n} w_i \cdot f\left(\frac{x-a}{2}\tau_i + \frac{x+a}{2}\right) + C_n \cdot L_{2n} \cdot (x-a)^{2n+1}.$$

Theorem 2 gives an indefinite integral as a linear combination of function evaluations plus an interval-valued error term. The attractiveness of equation (3.2) is its accuracy. An n-point Gaussian quadrature formula and a Taylor polynomial of degree $2n - 1$ both have error terms involving the same Lipschitz interval, L_{2n}, and the same factor $(x-a)^{2n+1}$. However, the Gaussian error coefficient is much smaller than the corresponding Taylor coefficient, so equation (3.2) usually yields a much narrower interval than equation (2.1). In addition, the widths of the rule portions of each formula are usually narrower for Gaussian quadrature.

The output from the pre-processing for the Gaussian quadrature approach consists of an algorithm for the evaluation of G as expressed in equation (3.2). This output may take several different forms.

The pre-processing can be designed to return the Gaussian weights, nodes, and error coefficient or to store them in an internal work area accessible to an evaluation routine. Since the weights and nodes are not machine numbers, they are returned as intervals whose endpoints differ by only one unit in the last place (ULP). Alternatively, weights and nodes can be given as rational numbers for use in symbolic systems which support exact rational arithmetic. An interval inclusion for L_{2n} is computed at pre-processing time for return to the client program. With this information, the client program can evaluate or manipulate G in any manner it desires.

Alternatively, the pre-processing can be designed to return or to print a character string containing the textual form of the expression for G. One can either assume that $f(t)$ is a function whose definition is known to the symbolic processor, or else one can textually substitute the expression for each point of evaluation for each occurrence of the character "t" in f. The resulting expression contains a combination of n evaluations of f plus an error term. This may be rather lengthy, but this is a common concern with symbolic processors. For the example given in equation (2.3), 3-point Gaussian quadrature gives the expression

$$G(x) = \frac{x}{2}\left[\frac{5}{9} \cdot f\left(-\sqrt{\frac{3}{5}} \cdot \frac{x}{2} + \frac{x}{2}\right) + \frac{8}{9} \cdot f\left(0 \cdot \frac{x}{2} + \frac{x}{2}\right) + \frac{5}{9} \cdot f\left(\sqrt{\frac{3}{5}} \cdot \frac{x}{2} + \frac{x}{2}\right)\right]$$

$$(3.3) \qquad\qquad + \frac{[-10.8138, 9.6729]}{2016000}x^7.$$

Finally, the pre-processing can be designed to write a subroutine which can be called at run-time to evaluate G.

We have illustrated the ideas of this section using a Gaussian quadrature formula, but any other quadrature formula similar to equation (3.1) could also be employed, provided only that an expression for the error is known in terms of derivatives of f. For example, Newton–Cotes or Gauss–Kronrod [12] formulae could be used, as could formulae with weight functions or formulae for special types of singularities.

4. Composite formulae. Often, the inclusion for G computed either by Taylor polynomial quadrature (Theorem 1) or by Gaussian quadrature (Theorem 2) is not sufficiently accurate in the sense that the width of the interval $[\underline{G}(x), \overline{G}(x)]$ is too wide for some values of x. When this happens, a composite formula is employed, either with fixed subintervals, or with subintervals which are determined adaptively by an algorithm such as SVALAQ [5].

Let $a = x_0 < x_1 < \ldots < x_K = b$ be a partition of $[a, b]$. Let $x \in [x_J, x_{J+1}]$. Then

$$g(x) = \sum_{j=0}^{J-1} \int_{x_j}^{x_{j+1}} f(t)\,dt + \int_{x_J}^{x} f(t)\,dt.$$

The composite formula for the anti-derivative consists of definite integrals for which inclusions are computed in advance, and the final indefinite integral is handled by either Theorem 1 or Theorem 2. The corresponding algorithm for G is a sum of J intervals plus the expression given either by equation (2.1) or by equation (3.2). $G(x)$ is continuous, provided that the algorithm used for the evaluation of $\int_{x_J}^{x} f(t)\,dt$ for $x \in (x_J, x_{J+1}]$ is the same as the one used to compute each $\int_{x_j}^{x_{j+1}} f(t)\,dt$.

The output from the preprocessing using composite rules is lengthier than before. The listing below shows portions of the output for the example given in equation (2.3) using 8 subintervals on $[0, 2]$ and an inclusion of f of degree 10.

```
If x is in [ 0.00000000000E+00,  2.50000000000E-01] then
g(x) is in  0.0
          + [ 9.99999999999E-01,  1.00000000002E+00] * (x -   0.0)^ 1
          + [ 0.00000000000E+00,  0.00000000000E+00] * (x -   0.0)^ 2
          + [-3.33333333340E-01, -3.33333333333E-01] * (x -   0.0)^ 3
              . . .
          + [ 0.00000000000E+00,  0.00000000000E+00] * (x -   0.0)^10
          + [-7.57575757597E-04, -2.38191714233E-04] * (x -   0.0)^11


Subinterval  2
If x is in [ 2.50000000000E-01,  5.00000000000E-01] then
g(x) is in  [ 2.44887887178E-01,  2.44887887308E-01]
          + [ 9.39413062813E-01,  9.39413062814E-01] * (x -   0.25)^ 1
```

```
    . . .
 + [-1.64301228451E-03, -1.64301228445E-03] * (x -  0.25)^10
 + [-5.24520727410E-04,  1.08208205213E-03] * (x -  0.25)^11
```

```
  . . .
```

```
Subinterval  8
If x is in [ 1.75000000000E+00,  2.00000000000E+00] then
g(x) is in  [ 8.74414994279E-01,  8.74415009146E-01]
      + [ 4.67706223839E-02,  4.67706223840E-02] * (x -  1.75)^ 1
          . . .
      + [-3.92377413881E-04, -3.92377413514E-04] * (x -  1.75)^10
      + [-1.00690524777E-02,  1.03470421381E-02] * (x -  1.75)^11
```

By taking the partition sufficiently fine, inclusions for $g(x)$ with differ by only a few units in the last place can be achieved.

The Pascal-SC program used to produce the listing above is available from the author.

5. Properties of G. The true anti-derivative is continuous. The computed anti-derivative should have the same property, especially when that function is being used as input for some other algorithm. Lyness [9] points out that when an adaptive quadrature routine is used to compute an (approximate) anti-derivative, the result is not a continuous function of x. This causes trouble when the result is used by an optimization routine.

The functions $\underline{G}(x)$ and $\overline{G}(x)$ computed by either Theorem 1 (Taylor polynomials) or Theorem 2 (Gaussian quadrature) are continuous. For composite formulae, G is continuous on each subinterval. G is continuous at each node in the partition by its dependence on $(x - x_J)$. Hence, anti-derivatives computed using the techniques of this paper are well suited for use an input for other algorithms.

Acknowledgment.

The author wishes to thank Professor Gary Krenz, with whom much of this work was done.

REFERENCES

[1] B. F. CAVINESS, *Computer algebra: Past and future*, J. of Symbolic Computation, 2 (1986), pp. 217–236.

[2] B. W. CHAR, G. J. FEE, K. O. GEDDES, G. H. GONNET, AND M. B. MONAGAN, *A tutorial introduction to Maple*, J. of Symbolic Computation, 2 (1986), pp. 179–200.

[3] G. W. CHERRY, *Integration in finite terms with special functions: The error function*, J. of Symbolic Computation, 1 (1985), pp. 283–302.

[4] GEORGE F. CORLISS AND GARY S. KRENZ, *Indefinite integration with validation*, ACM Trans. Math. Software (to appear) (1989).

[5]   GEORGE F. CORLISS AND LOUIS B. RALL, *Adaptive, self-validating numerical quadrature*, SIAM J. Scientific and Statistical Comput., 8 (1987), pp. 831–847.

[6]   J. FITCH, *Solving algebraic problems with REDUCE*, J. of Symbolic Computation, 1 (1985), pp. 211–227.

[7]   IBM CORP., *ACRITH High Accuracy Subroutine Library: General Information Manual*, IBM No. GC33 - 6163 - 02, 1986.

[8]   ULRICH KULISCH, (ED.), *Pascal-SC Manual and System Disks*, Wiley-Teubner, Stuttgart, 1986.

[9]   JAMES LYNESS AND J. J. KAGANOVE, *Comments on the nature of automatic quadrature routines*, ACM Trans. Math. Soft., 2 (1976), pp. 65–81.

[10] RAMON E. MOORE, *Interval Analysis*, Prentice-Hall, Englewood Cliffs, NJ, 1966.

[11] RAMON E. MOORE, *Methods and Applications of Interval Analysis*, SIAM, Philadelphia, PA, 1979.

[12] R. PIESSENS, E. DE DONCKER-KAPENGA, C. W. ÜBERHUBER, AND D. K. KAHANER, *QUAD-PACK: A Subroutine Package for Automatic Integration*, Springer Series in Computational Mathematics No. 1, New York, 1983.

[13] LOUIS B. RALL, *Automatic Differentiation: Techniques and Applications*, Springer Lecture Notes in Computer Science No. 120, New York, 1981.

[14] R. H. RAND, *Computer Algebra in Applied Mathematics: An Introduction to MACSYMA*, Pitman, Boston, 1983.

[15] R. H. RISCH, *The problem of integration in finite terms*, Trans. Amer. Math. Soc., 139 (1969), pp. 167–189.

[16] R. H. RISCH, *The solution of the problem of integration in finite terms*, Bull. Amer. Math. Soc., 76 (1970), pp. 605–608.

[17] R. S. SUTOR, *The Scratchpad II computer algebra language and system*, Proc. of Eurocal '85, vol II, edited by B. F. Caviness, Springer Lecture Notes in Computer Science No. 204, New York, 1985, pp. 32–33.

A TOOLBOX FOR NONLINEAR DYNAMICS

SHANNON COFFEY* [1], ANDRÉ DEPRIT† [2], ÉTIENNE DEPRIT*,
LIAM HEALY,‡ [3] AND BRUCE R. MILLER†

Abstract. Using the main problem of artificial satellite theory as an illustration, we review several developments which have had a significant impact on research in nonlinear dynamics. On the mathematical front, we point to the theory of Lie transformations; in the area of computational software, we explain how massively data parallel machines open the way for symbolic solution of large problems. Finally, we show how color graphics assist in the qualitative analysis of dynamical systems.

1. Introduction. The rapid progress in computers, software as well as hardware, has revolutionized research about dynamical systems. Prior to the computer age, mathematicians like Ch. Delaunay, G. W. Hill and E. W. Brown envisioned no other way for predicting the motion of a celestial object than by analytical theories. They encapsulated solutions to the differential equations in the form of power series into a body of formulas which they called a 'theory.' That phase once completed, they would turn to the task of breaking the theory into a sequence of steps, each one arranged to save as much of the intermediary calculations as possible. These arrangements which, today, we call 'flow charts', they referred to as the 'Tables' of a theory because they consisted for the main part in preparing partial results in the form of reference tables. Developing a Theory required years of effort, and so did the job of restructuring it to produce Tables. Mathematicians engaged in computational astronomy could not evade meeting head on the ultimate challenge: calculations in the almanac offices had to be done in 'real time.' Of what use would a lunar theory be if an average clerk could not predict the position of the moon from one place to the next in less time than it takes the moon to move from that place to the next?

With the advent of computers, numerical integration became the favored technique in orbit prediction for several reasons — modest use of processor time and memory, greater accuracy for shorter time periods, and ability to accommodate nonconservative forces too difficult to model analytically. More recently, however, analytic theories have enjoyed a resurgence owing to the increasing sophistication of algebraic software, the appearance of novel computational methods and the availability of color graphics. Nonlinear dynamics, at last, is finding a toolbox to serve its purposes.

Our line of research started many years ago, with the study of families of periodic orbits emanating from the triangular equilibria in the restricted problem of three bodies. In the neighborhood of the equilibria, the periodic orbits are functions of a small parameter to be expanded as Fourier series; coefficients in these series are obtained

*Naval Research Laboratory, Washington, D.C. 20375.

†National Institute of Standards and Technology, Gaithersburg, MD 20899. Partial support from the Computational and Applied Mathematics Program at the Defense Advanced Research Project Agency.

‡NAS/NRC Cooperative Research Associate at the Naval Research Laboratory.

as roots of recursive sets of linear equations [1,2]. These symbolic calculations led to devise general procedures for manipulating *en bloc* symbolic expressions of the form

$$(1) \qquad \sum_{i,j} C_{i,j} x_1^{i_1} x_2^{i_2} \ldots x_n^{i_n} \left\{ \begin{array}{c} \cos \\ \sin \end{array} \right\} (j_1 y_1 + j_2 y_2 + \ldots j_n y_n).$$

Expressions of that kind were dubbed Poisson series, and the name has stuck ever since [3].

During the same period, Imre Iszak [4], David Barton [5] and others were looking at ways of reproducing and expanding the grand literal theories upon which astronomers of the XIXth century had lavished so much time. Foremost among those stands the lunar theory. Rigorous to a fault, Delaunay had taken great pains to document his hand calculations, even taking the extraordinary precaution of reporting from one series to the next how each individual term arose. No wonder then that researchers wrapped up in algebraic processors eagerly tested the power and robustness of their systems against the achievements of Delaunay.

Armed with their own symbolic algebra system MAO (short for Mechanized Algebraic Operations), A. Deprit, J. Henrard and A. Rom took the challenge in steps. They first automated Birkhoff's normalization technique in the neighborhood of an equilibrium for a Hamiltonian system with two degrees of freedom [6], then expanded the main problem of artificial satellite theory in powers of the eccentricity [7] — a task that D. Brouwer speculated would be intractable with the computers available at the time.

Much has happened since MAO succeeded in checking Delaunay's theory, and extending it from order 8 to order 13 and even higher [8]. MAO's creators wavered for a while between assembler and Fortran implementations [9], eventually settling upon Fortran with a RATFOR preprocessing stage, yet, at the same moment, branching into PL/I where packaging the code into macros yielded more flexibility. The increasing need for flexibility in coding and ease of application to problems prompted a radical change of direction in 1984, when B. Miller transported MAO to a Lisp workstation to make programming 'objects' out of Poisson series. The Lisp version of MAO readily incorporated the new Lie algebraic methods devised to normalize Hamiltonian systems by canonical transformations near the identity mapping. Implementing the method of Lie transformations [10] in MAO stimulated mutual refinement and extensions between computational algorithms and their software implementation.

Over the years, applications extended from the three-body problem to other sectors of celestial mechanics, overflowing eventually into the general areas of classical mechanics. The latter applications include the Stoermer problem, i.e. the motion of a charged particle in a magnetic dipole, the dynamics of orbiting dust [11], the Hénon-Heiles oscillator, the quadratic Zeeman effect [12,13], and the Toda 3-point lattice [14]. All these problems pertain to the class of perturbed integrable Hamiltonian systems with a principal part that is either a Keplerian problem or a harmonic oscillator.

2. Symbolic algebra and Hamiltonian systems. In the 1960's and 70's, astronomers were not talking of normalizing Hamiltonians and their equations. Instead, they spoke of averaging Hamiltonians over fast variables, which they did by using one of Poincaré's *méthodes nouvelles*. In that context, averaging transformations

$$\chi : (q, Q) \rightarrow (p, P)$$

are represented by a system of implicit equations

$$P = \partial S/\partial p, \qquad q = \partial S/\partial Q$$

derived from a function in mixed variables $S \equiv S(p, Q)$.

Ideal for eliminating fast variables from a Hamiltonian, Poincaré's method is not well suited, though, to obtaining the averaging transformation in explicit form. For it necessitates solving simultaneous implicit equations for half of the system, and substituting the solutions into the other half. Astronomers were not yet equipped computationally to perform these operations beyond the first order. They were, of course, familiar with Lagrange's *inversion formula* for solving elementary implicit equations in one independent variable; beyond that, they knew only of the method of successive approximations. The theory of Lie transformations [15,10] evolved in the late 1960's to stave off these difficulties.

2.1. Normalization as an algebraic operation. We concern ourselves exclusively with canonical transformations depending on a small parameter ϵ of a particular type, namely those which are the general solution $(p(q, Q, \epsilon), P(q, Q, \epsilon))$ of a Hamiltonian system

(2)
$$\frac{dp}{d\epsilon} = \frac{\partial W}{\partial P}, \qquad \frac{dP}{d\epsilon} = -\frac{\partial W}{\partial p}$$

satisfying the initial condition

$$p(q, Q, 0) = q, \qquad P(q, Q, 0) = Q.$$

In regard to these objects, our major problem is a basic one: Given the power series

$$F(p, P, \epsilon) = \sum_{n \geq 0} \frac{\epsilon^n}{n!} F_n(p, P),$$

if W itself is a power series in ϵ, how do we obtain by machine the transformed function

$$F^*(q, Q, \epsilon) = F(p(q, Q, \epsilon), P(q, Q, \epsilon))$$

as a series

$$F^* = \sum_{n \geq 0} \frac{\epsilon^n}{n!} F_n^*(q, Q)$$

in powers of the small parameter? Well, we found that automated conversion works well when we recognize that the coefficients F_n^* are the values at $\epsilon = 0$ of the iterates $\mathcal{L}_W^n$ for the Lie derivative

(3)
$$\mathcal{L}_W : F \rightarrow (F; W)$$

of F in the direction of the vector field W. The notation $(F; W)$ in (3) represents the Poisson brackets of the two functions F and W.

Along those lines, normalizing a Hamiltonian that is a series in the powers of a small parameter ϵ

(4) $$\mathcal{H} = \mathcal{H}_0 + \sum_{n \geq 0} \frac{\epsilon^n}{n!} \mathcal{H}_n$$

means building a Lie transformation χ such that $(\mathcal{H}^*; \mathcal{H}_0^*) = 0$ for the transformed Hamiltonian $\mathcal{H}^* = \chi(\mathcal{H})$.

Actually, for most of the problems handled in the literature, normalization is merely an algebraic operation. Indeed, let $\mathcal{A}$ denote an algebra of functions containing the terms $\mathcal{H}_n$, and let $\mathcal{L}_0$ stand for the Lie derivative

$$\mathcal{L}_0 : F \to (F; \mathcal{H}_0).$$

In most cases, $\mathcal{L}_0$ acts as a semi-simple operator in $\mathcal{A}$, i.e. the vector space $\mathcal{A}$ may be decomposed into the direct sum

$$\mathcal{A} = \ker(\mathcal{L}_0|\mathcal{A}) \oplus \operatorname{im}(\mathcal{L}_0|\mathcal{A}).$$

Under that assumption, normalizing $\mathcal{H}$ means finding a Lie transformation which projects $\mathcal{H}$ into an element $\mathcal{H}^*$ in the kernel of the Lie derivative restricted to $\mathcal{A}$.

2.2. Automated normalizations. Without entering into details, recall that Lie transformations, and hence normalizations, are built by induction. Precisely this property makes normalizations most amenable to construction by computers. As the construction progresses at each order in ϵ, one encounters a partial differential identity of the form

(5) $$\mathcal{L}_0(W_n) + \mathcal{K}_{n,0} = \tilde{\mathcal{H}}_{n,0}$$

which we try to satisfy by choosing the two unknowns, namely the term W_n of order n in the Hamiltonian from which are derived the transformation equations (2) on the one hand, and the term $\mathcal{K}_{n,0}$, likewise of order n, in the transformed Hamiltonian of the system on the other.

As it happens most often, while $\tilde{\mathcal{H}}_{n,0}$ belongs to the Lie algebra $\mathcal{A}$, W_n belongs to an algebra $\mathcal{B}$ of functions such that $(f; g) \in \mathcal{A}$ for any f in $\mathcal{A}$ and any g in $\mathcal{B}$. In those circumstances, we choose $\mathcal{K}_{n,0}$ to be the part of $\tilde{\mathcal{H}}_{n,0}$ in the kernel of $\mathcal{L}_0$, which makes W_n a counter-image of $\tilde{\mathcal{H}}_{n,0} - \mathcal{K}_{n,0}$ in $\mathcal{B}$.

Solving (5) for W_n rarely proves trivial. Our favorite practice has been to select a standard representation for the algebra $\mathcal{A}$ and determine the action of $\mathcal{L}_0$ upon the appropriate expressions in $\mathcal{A}$. These rules of action indicate membership in the kernel, and tell us how to determine counter-images.

As a simple example, consider the elliptic oscillator

$$\mathcal{H}_0 = \tfrac{1}{2}\left(X^2 + x^2\right) + \tfrac{1}{2}\left(Y^2 + y^2\right).$$

In complex variables the Hamiltonian and Lie derivative take the form

$$\mathcal{H}_0 = i(uU + vV) \qquad \mathcal{L}_0 = i\left(u\frac{\partial}{\partial u} - U\frac{\partial}{\partial U} + v\frac{\partial}{\partial v} - V\frac{\partial}{\partial V}\right).$$

Assume that the perturbation belongs to the complex algebra $\mathcal{A}$ of multivariate polynomials in the variables u, v, U, V. The action of the Lie derivative on a given monomial is

$$\mathcal{L}_0(u^\alpha U^\beta v^\gamma V^\delta) = i(\alpha - \beta + \gamma - \delta)u^\alpha U^\beta v^\gamma V^\delta.$$

A monomial belongs to the kernel of $\mathcal{L}_0$ if and only if $\alpha - \beta + \gamma - \delta = 0$; otherwise, $u^\alpha U^\beta v^\gamma V^\delta$ is the image of the monomial

$$\frac{iu^\alpha U^\beta v^\gamma V^\delta}{(\beta - \alpha + \delta - \gamma)}.$$

Clearly, all algebraic computations involved in the normalization of an elliptic oscillator with polynomial perturbations are closed in the algebra $\mathcal{A}$. The successive differential equations of type (5) are solved automatically by syntactic matching of patterns and mappings. A similar argument holds when it is necessary to introduce dependent variables to reduce the complexity of expressions; the canonical forms aid in performing simplifications of these expressions.

2.3. Canonical simplifications. Not many of the systems one encounters in physics or astronomy belong to the class of polynomial Hamiltonians. Furthermore, given an algebra of perturbations, there exists no general procedure for building generators for the kernel of the Lie derivative restricted to that algebra.

Out of these difficulties grew a new line of research. Rather than going to the extreme of forcing the transformed Hamiltonian into the kernel of the Lie derivative $\mathcal{L}_0$, one conceives of converting the original Hamiltonian by Lie transformation into a 'simpler' function not necessarily in the kernel. So far, the few trials performed using this compromise solution have been very promising. In the next section, we mention how 'canonical simplifications' helped in normalizing the main problem of artificial satellite theory to the fourth order without resorting to developments in the powers of the eccentricity. As a result, the idea of canonical simplifications is now explored vigorously to see how it would contribute a lunar theory without expansions in the powers of the eccentricity of the sun, and also to study the attitude of a rigid body rotating about a fixed point.

Finally, we should question our fixation with canonical transformations. It is quite conceivable that normalization could be achieved by Lie transformations that are not canonical. Consider, for instance, a perturbed circular pendulum, the principal part being

$$\mathcal{H}_0 = \tfrac{1}{2}\Theta^2 - \omega^2 \cos\theta$$

when all terms $(\mathcal{H}_n)_{n>0}$ in the perturbation belong to the algebra $\mathcal{A}$ of periodic functions of the elongation θ with coefficients in the kernel of $\mathcal{L}_0$. Already at first order, a normalization by canonical Lie transformation introduces various types of elliptic integrals, thereby making it necessary for the software to extend $\mathcal{A}$ into a broader algebra mixing ordinary trigonometric functions with elliptic functions.

Some time ago, a canonical simplification was found to remove all periodic terms at the first and second order [16]. But one can achieve far more if one is willing to operate with Lie transformations that are not canonical. It is shown in [17] how to build by induction a non-canonical transformation $(\theta, \Theta) \to (\phi, \Phi)$ to convert not the Hamiltonian itself but the canonical equations derived from the perturbed Hamiltonian $\mathcal{H}$ into differential equations of the form

(6)
$$\frac{\partial \phi}{\partial \tau} = \Phi, \qquad \frac{\partial \Phi}{\partial \tau} = -\omega^2 \sin \phi.$$

Someone interested in performing the reduction by machine will appreciate the fact that elliptic functions and integrals do not show up in the literal developments. The construction in [17] takes place solely in the algebra of trigonometric functions in θ. There is, of course, a price to pay for using non-canonical transformations around Hamiltonian systems. The price here is a change of independent variable defined through an implicit equation. But, as we learn from K. R. Meyer in these proceedings, the symbolic solution of implicit equations is advancing rapidly.

3. Application: satellite theory. The discussion of Section can be made more concrete by examining the role of normalizations and canonical simplifications in the theory of an artificial satellite.

We assume the satellite can be taken as a point of negligible extent and consider the attraction by the Earth as the only force. Attach to the Earth an orthonormal frame $(\boldsymbol{b}_1, \boldsymbol{b}_2, \boldsymbol{b}_3)$, where $\boldsymbol{b}_3$ is the direction of the polar axis, and $\boldsymbol{b}_1$ lies on the Greenwich meridian. The position of the satellite in the Earth frame is given by the vector

$$\boldsymbol{x} = r \left[\boldsymbol{b}_3 \sin \beta + (\boldsymbol{b}_1 \cos \lambda + \boldsymbol{b}_2 \sin \lambda) \cos \beta \right],$$

with $r > 0$ representing the distance from the center of the Earth to the satellite, β the geographic latitude such that $-\pi/2 \le \beta \le \pi/2$, and λ the geographic longitude such that $0 \le \lambda < 2\pi$. Let μ stand for the Gaussian constant of the Earth, i.e., the product $\mu = k^2 m_\oplus$ of the Newtonian constant of universal attraction by the mass of the Earth; let also α denote the equatorial radius of the Earth. Then the gravity field of the Earth at the position occupied by the satellite is minus the gradient $\nabla_{\boldsymbol{x}} V$ of the potential

$$V = -\frac{\mu}{r} \sum_{n \ge 0} \left(\frac{\alpha}{r} \right)^n \left[\sum_{0 \le m \le n} (C_{n,m} \cos m\lambda + S_{n,m} \sin m\lambda) \, P_m^n(\sin \beta) \right].$$

The functions $P_m^n(w)$ are the associated Legendre polynomials

$$P_m^n(z) = (-1)^m \sqrt[m]{1 - z^2} \, \frac{d^m}{dz^m} P_n(z).$$

The field parameters $C_{n,m}$ and $S_{n,m}$ are dimensionless constants. By definition of the Gaussian constant, $C_{0,0} = 1$, and by setting the Earth-bound frame at the center of mass of the Earth, there follows that all three coefficients $C_{1,0}$, $C_{1,1}$ and $S_{1,1}$ are zero.

Let $(\boldsymbol{s}_1, \boldsymbol{s}_2, \boldsymbol{s}_3)$ be an orthonormal frame fixed in space set at the center of mass of the Earth. Assume that the polar axis of the Earth $\boldsymbol{b}_3$ is fixed in space, and that the Earth rotates at a constant angular velocity $n_\oplus$ about $\boldsymbol{b}_3$. Then we choose $\boldsymbol{s}_3 = \boldsymbol{b}_3$ and consider only the *zonal* form of the problem in which the potential reduces to the function

$$V = -\frac{\mu}{r} \left(1 + \sum_{n \ge 2} \left(\frac{\alpha}{r} \right)^n C_{n,0} P_n(\sin \beta) \right),$$

$P_n(w) = P_{n,0}(w)$ being the Legendre polynomial of degree n. By common usage, one sets $C_{n,0} = -J_n$. For the Earth, $J_2 \approx 10^{-3}$ whereas for $n \geq 3$, $|J_n| \approx 10^{-6} \approx J_2^2$. On account of these relative sizes, we retain only the first two terms. The dynamical system $\mathcal{H}_0 + J_2\mathcal{H}_1$, where

$$\mathcal{H}_0 = \frac{1}{2}\left(R^2 + \frac{\Theta^2}{r^2}\right) - \frac{\mu}{r},$$

$$\mathcal{H}_1 = \frac{\mu\alpha^2}{r^3}P_2(\sin\beta),$$

constitutes the main problem in artificial satellite theory.

The main problem being a perturbed Keplerian system, normalization amounts to eliminating the mean anomaly. After many trials, we discovered at last that the normalization arises most directly as the product of two canonical transformations [18]. The first step is only a canonical simplification whereby the Hamiltonian is stripped of all dependence on the mean anomaly except for terms in $(p/r)^2$. More precisely, $\mathcal{H}$ is converted into a series of the type

$$\mathcal{H} = \mathcal{H}_0 + \left(\frac{\alpha}{r}\right)^2 \sum_{n>0} J_2^n C_n$$

whose coefficients C_n belong to the kernel of $\mathcal{L}_0$. We dubbed this simplification the 'elimination of the parallax.' From the point of view of symbolic algebra, this simplification has the incomparable advantage of taking place entirely in the algebra of functions of the type

$$F = \left(\frac{\alpha}{r}\right)^2 \sum_{n\geq 0} C_n \left\{ \begin{array}{c} \cos \\ \sin \end{array} \right\} n\theta$$

with coefficients C_n in the kernel of $\mathcal{L}_0$. More significantly, elimination of the parallax removed most of the difficult trigonometric functions in θ which cause difficulty in an ordinary Delaunay normalization, allowing us to eliminate the mean anomaly to the fourth power of J_2 by machine without resorting to developments in the powers of the eccentricity. The latter alternative is no longer attractive with today's aerospace engineers launching satellites at eccentricities greater than 0.1. This achievement paved the way for development of a complete third order theory for the main problem of an artificial satellite.

Following the elimination of the parallax, a Delaunay normalization eliminated the remaining short period terms for the main problem. To complete the theory, a long period transformation, once again a Lie transformation, was constructed to produce the secular Hamiltonian [19].

Our first attack on the satellite problem produced the third order theory mentioned above and a second order theory which included zonals through J_7. The algebra was first performed on the early PL/I versions of MAO and later on Dasenbrock's Fortran processor. Integer overflows, however, caused by the 32 bit word length on the Texas Instruments ASC vector computer limited the number of zonals that could be included in the potential to J_7. Recently, in exercising the Lisp version of MAO, we revisited the satellite problem. Given the extended precision integer arithmetic of the Lisp machine, we were able to produce the transformations to second order for the combined potential of J_2 through J_{10}. We mention this detail to underscore the fact that integer arithmetic with indefinite length has become an indispensable instrument in the toolbox of nonlinear mechanics.

4. Symbolic algebra systems. In principle, general purpose symbolic manipulation systems, such as Macsyma, Reduce and Mathematica, provide all the functions necessary to perform the desired normalizations and simplifications. These general purpose processors, however, suffer from the "free-form" specification of expressions which ignores the underlying algebraic structure. Specifying the exact operation desired or producing results in the precise form required often proves quite difficult. Moreover, by handling mathematical expressions of arbitrary structure, these systems cannot optimize their storage schemes or their algebraic algorithms for expressions of a fixed structure. Our experience with both general and special purpose processors clearly demonstrates that our research problems must be attacked with software specifically designed for the task at hand. In this paper, therefore, we concentrate on special purpose algebraic systems.

4.1. Serial processing. The early symbolic processors were special purpose programs, most of them, if not all, developed by individuals interested in solving specific classes of problems. Although each processor grew in capabilities with the research interests of its creator, nonetheless, the processor never departed from the algebraic structure characterizing the class of problems for which it was originally designed.

The Echeloned Series Processor (ESP) was developed in assembler on the IBM 360/44 computer for Delaunay's theory [20]. The ESP turned out to be barely adequate for the lunar theory, but spurred development of symbolic codes along the same pattern. In 1973, R. Dasenbrock at the Naval Research Laboratory developed a processor in Fortran [21]. This processor finds continued use at NRL [22]; portable versions were made at the University of Zaragoza for IBM/RT's, the CNES in Toulouse for Sun's, and the University of Cincinnati for PC's.

Developing the artificial satellite theory on Dasenbrock's processor exposed several limitations of the compilers and hardware of the time. The integer length of 32 bits limited the size of the rational coefficients. Memory limitations always posed a serious problem, partially as a result of scrimped core and no virtual memory, and partially as a result of the flat data representation for the expressions. Long run times of several hours on the ASC advanced array processor were common for large problems. Indeed, the ASC proved to be of little advantage for these types of calculations. Algebraic manipulations require tremendous amounts of localized operations like sorting, combining terms, and other operations to maintain the canonical form of the expressions. Such local operations prove difficult to vectorize, and we ended up using the ASC as a fast scalar machine. In sum, while the software could exploit very efficiently the algebraic structures inherent to the canonical representations, it made it very inconvenient to fit the problem at hand.

In 1984, Miller reincarnated MAO on a Symbolics workstation using an object-oriented dialect of Lisp. This version of MAO introduced a number of new features, most notably the polymorphism of functions and advanced mathematical typography on a bitmapped screen.

In the current version of MAO, both mathematical functions (such as multiplication) and operational functions (such as printing) are polymorphic. Each class of algebraic object knows how to handle addition, multiplication, printing, etc. in the appropriate way. For example, the product of two algebraic objects may be a multiplication or a scaling depending on the types of the operands. By using these generic operations, the code which implements polynomial multiplication may be optimized quite readily, since this routine concerns itself only with manipulating the two expressions at the polynomial level. The types of the coefficients and even the number of variables prove irrelevant, as long as generic routines exist to multiply and add terms.

The polymorphism of functions also provides a user of MAO with great freedom in structuring the solutions to his problem. By combining modules defined for manipulating polynomials, series and Fourier sums, the user constructs expressions matching exactly the algebra of his computational methods. For example, one might choose to work with series in the small parameter ϵ, whose terms are polynomials in L and n, whose coefficients are Fourier sums in the angles ℓ and g, whose coefficients are polynomials in e, whose coefficients are rational numbers (see Figure 1). Having this hierarchy of algebras, MAO keeps all expressions in this form, ready for any simplifications or other operations requiring the results to be in canonical form.

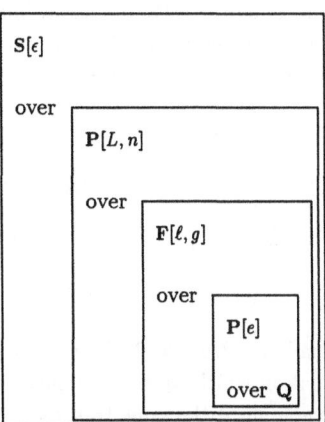

Figure 1: Hierarchy of algebras

On the whole, MAO has turned into a very handy tool for solving medium size problems. Yet, in spite of all provisions to ensure speed and efficiency, MAO appears too slow to handle very large problems. We do not invent these problems for the sake of pushing the equipment to the limit; they are out there, begging for a solution. The main problem of lunar theory was solved to an accuracy of 50m in the radius. Presently, the U.S. Naval Observatory time service requires a solution accurate to a

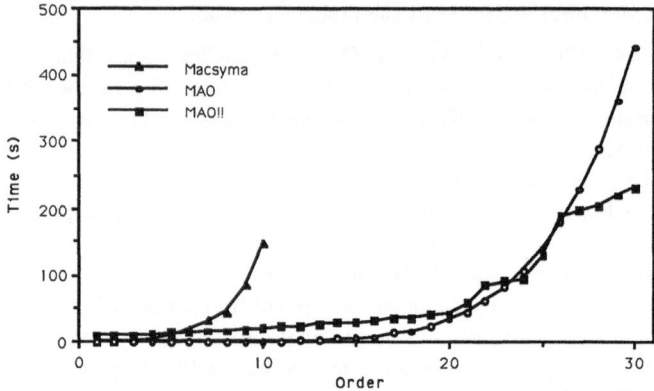

Figure 2: $(a/r)^5$ with rational coefficients

few centimeters. Most likely, so stringent a precision will tax the Symbolics environment beyond its present capabilities in both speed and memory. Thus, we turn once again to rebuilding MAO on a new type of machine, this time on a massively parallel processor — the Connection Machine (CM).

4.2. Parallel processing. As suggested in Figure 2, a massively parallel processor constitutes a powerful tool for manipulating large Poisson series. In the benchmark picture, we started with the ratio a/r as a Fourier series in the mean anomaly ℓ with polynomial coefficients in the eccentricity e over the rationals. Textbooks in celestial mechanics give only the first few terms of the series a/r:

$$
\frac{a}{r} \;=\; 1 \quad + e\cos\ell\left(1 - \tfrac{1}{8}e^2 + \tfrac{1}{192}e^4 + \ldots\right) + e^2\cos 2\ell\left(1 - \tfrac{1}{3}e^2 + \ldots\right)
$$
$$
+\, e^3\cos 3\ell\left(\tfrac{9}{8} - \tfrac{81}{128}e^2 + \ldots\right) + e^4\cos 4\ell\left(\tfrac{4}{3} + \tfrac{2}{5}e^2 + \ldots\right)
$$
$$
+\, e^5\cos 5\ell\left(\tfrac{625}{384} + \ldots\right) + \ldots.
$$

Acting upon a suggestion made by Prof. K. R. Meyer, we solved Kepler's equation by means of a Lie transformation which we then applied to obtain the series a/r to power 30 in e. We could have applied the same transformation to the power $(a/r)^5$, but we did not. Our purpose here is to produce an expression long enough that, by raising it to the fifth power, would exercise to the fullest the capabilities of the Connection Machine.

In one calculation of Figure 2, we use Macsyma; in another, we employ MAO on a Symbolics Lisp workstation. In the third, we compute the product on a Connection Machine, for which we have developed a package of procedures (MAO!!) written in *Lisp. In each run, we timed the operations for increasing orders in e. At order $n = 30$, each factor contains 256 terms. Macsyma is desperately slow—it took almost two and a half minutes to compute the result to order 10. MAO reached order 30 in approximately seven and a half minutes, while MAO!! did the same nearly twice as fast. The shape of the curve representing timings on the CM exhibits several interesting features. Up to order 22, the curve is practically linear; this is easily explained by the small increase in overhead due to the incremental growth in the length of the series from one order to the next. Then, the timing curve jumps to another linear slope slightly steeper than the initial one, and the phenomenon reproduces itself periodically at higher orders. At order 22, the jump results from the program having exhausted the available pool of physical processors and reconfiguring the real machine to behave as a logical machine with twice as many virtual processors. At order 26, the configuration changed again, again redoubling the number of virtual processors.

The tool developed on the Connection Machine has been dubbed MAO!! [24] — the suffix '!!' follows the CM programmer's convention for denoting parallelism. MAO!! achieves its gains over MAO by spreading Poisson series over thousands of processors. Each processor in the CM holds a single term of a Poisson series. This distribution provides a simple resource allocation scheme flexible enough to deal with the constant explosion and implosion of partial results so typical of symbolic algebra. In addition, the scheme permits many series to remain active in the CM memory.

From the standpoint of parallelism, algebraic operations fall into two classes. Multiplication by a monomial, partial differentiation and integration are local operations, requiring only isolated computation in each processor. On the other hand, multiplication and simplification are global operations in the sense that processors representing terms of the series must communicate among themselves. Global operations bring forth the real power of the CM.

Among the global operations, we concentrated on two problems: the simplification of like terms, and the multiplication of Poisson series. In both cases, we succeeded in introducing a high degree of parallelism. The secret was to take advantage of the possibility of restructuring the CM as a grid on which global patterns of communication act like translations in n-space. In this regard, combination of like terms turns into a sorting to put all like terms next to one another along intervals of a one-dimensional grid; after sorting comes a scanning to sum the like terms in the processor at the end of each segment.

MAO!! multiplies Poisson series by replicating the factors and forming all partial products at the same time. The code is best understood by looking at a simple example. To multiply a second degree polynomial in one variable $a + bx + cx^2$ by a polynomial $A + Bx$, we arrange the machine so that the first six processors on the one-dimensional grid contain the following quantities.

a	b	c	a	b	c
x^0	x^1	x^2	x^0	x^1	x^2
A	A	A	B	B	B
x^0	x^0	x^0	x^1	x^1	x^1

Then, all partial products are computed in parallel so that the one-dimensional grid now holds the quantities below.

a	b	c	a	b	c
x^0	x^1	x^2	x^0	x^1	x^2
A	A	A	B	B	B
x^0	x^0	x^0	x^1	x^1	x^1
aA	bA	cA	aB	bB	cB
x^0	x^1	x^2	x^1	x^2	x^3

There remains to pass the terms to the simplification routine and store away the remaining terms. The dynamic virtualization mechanism of the CM makes the multiplication of large series possible. For instance, when multiplying two Poisson series of 256 terms each, the intermediate result will have 2^{17} terms, while our CM provides only 2^{14} physical processors. Nevertheless, the CM may be configured as if it had 2^{17} virtual processors, where each physical processor emulates eight virtual ones.

In our opinion, massive parallelism presents a viable option for processing very large Poisson series. We intend to continue the development of MAO!! by introducing hierarchical structures like those in MAO, as well as refining the parallel and numerical algorithms. In addition, the introduction of the Data Vault — a parallel mass-storage device connected directly to the CM — offers the possibility of caching large numbers of series, constituting in effect a virtual memory.

5. Graphical studies of flows.

Although a normalization may vastly simplify a system, there remains much to do to understand its global behavior. By virtue of the normalization described in Section, the system has been reduced from one describing the evolution of the position of the satellite to one describing the evolution of the 'state' of the orbit, the instantaneous ellipses upon which the satellite moves. The stable equilibria of the reduced system indicate which orbits are 'safe'; a satellite would remain on or near such an orbit for a long time. The purpose of performing the normalization is to better understand the dynamics of the system. Once the algebraic manipulations are done and we have the normalized Hamiltonian, we still need to extract the qualitative features of the dynamics.

We are facing, from an algebraic point of view, a comparatively unstructured problem: whereas previously we were able to identify algebraic structures amenable to automated manipulations, we are now dealing with Hamiltonians usually so short that we can afford to process them by general purpose systems for symbolic and algebraic manipulations. Our guides at this point are insight and experience. For we are now trying to capture the major features in the global flow determined by normalized Hamiltonian equations.

As expected we start by locating the singularities in the system. In particular we are especially attentive to the creation and annihilation of equilibria as the parameters of the system are changing. One can envision several ways of learning these things, and, in fact, we use no single tool but a combination of numeric, symbolic and graphical techniques.

5.1. Topology to the rescue.

Carried away by tradition, the first people to study the satellite problem assimilated the phase space (g, G) to a cylinder, in part because they felt comfortable in looking at the system as a simple pendulum under perturbations. The fact is that the model of the pendulum is totally misleading. It misses those exceptional situations at $G = L$ where, the eccentricity being zero, the argument of perigee g has no meaning. Furthermore, it excludes the whole manifold of equatorial orbits because when the inclination I is zero, i.e., at $G = H$, the longitude h of the ascending node has no meaning.

However, as Elie Cartan showed some eighty years ago, on each manifold $L =$ constant, the set of bounded orbits that one gets from making constant all Delaunay elements but the mean anomaly ℓ consists of a pair of two-dimensional spheres. One can see that easily: indeed the vector functions

$$G = x \times X \quad \text{and} \quad A = (L/\mu)(X \times G - \mu x/r)$$

are independent of ℓ, and so are $\sigma = \frac{1}{2}(G + A)$ and $\delta = \frac{1}{2}(G - A)$. The identities $\|\sigma\|^2 = \|\delta\|^2 = L^2/4$ define the two spheres recognized by Cartan.

As we are interested mainly in perturbed Keplerian problems where the longitude of the ascending node is ignorable we found convenient to use coordinates other than the vectors σ and δ. We take

$$\eta_1 = \sqrt{L^2 - H^2} \cos h, \qquad \xi_1 = LG \sin I \, e \cos g,$$
$$\eta_2 = \sqrt{L^2 - H^2} \sin h, \qquad \xi_2 = LG \sin I \, e \sin g,$$
$$\eta_3 = H, \qquad \xi_3 = G^2 - \frac{1}{2}(L^2 + H^2).$$

In these coordinates, the orbital space is represented by the two spheres

$$\eta_1^2 + \eta_2^2 + \eta_3^2 = L^2, \qquad \xi_1^2 + \xi_2^2 + \xi_3^2 = \frac{1}{4}(L^2 - H^2)^2.$$

So, for a perturbed Keplerian system in which the longitude h is ignorable, the reduction by the group SO(2) identifies the orbital space to the unique sphere in the Euclidean space based on the coordinates (ξ_1, ξ_2, ξ_3).

Once the Hamiltonian is expressed in these coordinates, the equations of motion are

(7) $$\dot{\xi_i} = (\xi_i; \mathcal{H}) \quad \text{for} \quad i = 1, 2, 3.$$

These equations cover all possible motions on the spheres, including those in the neighborhood of the north pole $(0, 0, \frac{1}{2}\sqrt{L^2 - H^2})$ and the south pole $(0, 0, -\frac{1}{2}\sqrt{L^2 - H^2})$. At the north pole, $G = L$, hence that point represents circular orbits with an inclination such that $\cos I = H/L$; at the south pole, since G is equal there to H, we find the class of equatorial orbits with eccentricity $e = \sqrt{1 - H^2/L^2}$. The coordinates (g, G) amount to a Mercator projection of the sphere, which excludes the north and south poles. Curves on a sphere are easy to draw on the screen of a terminal. In an orthographic projection, there is no need for a special algorithm to decide whether a point is visible on the screen.

5.2. Phase flows by numerical integration. In the main problem of artificial satellite theory, the global equations (7) tell that the south pole is an equilibrium, call it $E_\dagger$, and that this equilibrium is stable for any H. The north pole is also an equilibrium, call it E_0, but sometimes stable, at other times unstable. This change of stability stems from two bifurcations. These facts are established by using a Newton-Raphson iteration to solve analytically the equilibria equations

$$(8) \qquad \partial\mathcal{H}/\partial g = 0, \qquad \partial\mathcal{H}/\partial G = 0$$

in the neighborhood of $H = L/\sqrt{5}$. Details of the calculations have been published elsewhere [25]. This was accomplished with Macsyma on a Lisp workstation, and, later, Mathematica [26] on a Macintosh.

Once the equilibria were located and their stability character established, we drew a sample of averaged orbits by integrating numerically the differential equations

$$(9) \qquad \dot{G} = \partial\mathcal{H}/\partial g, \qquad \dot{g} = -\partial\mathcal{H}/\partial G$$

Sampling was first accomplished at the terminal by selecting what we thought would be the appropriate initial conditions. After a long while, we came to an algorithm for selecting automatically those initial conditions that would produce the most telling image of the phase flow.

Our tool for doing the plots, which we named the "Doodler," operates somewhat like a hybrid of a plotting package and an object-oriented drawing program. Lines, curves, and figures can be drawn under program control. In addition, one can use the mouse to annotate or manipulate the images. For instance, one can superimpose the plots to make a collage of the kind that one sees in [25, Figures 2 and 5].

General purpose algebraic systems like Macsyma and Mathematica, on the one hand, and interactive graphic software like the Doodler proved effective in attacking more difficult systems. On account of the interest engineers have for "frozen orbits," we enlarged the main problem to involve a few more zonal harmonics, namely those of degree 4 and 6.

Save for a rotation by 90° the harmonic of degree 4 introduces no qualitative changes. The same, however, cannot be said for the harmonic of degree 6. Eliminating G between the equations (8) produces a resultant that is quadratic in $\sin g$, it opens the prospect of equilibria appearing at values of g away from multiples of $\pi/2$ whose appearance and annihilation are the effect of additional pitchfork bifurcations. At this stage of our analysis we began to feel very painfully how much we were straining the capabilities of our equipment. We had reached the point where a diagram of the phase flow produced by numerical integration required 60 uninterrupted hours on the Lisp workstation.

5.3. Phase flows by color painting. Investigation of the critical inclination in the main problem of satellite theory was initially limited to a neighborhood of $H/L = 1/\sqrt{5}$. As the value of H/L decreases, what becomes of the global dynamics? Answering the question analytically by iterating the Newton-Raphson procedure is out of the question. The result would be so complex as to preclude any significant interpretation. The only way then is to proceed numerically. The function Solve of Mathematica proves eminently suitable in that direction. To our surprise we found that, close to $H/L = 0.0306$ for $J_2\alpha^2/a^2 = 0.001$, there appear four more roots in G for equations (8) when g equals 0 or π.

This discovery made us wish to be able to see snapshots of the phase flow as the ratio H/L runs from $1/\sqrt{5}$ to 0. Numerical integration will not serve this purpose. Let us approach the problem from a different perspective. Since we are dealing with conservative Hamiltonians having only one degree of freedom, the average orbits are the level curves of the Hamiltonian over the orbital spheres. Initially, we experimented with the contouring features of Mathematica; we found them wanting. Not all interesting features could be shown simultaneously due to insensitivity to changes of scale. Furthermore, numerical roundoff gave poorly defined trajectories.

There is an alternative that solves all these problems and furthermore is ideally suited to the data-parallel architecture of the Connection Machine. At each point in the phase space we compute the value of the Hamiltonian; we convert that value into a color code; we assign that color code to the pixel representing the point. This is what we call "painting the Hamiltonian onto phase space." These computations are performed for each point independently of the computations for any other points, and thus they can proceed in parallel. On the screen, we see strips of different colors; the boundaries between adjoining strips are a substitute for contour levels of the Hamiltonian.

A straightforward painting of the Hamiltonian would suffer from the same problem as contour drawing in that it would be insensitive to changes in scale. The algorithm mapping value onto color must be selected carefully, for there is a difference between the map we seek and a cartographer's contour map. The cartographer wants to represent altitudes uniformly throughout the map. For us, however, comparison of altitudes is not the issue. The altitude of a point over the sphere is determined by the value of the Hamiltonian at that point. We are not interested in reading values of the Hamiltonian in the map; we only want to relate points if they are of the same value and to contrast them if they are not. In particular, we care about marking peaks, hollows, and passes; we are especially attentive to the contours around these singularities. These are the features that we want to emphasize even if it means losing all information about their relative heights.

The new equilibria just mentioned have a Hamiltonian value which is about 10^4 times smaller than that of the critical inclination points. In a uniform height scale, it would take an enormous number of colors (or, for a contour plot, an enormous number of contour lines) to enhance these very low features, so many in fact that the interesting areas would have only slight color gradation. To a cartographer, the problem is one of separating a huge mountain from a shallow lake at its foot, and furthermore of showing the details in the lake and around the summit. A cartographer might solve the problem by printing the heights of the mountain and the lake on the map. We do not care so much about the heights as we do about showing which points in the neighborhood of the mountain and the lake are of the same height and about contrasting points at the same elevation with nearby points at slightly different altitudes.

One way of achieving this result is by covering the range of Hamiltonian values not by a single spectrum, but by covering with several spectra. This will effectively show finer detail. The procedure, however, has its limits. Too many bands of color would make the picture confusing.

Figure 3: The phase flow slightly below $H/L = 1/\sqrt{5}$. The region depicted is an orthographic projection of the northern hemisphere upon the equatorial plane $\xi_3 = 0$. The north pole is at the center of the figure; the positive horizontal axis corresponds to $g = 0$.

Another way is to give up a uniform height scale. Whether shallow or high, the interesting points and their neighboring flows generally occupy about the same area in phase space. We draw a very fine scale around the stationary points, and take longer strides along the mountain sides and other less interesting areas. Thus we shall weigh the assignment of colors by the distribution of values, so that any strip of a given color has approximately the same number of points as any other strip. Our shallow lake, the newly discovered equilibria, occupies about the same area in phase space as the mountain of critical inclination, so it will now appear with as many color strips, even though the range of values of the Hamiltonian are dramatically different.

Calculating the Hamiltonian at each point and mapping its value to a color are operations that can be performed in parallel. On the Connection Machine (16K processors with floating point accelerator), the rendering of a 512×512 plot takes about one second; contrast this to the Symbolics where it takes about 40 minutes.

The first illustration of the painting technique, Figure 3, reproduces in black and white a color phase diagram on the display of the Connection Machine [27]. It shows the flow of the averaged equations in a neighborhood of the critical inclination after the two bifurcations which have alternatively changed the stability of the north pole from stable to unstable and back to stable again. Figure 4 shows the newly discovered equilibria. These equilibria appear by a saddle node bifurcation. The ability of the Connection Machine to display rapidly the phase flow as it evolves through the bifurcation is very effective in animating the mechanism by which these equilibria come to existence.

At the speed of the Connection Machine, one can make a movie of phase flow as the ratio H/L is varied. Pictures are produced fast enough that one's mind re-

Figure 4: The eyes of the hippopotamus. A view from the south pole.

tains knowledge of the behavior, so that creative exploration with different values of parameters, or zooming in on some section of phase space yields a solid feel of the behavior, sometimes even new knowledge as well. In fact, a suspicious looking pattern at the previously unstable points of $g = \pi/2$ and $g = 3\pi/2$ at low values of H turned out to be, on closer examination, the aftermath of a pitchfork bifurcation into one stable and two unstable equilibria, the new unstable points being at slightly different values of g. This bifurcation had not been anticipated, although examination of the equilibria equations after the fact showed how they occur. Figure 5 is a zoom of the region around the equilibrium $g = \pi/2$ where the unanticipated bifurcation occurred.

6. Conclusions.

Most recently, visualization has been added to our collection of tools to provide insight into the global behavior of dynamical systems. The appeal of color graphics as an exploratory tool is not to be turned down. For it is at the start of a qualitative analysis rather than in the middle of writing the research report that global pictures should be sought.

Unlike other disciplines identified with pure geometry, nonlinear dynamics should not and never shall be one of those lofty towers where individuals feel compelled to live up to some intellectual asceticism away from machines and gadgets. The discipline instead must try with tenacity to keep pace with computational technology and make room for its innovations the same way. The challenge thus is endless, for each generation of mathematical physicists needs to keep abreast of techniques relentlessly emerging from the engineering shops. There is excusable dabbling and playfulness in shopping for a tool.

114

Figure 5: Zooming onto a surprise bifurcation in Figure 4.

References

[1] A. DEPRIT AND A. DELIE, *Trojan Orbits I. D'Alembert Series at L_4*, Icarus, **4** (1965), pp. 242–266.

[2] A. DEPRIT, *Limiting Orbits at the Equilateral Centers of Libration*, Astron. J., **71** (1966), pp. 77–87.

[3] A. DEPRIT, J. M. A. DANBY AND A. ROM, *The Symbolic Manipulation of Poisson Series*, Boeing Document D1–82–0481 (1965).

[4] J. M. GERARD, I. ISZAK AND M. P. BARNETT, *Mechanization of Tedious Algebra: The Newcomb Operators of Planetary Theory*, Comm. ACM, **8** (1965), pp. 27–32.

[5] D. BARTON, *A scheme for manipulative algebra on a computer*, Computer J., **9** (1967), pp. 340–344.

[6] A. DEPRIT, J. HENRARD AND A. ROM, *Trojan Orbits II. Birkhoff's Normalization*, Icarus, **6** (1967), pp. 381–406.

[7] A. DEPRIT AND A. ROM, *The Main Problem of Artificial Satellite Theory for Small and Moderate Eccentricities*, Celest. Mech., **2** (1970), pp. 166–206.

[8] A. DEPRIT, J. HENRARD AND A. ROM, *Lunar Ephemeris: Delaunay's Theory revisited*, Science, **168** (1970), pp. 1569–1570.

[9] A. ROM, *Mechanized Algebraic Operations (MAO)*, Celest. Mech., **1** (1970), pp. 301–319.

[10] A. DEPRIT, *Canonical Transformations Depending on a Small Parameter*, Celest. Mech., **1** (1969), pp. 12–31.

[11] A. DEPRIT, *Dynamics of Orbiting Dust under Radiation Pressure*, The Big Bang and Georges Lemaître, ed A. Berger, The Reidel Publishing Company, Dordrecht, (1984), pp. 151–180.

[12] S. L. COFFEY, A. DEPRIT AND B. R. MILLER, *The Quadratic Zeeman Effect in Moderately Strong Magnetic Fields*, Ann. New York Acad. Sc., **497** (1987), pp. 22–36.

[13] A. DEPRIT, S. FERRER, M.-J. MARCO AND J. PALACIÁN, *The Hydrogen Atom in Parallel Electric and Magnetic Fields (m=0)*, Physics B, submitted for publication.

[14] A. DEPRIT AND B. R. MILLER, *Normalization in the Face of Integrability*, Ann. New York Acad. Sc., **536** (1988), pp. 101–126.

[15] G.-I. HORI, *Theory of General Perturbations with Unspecified Canonical Variables*, Proc. Astron. Soc. Japan, **18** (1969), pp. 1287–1296.

[16] A. DEPRIT, *The Ideal Resonance Problem at First Order*, Advances in the Astronautical Sciences, **46** (1982), pp. 521–526.

[17] J. HENRARD AND P. WAUTHIER, *A Geometric Approach to the Ideal Resonance Problem*, Celest. Mech., **44** (1989), pp. 227–238..

[18] A. DEPRIT, *The Elimination of the Parallax in Satellite Theory*, Celest. Mech., **24** (1981), pp. 111–153.

[19] S. L. COFFEY AND A. DEPRIT, *A Third Order Solution to the Main Problem in Satellite Theory*, J. Guidance, Control, and Dynamics, **5** (1982), pp. 366–371.

[20] A. ROM, *Echeloned Series Processor (MAO)*, Celest. Mech., **3** (1971), pp. 331–345.

[21] R. R. DASENBROCK, *Algebraic Manipulation by Computer*, NRL Report **7564** (1973).

[22] ————————— *A FORTRAN-Based Program for Computerized Algebraic Manipulation*, NRL Report **8611** (1982).

[23] A. DEPRIT AND E. DEPRIT, *Massively Parallel Symbolic Computation*, Proceedings of the ACM-SIGSAM 1989 International Symposium on Symbolic and Algebraic Computation, ACM Press, New York, (1989), pp. 308–316.

[24] ————————————— *Processing Poisson Series in Parallel*, J. Symbolic Computation, submitted for publication.

[25] S. L. COFFEY, A. DEPRIT AND B. R. MILLER, *The Critical Inclination in Artificial Satellite Theory*, Celest. Mech., **39** (1987), pp. 365-406.

[26] S. WOLFRAM, *Mathematica. A System for Doing Mathematics by Computer*, Addison-Wesley, Reading, MA, 1988.

[27] S. L. COFFEY, A. DEPRIT, E. DEPRIT AND L. HEALY, *Visualization of Phase Flows*, submitted for publication.

COMPUTER ASSISTED PROOFS OF STABILITY OF MATTER

R. DE LA LLAVE*

Abstract. We review some recent progress in the study of Schrödinger equations for arbitrarily many fermions interacting via Coulomb forces. The goal is to prove lower bounds for the infimum of the spectrum which are reasonably close to optimal. Some of the key estimates are established with the help of a computer.

Key words. N–body problem, Schrödinger equation, Taylor methods, Interval arithmetic, Ground states.

AMS(MOS) subject classifications. 81–08, 81C99, 82–03, 2A15.

1. Description of the problem and motivation.

The problem we have considered is to obtain lower bounds for the so–called N–body problem in quantum mechanics.

The most standard version of the problem can be formulated mathematically as:

Problem 1.1. *For any* $N, M \in \mathbf{N}$, $Z \in \mathbf{R}$, *choose* M *points* $y_1, \ldots, y_M \in \mathbf{R}^3$ *and consider the operator:*

$$(1.1) \quad H^{N,M} \equiv \sum_{i=1}^{N} -\Delta_{x_i} + \alpha \left[\sum_{\substack{1 \le j \le N \\ 1 \le k \le N \\ j < k}} \frac{1}{|x_j - x_k|} + \sum_{\substack{1 \le j \le M \\ 1 \le k \le M \\ j < k}} \frac{Z^2}{|y_j - y_k|} - \sum_{\substack{1 \le j \le N \\ 1 \le k \le M}} \frac{Z}{|x_j - y_k|} \right]$$

acting on the Hilbert space L^2_{antisym}, *the space of complex valued functions* Ψ *of* N *variables in* $\mathbf{R}^3$ *whose square is integrable and which satisfy the antisymmetry condition:*

$$(1.2) \quad \Psi(x_1, \ldots, x_i, \ldots, x_j, \ldots, x_N) = -\Psi(x_1, \ldots, x_j, \ldots, x_i, \ldots, x_N)$$

Define

$$(1.3) \quad E_{N,M} = \inf_{y_1, \ldots, y_M \in \mathbf{R}^3} \inf_{||\Psi|| = 1} \left\langle \Psi, H^{N,M} \Psi \right\rangle$$

*Dept. of Math., University of Texas, Austin, TX 78712. Supported in part by N.S.F. grants

Show that:

(1.4) $$E_{N,M} \geq -C(Z)(N+M)$$

The physical interpretation can be found in any textbook in quantum mechanics. Let us recall it briefly:

The setup describes a system of N electrons interacting among themselves and with M nuclei of charge Z – in physical applications it is an integer – via electric forces described by Coulomb's law. In appropriate units, the strength of the electrostatic force is measured by α. The nuclei are assumed to be at fixed positions $y_1, \ldots, y_M$. The "wave function" $\Psi(x_1, \ldots, x_N)$ has the meaning that $|\Psi(x_1, \ldots, x_N)|^2$ is the probability density of finding one electron at x_1, another one at x_2, etc. The meaning of $\langle \Psi, H\Psi \rangle$ is the "*expected energy*" of the system. The antisymmetry requirement for the wave function is the so–called *Pauli exclusion principle*, which implies that electrons, even if undistinguishable cannot occupy the same state.

Remark. The model as it was described, does not take into account the "*spin*" of the electrons. To include it, we just consider the variables x_i entering in Ψ to range over $\mathbf{R}^3 \times \{-1, 1\}$ rather than $\mathbf{R}^3$. This modification affects all the numbers coming from the theory.

For physical systems at relatively low temperature, one expects that they will get rid of as much as possible of the energy and will be described most of the time by the states where the energy is close to the minimum $E_{N,M}$.

Lower bounds for $E_{N,M}$ can be interpreted as statements that a system cannot give off much energy by rearranging itself in a particularly favorable state (these arrangements would happen spontaneously, so they, probably, should have been observed if they were possible). We point out that to show that $E_{N,M} > -\infty$ for N or $M \geq 2$ is non trivial and was proved for the first time by Kato in [Ka]. For the corresponding model based on classical mechanics, indeed $E_{N,M} = -\infty$ except in the trivial case $N = 0$ or $M = 0$.

The fact that $E_{N,M}$ does not grow faster than linearly with the total number of particles has important physical consequences. Once one proves the bound (1.4) it is comparatively easy to show that $\lim_{N \to \infty} E_{N,N}/N$ exists. This shows that, when N is macroscopically large $E_{2N,2N} \approx 2E_{N,N}$. In physical terms, when we put together two equal large systems, the energy of the system is approximately twice the energy of each one. We do not gain energy by making a bigger out of two smaller ones. If, instead, we had $E_{N,N} \approx N^\tau$, then, E_{2N} would be $\approx 2^\tau E_N$ so that putting together two masses of N particles, would give off $(2^\tau - 2)E_N$. If $\tau > 1$ even by a tiny amount, systems large enough for the approximation to hold, would tend to absorb neighboring systems and give off energy. This situation is described as *collapse*, and the fact that there is no collapse, ((1.4)) is described as *stability of matter*.

We should stress that the behavior of bulk matter can be described by different models depending on which physical effects we decide to model. Since there is no all-encompassing physical theory, such models involve, of necessity approximations that can only be valid for certain ranges of parameters. We will not go into the physical discussion of limits of validity of models, but rather prove rigorous theorems about the models themselves. When we use the abbreviation *matter collapses* or *matter is stable* it should be understood as a mathematical statement about the model being discussed at the moment. We just point out that, even if we will not argue for it, the theorems we prove are considered to apply for physically relevant parameters.

Stability of matter, even if part of every day experience, is surprising because, after all, the expression we are bounding has N^2 terms. If stability is to be true at all, it has to depend on very delicate cancellations of the N^2 terms entering in the potential. The fact that these cancellations are subtle is illustrated by the fact that, if we forget about the antisymmetry requirements, the result is false [Li2]. The form of the potential is also crucial since apparently tamer potentials, finite and exponentially decreasing can lead to collapse ([Th] p. 258.) Gravitation is an example of a physical model which predicts collapse. For sufficiently large N –a few times the mass of the sun– E_N grows faster than linear and matter indeed collapses. Those are the *neutron stars*. (See [Th] Chapter 4.2, [LT2].)

It is worth pointing out that since energy controls the statistical properties of a system in contact with a heat bath, a problem closely related to stability is to show that thermodynamic properties such as entropy, etc. are asymptotically proportional to the number of particles and that they satisfy some qualitative properties such as convexity. We refer to [Li1], [Li2], [Th] for a review of implications of stability of matter and related subjects. [Th] contains a wealth of material on the problem of deriving properties of bulk matter from first principles using quantum mechanics.

Since failure to account for the bulk properties of matter was one of the main reasons the classical theory of matter was abandoned, it is reassuring to be able to show that quantum mechanics can indeed account for these properties.

Stability of matter was proved first by Dyson and Lenard [DL1], [DL2]. Another much simpler proof was discovered by Lieb and Thirring [LT1]. A few years ago, C. Fefferman showed in [Fe1] that, assuming that (1.4) holds with a constant C close to the one for a configuration describing separate atoms, it follows from first principles that, for certain ranges of temperature and density, matter formed of protons and electrons can be described as a plasma – almost homogeneous density of protons and electrons and almost no correlations –. For other ranges of temperature and density, matter can be described as a gas of hydrogen atoms – the correlation between protons and electrons are almost those of a hydrogen atom and the correlation between different atoms is very small–.

Clearly, in view of this result, proving good lower bounds of the constant C in (1.4) became an interesting problem and provided motivation for the search of new methods that could provide better bounds.

What we are going to describe in this paper is part of a program initiated by C. Fefferman to obtain lower bounds for the N–body problem in quantum mechanics.

We are going to describe only the part of the program in which computer assisted proofs play an important role and in which the author has been personally involved. Both restrictions are non–trivial: there are very important results that follow from the same techniques without using the computer and there are computer–assisted results obtained by others, notably H. Trotter, L. Seco and C. Falcolini. The work in progress described in section 4, has been done in collaboration with C. Fefferman, H. Trotter and C. Falcolini.

2. Relativistic stability of matter.

The problem considered here was not the standard N–body problem in quantum mechanics but rather a modification in which the kinetic energy is supposed to be $|p|$ rather than the usual $p^2/2m$ (we take units in which $m = 1/2$). Since $|p| \leq p^2 - 1/4$, lower bounds for this problem imply lower bounds for the standard problem. In particular, the following theorem proved by computer–assisted methods in [FL] can be used to prove (1.4) for $Z = 1$.

Theorem 2.1. *If $\alpha \leq 70/144\pi$, then the operator*

$$(2.1) \quad H \equiv \sum_{i=1}^{N} (-\Delta_{x_i})^{1/2} + \alpha \left[\sum_{j<k} \frac{1}{|x_j - x_k|} + \sum_{j<k} \frac{1}{|y_j - y_k|} - \sum_{j,k} \frac{1}{|x_j - y_k|} \right]$$

satisfies $\langle \Psi, H\Psi \rangle \geq 0$ for any $\Psi \in L^2_{\text{antisym}}$.

Remark. The fact that we have the square root of the Laplacian in place of the Laplacian makes all the terms in the Hamiltonian have units of inverse of length. This has as a consequence that if there is a configuration with negative energy, by shrinking the positions of the protons and electrons, we get another configuration of smaller energy. If there indeed was a configuration with negative expected energy, we could obtain an unlimited amount of energy form it by shrinking indefinitely. This is the relativistic collapse. Using trial functions, it is possible to show that this collapse indeed happens when $\alpha \geq 2/\pi$ [He] [We]. Notice that then, the conclusions of Kato's theorem do not hold for this model. This is not surprising since Kato's theorem uses essentially that, since the Laplacian has units of Length^{-2} and the potential of Length^{-1}, for phenomena happening in small scales, the Laplacian is the dominant term.

The scale invariance of (2.1) also makes it impossible to use the methods based on statistical methods that we mentioned before.

Some landmark papers in the study of this problem are: [We], [He] studied the problem of one nucleus and one electron – it can be solved using the Mellin transform –, [DL3] studied the problem of one electron and many nuclei, and [Co] proved stability of matter for this model provided that α, $Z\alpha$ was small enough – the estimate quoted in the paper was 10^{-200} –.

In this section, we will describe the computer–assisted method used in [FL] to prove Theorem 2.1 Further developments along similar lines, can be found in [Fe3], [LY].

The basic strategy in the proof of stability of matter of [FL] – and some others – is to bound the problem by another one that decomposes into as many independent pieces as nuclei. Bounding from below the independent pieces is very similar to bounding from below the energy of atoms. The caricature of atoms can be bounded from below by a problem describing independent electrons.

The main analytical device introduced in [FL] to achieve this decompositions is to rewrite the expected value of the Hamiltonian as an integral over the space of balls. If we denote by $B(z,R)$ the ball of center $z \in \mathbf{R}^3$ and radius $R \in \mathbf{R}^+$ we have:

$$(2.2) \qquad \frac{1}{|x-x'|} = \frac{1}{\pi} \int_{z \in \mathbf{R}} \int_{z \in \mathbf{R}^+} \chi_{x,x' \in B(z,R)} \frac{dzdR}{R^5}$$

$$(2.3) \qquad \int_{z \in \mathbf{R}^3} \int_{R \in \mathbf{R}^+} \int_{x,y \in B(z,R)} |u(x)-u(y)|^2 \, dx dy \frac{dzdR}{R^8}$$
$$= \frac{16\pi}{35} \int \frac{|u(x)-u(y)|^2}{|x-y|^4} \, dx dy = \frac{32\pi^2}{35} \left\langle u, (-\Delta)^{1/2} u \right\rangle$$

Except for the value of the constants, one can convince oneself quickly that the formulas are correct observing the invariance under rotation and the properties of both sides under scaling.

These formulas have immediate consequences. Calling

$$\mathcal{N}(z,R,x_1,\ldots,x_N) = \sum_{\substack{i<j}}^{N} \chi_{x_i,x_j \in B(z,R)}$$

$$\mathcal{M}(z,R,y_1,\ldots,y_M) = \sum_{\substack{i<j}}^{M} \chi_{y_i,y_j \in B(z,R)}$$

the number of electrons and of protons in a ball, we have:

(2.4)
$$V(x_1,\ldots,x_N;y_1,\ldots,y_M) =$$
$$\frac{1}{\pi} \int_{z \in \mathbf{R}^3} \int_{R \in \mathbf{R}^+} (\mathcal{N}(\mathcal{N}-1)/2 + \mathcal{M}(\mathcal{M}-1)/2 - \mathcal{N}\mathcal{M}) \frac{dzdR}{R^5} =$$
$$\frac{1}{\pi} \int_{z \in \mathbf{R}^3} \int_{R \in \mathbf{R}^+} \left(\frac{1}{2}(\mathcal{N}-\mathcal{M})^2 - \mathcal{N}/2 - \mathcal{M}/2 \right) \frac{dzdR}{R^5}$$

Likewise, the kinetic energy can be rewritten:

$$(2.5) \quad \sum_{k=1}^{N} \left\langle \Psi, (-\Delta_{x_k})^{1/2} \Psi \right\rangle = \frac{35}{32\pi^3} \int_{z \in \mathbf{R}^3} \int_{R \in \mathbf{R}^+} \int_{x_k \in (\mathbf{R}^3)^{N-1}} T_k(z,R,x_k) \frac{dzdR}{R^5}$$

where $T_k(z, R; x_k) = \int_{x', x'' \in B(z,R)} |\Psi(x', x_k) - \Psi(x'', x_k)|^2 dx' dx''$.

The formulas (2.4), (2.5) are interesting because they write the energy as integral of local quantities. Notice also that (2.4) shows that, unless the charge in a ball is more or less balanced, the contribution of this ball is largely positive. This agrees with the experimental fact that, in states with low energy, the electrostatic charge is almost exactly balanced.

In this representation, it is also possible to take into account the effect of antisymmetry. Due to the antisymmetry, two electrons in the same ball cannot both be in a state that minimizes kinetic energy. This raising of the energy can be interpreted as an extra repulsive term. As we mentioned before, this extra term is crucial to the proof of the result.

Reducing the original problem to a set of independent problems for each nuclei can be achieved by dividing the contribution of the balls among the nuclei. The contribution of a ball is divided equally among all the nuclei it contains. If it contains none, its contribution is assigned to the nucleus nearest to the center.

The problem for each nucleus can be further estimated to show it is bounded from below if for any choice of $0 < R_1 < R_2 < 1$, the quadratic form $\mathcal{Q}(\Psi) = \int_0^1 \bar{\Psi}(s) A[\Psi](s) s^2 ds$ with

(2.6)
$$A[\Psi](s) = W(s)\Psi(s) - 2\pi \int_0^1 K(s,t)\Psi(t)t^2 dt$$

$$W(s) = -1/s + \gamma \int_s^1 \int_0^R t^2 \frac{dt dR}{R^5 M(R)}$$

$$K(s,t) = \gamma' \int_{\max(s,t)}^1 \frac{dR}{R^5 M(R)}$$

$$M(R) = 1 + \chi_{[R_1, 1]} + \chi_{[R_2, 1]}$$

is positive for all Ψ in an appropriate domain that ensures that the form makes sense. γ and γ' are constants related to the original α.

We omit the details which can be found in [FL], but we point out that the variational equations for this problem are, formally, O.D.E's and the conditions on the domain can be interpreted as boundary conditions. It is furthermore possible to show that, for certain ranges of parameters, the minimizing solution exists and indeed satisfies the O.D.E.

It is possible to show analytically that the quadratic form is positive for certain values of $\alpha \leq \alpha^*$. If we can show that for $\alpha \in [\alpha^*, \alpha^{**}]$ there are no solution of the eigenvalue problem with zero eigenvalue, then the form will be positive when $\alpha \leq \alpha^{**}$. It is the later problem that will be tackled with the computer.

3. Computer assisted lower bounds

3.1. Interval arithmetic. Interval arithmetic, introduced by R. Moore in the 60's provides a very convenient framework in which to perform rigorous calculations with the computer. We refer to [Mo],[KM] for more details.

Usually, computers come equipped with an approximate arithmetic. Arithmetic operations are only defined on finite set $\mathcal{R} \subset \mathbf{Q}$, – the representable numbers – which the machine can manipulate very effectively. The manufacturer supplies arithmetic operations acting on $\mathcal{R}$ which produce either another representable which is close to the right answer or raises an exception indicating that no such a number can be found – e.g. an overflow or division by zero –. Since approximate of approximate may not be approximate at all, it is very difficult to asses the reliability of the final result of a large number of arithmetic operations. The situation is even worse when, as happens in practice, the arithmetic operations are meant to be an approximation to an analytic operations which is our real interest.

The basic idea of interval arithmetic is to introduce arithmetic operations that produce representable numbers which are upper (resp. lower) bounds for the true answer. Using them systematically, it is possible to obtain upper and lower bounds of arithmetic expressions given upper and lower bounds of the variables entering in them.

A convenient way of organizing this idea is to introduce an structure *"interval"* whose boundaries are representable numbers. We take it to mean a set of possible values which an expression can take.

It is possible to write routines that, given intervals, can produce intervals which are guaranteed to contain the true result of the mathematical operation performed on any number contained in the input intervals.

The result of translating an algebraic operation in its interval counterparts is guaranteed to contain the result of the mathematical operation whenever the variables are contained in the input intervals.

It is well known that interval arithmetic results can be very pessimistic because they assume worst case estimates at each stage of the computation.

Two main effects are *"subdistributivity"*: For any three intervals I, J, L, we have denoting by $+_v$ and $*_v$ the interval operations.

$$(\mathcal{I} +_v \mathcal{J}) *_v \mathcal{L} \subset (\mathcal{I} *_v \mathcal{L}) +_v (\mathcal{J} *_v \mathcal{L})$$

and *"coherence"* effects. If two expressions e_1, e_2 involving the variable x are negatively correlated (e. g. x and $-x$), and $\mathcal{E}_1$, $\mathcal{E}_2$ are intervals guaranteed to contain e_1, e_2, then $\mathcal{E}_1 +_v \mathcal{E}_2$ will be a pessimistic estimate of $e_1 + e_2$ even if $e_1 \subset \mathcal{E}_1$ $e_2 \subset \mathcal{E}_2$ are sharp bounds.

Even if these effects are always a theoretical possibility, there are many applications where the intervals that we need to consider have a width which is negligible compared with the value they represent, so that these effects are not worrisome. Very often, if the subdistributivity and coherence effects are serious, it should be interpreted as a symptom that the algorithm used has been poorly chosen or that the implementation should be improved.

3.2. Non–existence of solutions of boundary value problems. Most of the computer assisted proofs based on interval arithmetic had been geared to "*validate*" a numerical calculation –an exception is [MP]–. In our case, we want to use the interval arithmetic to "*exclude*" the existence of solutions of a family of boundary value problems for certain range of parameters.

The basic idea is to reduce the problem of existence of solutions to whether a certain function takes the value zero or not. If we compute an interval which is guaranteed to contain all possible values of the function but does not contain zero, we are sure that our function does not take the value zero and, hence, the problem does not have a solution.

Remark. We believe that in these exclusion problems, interval arithmetic presents notable advantages – besides that of rigor – over traditional methods. Within the philosophy of traditional numerical analysis, the most natural way of proceeding would be to compute the function on a grid and check whether the values obtained are safely away from the value zero. To get convincing results one should make estimates of the derivative of the function. Either if we obtain them analytically or numerically, it is likely that we will have to use a very fine grid. The need of computing in a fine grid can easily overcome the overhead involved with the use of intervals.

For a second order boundary value problem, the natural function to consider is the value of the Wronskian of two solutions, each of which satisfies the boundary conditions at one end. It is well known that if the Wronskian does not vanish identically, the two solutions satisfying the boundary conditions are linearly independent and it is possible to obtain a Green function so that the only solution of the homogeneous problem is the zero solution.

A possible way of organizing the calculation, would be to write an O.D.E. solver that given the values of the solution and the derivative at a certain point, can compute them at a neighboring point. Then, starting from the boundaries, it should be possible to compute the values of the function and the derivatives at an intermediate point.

In our case, some inessential complications arise from the fact that there are points in which the O.D.E. has discontinuities and because it is singular at the boundaries. This requires that the solver uses different algorithms. Other mathematical complications such as certain values that have to be treated specially or the need to exclude some degenerate ranges of parameters can be dealt with with the same methods.

It is interesting to point out that the most popular algorithms to solve O.D.E's – Runge–Kutta, predictor–corrector, Stoerr–Burlish – are not easy to make rigorous. Theoretical error analysis are available for the first two, but they seem too cumbersome to implement in practice.

An alternative well suited for interval arithmetic are the so–called "*Taylor meth-*

ods". The basic idea is to use the fact that, by matching powers in the O.D.E., it is possible to determine the Taylor expansion of the solution. The remainder can be estimated by using a majorant method (an exposition of the method can be found in [Po] §20 – §27). We have found it more convenient to use the contraction mapping principle to estimate the errors because of reasons that will be clarified later. The basic idea is that a solution of $y' = f(y)$ is a solution of $y(t) = y_0 + \int_0^t f(y(s)) \, ds \equiv \mathcal{T}[y](t)$. This is a fixed point equation for the operator $\mathcal{T}$. It is possible to show that $\mathcal{T}$ is a contraction in appropriate spaces. If y^* is a polynomial for which we can show $\|y^* = \mathcal{T}[y^*]\| \leq \epsilon$, then, there will be a true solution y satisfying $\|y - y^*\| \leq \epsilon/(1 - \|D\mathcal{T}\|)$.

We refer to [Mo] for a discussion of Taylor methods with error bounds. We also refer to [Ch] for a program that given a differential equation, generates a heuristic FORTRAN program implementing a Taylor method. Other alternatives, reducing to integral equations are mentioned in [KM].

We also point out that if the equation contains parameters, translating all the algebraic operations into interval arithmetic, it is possible to obtain bounds which are uniform when the parameters range on the intervals we assign them.

3.3. Implementation and practical considerations. We wrote an interval arithmetic package in standard PASCAL. Using these routines, we implemented the O.D.E. solvers needed. Originally, the routines were written for the 4.1 BSD PASCAL compiler on a VAX 11/750. We checked that the program run without any difference under the ULTRIX compiler for the same machine. Later, it proved very easy to get the program running under SUN's or MS-DOS machines with TURBO or Microsoft PASCAL.

The O.D.E. solvers we constructed were as conservative as possible. Given intervals containing initial time, the final time, the values of the function and its derivative and the parameters, they produced intervals guaranteed to contain the solution and its derivative at the final time for any data contained in the original intervals.

The O.D.E. solvers we constructed used very heavily the form of the equations being considered. Somewhat more general O.D.E. solvers have been constructed in [Se], but for exclusion problems, it seems counterproductive to attempt generality.

The main difficulty in exclusion problems seems to be that since we want to obtain uniform bounds over a macroscopic range of parameters, we have to work with intervals whose width is much bigger than the round off error. In those circumstances, the coherence and subadditivity properties play a very important role. It becomes important to organize the calculations in such a way that the expressions have as many built in cancellations as possible. For example, using recursive evaluation of the coefficients of a Taylor expansion, even if it works very well with sharp intervals, leads to catastrophic growth if the intervals are "*fat*". This is why, to estimate the truncation error in the O.D.E. solver we use a contraction mapping principle.

Remark. It is quite possible that, for the study of O.D.E.'s, keeping bounds on the derivative and the function is a bad idea. We can think of a second order O.D.E. as a flow on two dimensional space. Typically, a rectangle, will evolve into a much more complicated figure. If we require bounds of the form of a rectangle, we may be overestimating grossly the region. For linear second order O.D.E.'s , a possibility is to perform a Ricatti transformation $\Psi = u'/u$ and study the equation for Ψ. A variant of this approach is used in [Se]. Other possibility is to study more general data structures to bound the region in space (see [MP] or the contribution by Muldoon to this volume). Other methods are considered in [LR].

In the case considered in [FL], we decided to postpone tackling general issues till we got more experience. The way of defeating the growth was to rewrite the program till a substantial amount of coherence and subadditivity was taken out. The places where those effects were more harmful were decided empirically. In the most delicate cases, we decided to evaluate in different ways and take intersections.

The estimates obtained are more conservative the wider the intervals which bound the parameters are. It could happen that the interval obtained by applying the routine to the interval we are interested in is much bigger than the union of the results obtained running the program on subintervals that cover the original interval. Clearly, both of them are valid estimates. A convenient way of organizing the program was to write it in two different levels. A first level consists of a routine called "excluded " which propagates the solution and checks whether the resulting interval for the Wronskian contains zero. If it does not, excluded reports success. The second level is a routine superexcluded which calls excluded and reports success if excluded succeeded. If excluded failed, superexcluded divides the interval in parameters space into pieces and calls itself on each on of the pieces. If all of them succeed, then it reports success. Notice each one of the calls to superexcluded has the capability of generating new calls if the subsequent evaluation fails to show that there is no solution. The final result is an adaptative method that considers finer subdivisions where they are needed. Many other adaptative methods can be organized in the same way.

4. Current work

It turns out that many of the ideas described before can be used for the canonical model of matter Problem 1.1 including spin, even though the values of Z have to be suitably restricted.

It is possible to find representations of the kinetic energy as an integral over balls, and use the antisymmetry. The resulting formula can be divided into independent problems by estimating the contribution of certain balls that contain sufficiently many particles. Again, it is possible to reduce the study of the problem to a quadratic form on one–dimensional functions by using only radial estimates.

We are lead to the problem of estimating from below the sum of negative eigenvalues of a quadratic form very similar to the one considered in (2.6). There is a potential term and integral kernels which are piecewise the integrals of rational functions.

For this problem, it does not seem to be practical to reduce to a differential equation. Moreover, the fact that the Hamiltonian is not scale invariant and that we want to introduce the parameter Z makes us to have to consider several problems. The amount of work can be cut because convexity properties make it sufficient to study a finite number of points.

The line of attack we are currently pursuing is to compute the kernels using symbolic manipulators and then produce estimates using the methods based on interval arithmetic in Banach spaces of analytic functions as described e.g. in [La], [EKW], [EK]

REFERENCES

[Ch] Y. F. Chang: The ATOMCC toolbox. Byte **11,4**, 215–226 (1986).

[Co] J. Conlon: The ground state energy for a classical gas. Comm. Math. Phys. **94**, 439–XX (1984).

[DL1] F. Dyson, A. Lenard: Stability of matter I. Jour. Math. Phys. **8**, 423–434 (1967).

[DL2] F. Dyson, A. Lenard: Stability of matter II. Jour. Math. Phys. **9**, 698–711 (1968).

[DL3] I. Daubechies, E. H. Lieb : One–electron relativistic molecules with Coulomb interaction. Comm. Math. Phys. **90**, 497–510 (1983).

[EKW] J-P. Eckmann, H. Koch and P. Wittwer: A computer assisted proof of universality in area preserving maps. Mem. of the A.M.S. **289**, (1984).

[EW] J-P. Eckmann, P. Wittwer: " *Computer methods and Borel summability applied to Feigenbaum's equation*", Springer Verlag, NY (1985).

[FL] C. Fefferman, R. de la Llave: Relativistic stability of matter I. Rev. Mat. Iber. **2**, 119–213 (1986).

[Fe1] C. Fefferman: The atomic and molecular nature of matter. Rev. Mat. Iber. **1**, 1–44 (1985).

[Fe2] C. Fefferman: The N–body problem in quantum mechanics. Comm. Pure and Appl. Math. **39**, S67–S109 (1986).

[Fe3] C. Fefferman: Graduate courses taught at Princeton 1986–1987,1987–1988.

[He] I. Herbst: Spectral theory of the operator $\left(p^2 + m^2\right)^{1/2} - Ze^2/r$. Comm. Math. Phys. **53**, 285–294 (1977).

[KM] E. W. Kaucher, W. Miranker: "*Self-Validating numerics for function space problems*", Academic Press, Orlando (1984).

[Ka] T. Kato: Fundamental properties of Hamiltonian operators of Schrödinger type. Trans. Am. Math. Soc. **70**, 195–211 (1951).

[LL] J. Lebowitz, E. H. Lieb: The constitution of matter: Existence of thermodynamics for systems composed of electrons and nuclei. Adv. in Math **9**, 316–398 (1972).

[LR] R. de la Llave, D. Rana: Algorithms for the proof of existence of special orbits. preprint

[LT1] E. H. Lieb, W. Thirring: Bound for the kinetic energy of fermions which proves the stability of matter. Phys. Rev. Lett. **35**, 687–689 (1975).

[LT2] E. H. Lieb, W. Thirring: Gravitational collapse in quantum mechanics with relativistic kinetic energy. Ann. of Phys. **155**, 494–512 (1984).

[LY] E. H. Lieb, H.–T. Yau: The stability and instability of relativistic matter. Comm. Math. Phys. **118**, 177–213 (1984).

[La] O.E. Lanford III: Computer assisted proofs in analysis. Physica **124A**, 465-470 (1984).

[Li1] E. H. Lieb: Stability of matter. Rev. Mod. Phys. **48**, 553–569 (1976).

[Li2] E. H. Lieb: Thomas–Fermi and related theories of atoms and molecules. Rev. Mod. Phys. **53**, 603–642 (1981).

[MP] R. MacKay, I.C. Percival: Converse K.A.M.: theory and practice. Comm. Math. Phys. **98**, 469-512 (1985).

[Mo] R.E. Moore: "*Methods and applications of interval analysis*", S.I.A.M. Philadelphia (1979).

[Po] H. Poincaré: "*Les methodes nouvelles de la méchanique céleste*", Gauthier Villars, Paris (1891–1899).

[Th] W. Thirring: "*A course in mathematical Physics IV, Quantum mechanics of large systems*", Springer Verlag (1981).

[We] R. Weder: Spectral analysis of pseudodifferential operators. Jour. Funct. Anal. **20**, 319-377 (1975).

ACCURATE STRATEGIES FOR K.A.M. BOUNDS AND THEIR IMPLEMENTATION

R. DE LA LLAVE* AND D. RANA†

Abstract. We study perturbative expansions for quasi–periodic solutions of non–linear systems. We describe how to construct and implement algorithms that prove convergence of these expansions for values of the perturbation parameter as close to optimal as desired. The method is based on a constructive form of K.A.M. theory and implemented using interval arithmetic. For some cases, the algorithms have been run on a computer yielding results better than 90% of optimal.

Key words. K.A.M. theory, Interval arithmetic, Perturbation methods, Stability bounds.

AMS(MOS) subject classifications. 39–04, 39B99, 70K50, 58F30, 58F27, 58F10, 65J15, 30D05

1. Description of the problem

1.1. Introduction. Many problems of interest in celestial mechanics are *"close"* to systems that can be solved exactly. In such cases it is natural to consider perturbation expansions that allow us to understand features of the system of interest in terms of those of the exactly solvable one. Typically, one can compute all the terms in these expansions recursively, but frequently this computation involves expressions whose denominators become arbitrarily small in the course of the calculation.(See [Po] §126, §146 ff.) The convergence of these expansions is difficult to settle in general and it has been known for a long time that they sometimes diverge. In the 60's Kolmogorov, Arnol'd and Moser found a systematic way of establishing convergence of these expansions in many cases. These techniques, usually known as K.A.M. theory, have some limitations from the point of view of physical applications. They are technically complicated and the range of validity that can be established in some typical problems is much smaller than the values that would be relevant for applications. (See [Mo1] for a discussion of some astronomical problems.)

We have considered the problem of systematically improving the values yielded by K.A.M. theory. We show that there are algorithms that can make the K.A.M. theory produce results arbitrarily close to the true ones. In two applications of the theory, we have actually implemented the algorithms on a computer, paying attention to round–off errors so that the results of the programs can be considered as rigorous theorems. For those cases, we have also obtained – or found in the literature – values for which the conclusions of the theorem are known to be false. In both cases, the lower bounds for the range of validity we prove are 90% of the values for which the conclusions are known to be false.

*Dept. of Math., University of Texas, Austin, TX 78712. Supported in part by N.S.F. grants
†Math. Dept. Columbia Univ. Supported in part by N.S.F. grants

1.2. Classical results. The K.A.M. techniques can be applied to many problems in dynamics. Two problems that involve all the essential difficulties of the theory are the *Siegel center theorem* (originally proved by Siegel [Si] using other methods) and the *twist mapping theorem* (originally proved by Moser [Mo2].)

Theorem 1.1. *Given a family of analytic functions on* **C**

$$f_\epsilon(z) = az + \frac{1}{\epsilon}\hat{f}(\epsilon z), \qquad f_0(z) = az$$

where $\hat{f}(z) = \mathcal{O}(z^2)$ *and* a *satisfies, for some* $C, \nu > 0$

(1.1) $$|a^n - 1|^{-1} \le Cn^\nu, \quad n \in \mathbf{N}$$

Then, there exists an $\epsilon_0 > 0$ *such that, if* $|\epsilon| \le \epsilon_0$ *there is a conformal change of variables* Ψ *defined on the unit disk and satisfying*
  i) $\Psi(0) = 0$
 ii) $\Psi'(0) = 1$
iii) $\Psi^{-1} \circ f_\epsilon \circ \Psi = az$

Remark. The above theorem can be interpreted as saying that for small enough $|\epsilon|$ the perturbed dynamics – f_ϵ – is just the unperturbed dynamics $-f_0-$ in a distorted system of coordinates. This is very similar to the Hamilton–Jacobi method of perturbation theory, whose basic idea is to try to reduce the perturbed system to the unperturbed one by making a suitable change of variables.

Remark. Observe that f_ϵ is just a scaled version of f_1, $f_\epsilon(z) \equiv \frac{1}{\epsilon}f_1(\epsilon z)$. As stated, the theorem produces an ϵ for which the domain of the conjugating function, Ψ contains a ball of radius 1. Alternatively, we can use the above scaling to find the biggest radius on which we can define a Ψ linearizing f_1. That radius is then $\le \epsilon_0$. Since switching from one formulation to another is so easy, we will use either one, depending which one is more natural for the aspect of the problem we are discussing. In most other problems, e.g. the twist mapping theorem that we will discuss below, the size of the perturbation and the domain of definition of the solutions are described by different parameters.

Theorem 1.2. *Let* F_ϵ *be a* C^k *family of* C^k *diffeomorphisms of* $\mathbf{R} \times \mathbf{S}^1$:

$$F_\epsilon(x, \theta) = (f_\epsilon(x, \theta), \phi_\epsilon(x, \theta))$$

for which:
  i) *the area* $dx \wedge d\theta$ *is preserved*
 ii) $\phi_x(x, \theta) \ge c > 0$, *where* c *is a constant*
iii) $\int (f(x, \theta) - x)d\theta = 0$
iv) $F_0(x, \theta) = (x, (\theta + x)(\mathrm{mod}1))$

For some $C > 0$, $0 < \nu < k/3$, let $\omega \in \mathbf{R}$ satisfy:

$$|n\omega - m|^{-1} \leq Cn^\nu \quad n, m \in \mathbf{Z}, n \neq 0$$

Then, for any $k' < k - 3\nu$ there exists an $\epsilon_0 > 0$ such that if $|\epsilon| < \epsilon_0$ there is a $C^{k'}$ map, $K : \mathbf{S}^1 \mapsto \mathbf{R} \times \mathbf{S}^1$, whose image is a topologically non–trivial circle, and for which:

(1.2)
$$F_\epsilon \circ K = K \circ T_\omega = 0,$$

where $T_\omega(\theta) = (\theta + \omega)(\mathrm{mod}\ 1)$. K_ω is invertible on its range and the inverse is differentiable.

Remark. Notice that all the orbits of the unperturbed system F_0 move in circles winding around the cylinder, so that one can find a K_ω for every ω. The theorem states that some of these unperturbed motions survive, up to a change of variables, in the perturbed system.

Remark. This problem appears frequently in celestial mechanics. For example, it appears in the study of an integrable system of one degree of freedom subject to periodic forcing, in the three body problem, in the motion of a satellite around an oblate planet and, with suitable modification, in the study of the motion of a billiard ball on a convex planar table, etc.

Remark. In the application mentioned above, the conclusions of the theorem have dramatic importance for the long time behavior of the system. The theorem asserts the existence of invariant circles going around the cylinder. These circles are clearly barriers for global diffusion. One orbit that starts on one side of the circle cannot contain points on the other side. It has been shown recently [Ma] that circles winding around the cylinder are the only obstruction for global diffusion, though the circles may not be as differentiable or have winding numbers with Diophantine properties as those circles considered in the K.A.M. theory.

Applying most of the versions of these theorems in the literature to concrete examples e.g.

(1.3)
$$f_\epsilon = e^{2\pi i \gamma} z + \epsilon z^2, \quad \gamma = \frac{\sqrt{5} - 1}{2}$$

for Theorem 1.1 or the "standard map":

(1.4)
$$F_\epsilon(x, \theta) = \left(x + \frac{\epsilon}{2\pi} \sin(2\pi\theta), (x + \theta + \frac{\epsilon}{2\pi} \sin(2\pi\theta))(\mathrm{mod}\ 1) \right).$$

for Theorem 1.2, we find that the ϵ for which smallness conditions of the theorem, also known as K.A.M. bounds, apply, are several orders of magnitude smaller than those relevant for physical situations.

We have considered the problem of finding systematic methods of improving the K.A.M. bounds. We find that there are systematic optimizations of some K.A.M. proofs which yield bounds converging to optimal. Other proofs have intrinsic limitations so that, no matter how optimized in the free parameters, they will yield results that remain a finite distance from the optimal values.

2. Theory

2.1. Definitions. The proofs of K.A.M. theorems involve many, in fact infinitely many, arbitrary choices. For example, at each step one has to choose a decreasing domain of analyticity in which to compute bounds, one has a choice of which norms to use, etc. At the expense of computational effort, a proof of a theorem can yield improving K.A.M. bounds by optimizing the choices, though the choices which are optimal for one system could be far from optimal for another one.

Definition 2.1. *A strategy for K.A.M. bounds is an algorithm which, for a family of problems $\mathcal{F}_\epsilon$ (f_ϵ for Theorem 1.1, F_ϵ for Theorem 1.2) and any positive integer N, produces a positive number ϵ_N, where ϵ_N can be computed out of $\mathcal{F}_\epsilon$ using less than N operations, and such that if $|\epsilon| < \epsilon_N$ the conclusions of the K.A.M. theorem hold.*

By operations we mean either arithmetic operations between real numbers or comparisons between them. The algorithm may have *branches*, that is the operations to be performed may depend on the result of a comparison.

Some strategies are optimal in the following sense:

Definition 2.2. *We say that an strategy for K.A.M. bounds is accurate if, whenever we have a family $\mathcal{F}_\epsilon$ for which $\lim \epsilon_N = \epsilon^*$, then for any $\epsilon^{**} > \epsilon^*$, the conclusions of the theorem are false for some $\mathcal{F}_{\epsilon'}$, $\epsilon^* \leq \epsilon' \leq \epsilon^{**}$.*

Analogous definitions can be made for the case in which the theorem we prove states that the conclusions of K.A.M theory do not hold. In that case, we will speak of *"(accurate) strategies for converse K.A.M. bounds"*.

The concept of accuracy is related to the *"finite computability hypothesis"* of [MP], established in [St]. In our language, the finite computability hypothesis would have been formulated saying that the algorithm of [MP] is an accurate strategy for the non–existence of invariant circles.

2.2. Some accurate strategies for K.A.M. bounds. From the analytic point of view, K.A.M. theorems can be formulated as assertions that certain functional equations have solutions, and the theorem with its proof can be interpreted as an implicit function theorem. The functionals considered usually have derivatives (understood in some appropriately weak sense), and the existence of a formal perturbation theory (up to first order) can usually be formulated in terms of the inverse of this derivative. The difficulty of small divisors implies that the inverse of the derivative is not bounded. In such cases it is impossible to apply the usual implicit function theorem between Banach spaces, and indeed, in this generality, the result of an implicit function theorem is false. Even though the usual algorithms for the proof of an implicit function theorem (e.g. those in [Di]) do not converge, sometimes a variation of the Newton method can be used for our problems. This is usually called a *"hard"* implicit function theorem because the proof uses the detailed structure of the functional, delicate properties of the spaces where the functional is defined, and assumptions on the small divisors.

For a systematic review of how to formulate many small divisor problems as consequences of a general implicit function theorem see [Ze1], [Ze2], [Bo]. Many other problems not treated in these papers, such as Siegel's center theorem, can also be fitted into this framework. The application of a general implicit function theorem to K.A.M. theory has shortcomings. Some well known theorems do not fit easily. Frequently it is also possible to obtain sharper differentiability conclusions by walking trough the steps of the proof and optimizing for the particular case being considered.

On the other hand, for the problems considered here the systematization introduced from this point of view is essential. By performing the proof of the implicit function theorem carefully, it is possible to find a finite set of explicit conditions that will guarantee that a *"Newton method"* started on a sufficiently approximate solution will converge to a true solution.

The prototype of such a theorem is:

Theorem 2.3. *For certain choices of Banach spaces $X_0 \subset X_1$ and certain functionals $T : X_0 \mapsto X_0$ there exist a computable function $g : \mathbf{R}^{+n} \times \mathbf{R}^+ \mapsto \mathbf{R}^+$ satisfying:*

$$\lim_{\mu \to 0} g(M_1, \ldots, M_n; \mu) = \infty \quad \forall M_1, \ldots, M_n$$

and computable functionals $\mathcal{L}_1, \ldots, \mathcal{L}_n : X_0 \mapsto \mathbf{R}$ with the following property. Suppose that $h^* \in X_0$ satisfies:

i) $\|T(h^*)\|_{X_0} \leq \mu$

ii) $\mathcal{L}_1(h^*) \leq M_1, \ldots, \mathcal{L}_n(h^*) \leq M_n$

iii) $g(M_1, \ldots, M_n; \mu) \geq 0$

Then, $\exists\, h \in X_1$ s.t. $T(h) = 0$ and $\|h - h^*\|_{X_1} = \mathcal{O}(\mu)$.

Remark. With a theorem such as the one described above, by checking a finite number of conditions on a function which approximately solves the functional equation, we are assured that there is a true solution nearby. Constructive implicit function theorems have played an important role in the *"validation"* of numerical computations. If the approximate solution, h^* is the result of an heuristic numerical algorithm, by verifying the hypothesis of the theorem we are assured that there is a true solution close to the computed one. In most of the applications considered so far, (e.g. most of those in [Mo3], [KM], [La]) it sufficed to use a version of the implicit function theorem patterned on the usual implicit function theorem in Banach spaces rather than on Nash–Moser implicit function theorems.

Remark. Notice that to conclude the existence of a true solution we do not need to analyze the algorithm used to produce the approximate solution. It suffices to verify rigorously that the conditions of the theorem are met.

Remark. Rather that considering a functional, if we consider a continuous family of functionals T_ϵ, as in the Siegel center theorem or in the twist mapping theorem, the hypothesis of Theorem 2.3 will be verified for a nontrivial interval. By applying the constructive implicit theorem to, a guess (or possibly several guesses) h^*, it is possible to prove the K.A.M. theorem for a range of parameters.

Remark. In most of the applications in the mathematical literature, h^* is taken as the solution of the unperturbed system. This is usually necessary because the integrable system is the only system in the family for which it is possible to compute the solution to the functional equation in closed form, and more importantly, because the integrable system is the one for which it is possible to compute a perturbation theory. For our applications it is quite important that the Newton method starts on an arbitrary approximate solution. Some versions of Nash–Moser implicit function theorems, notably those of [Ze1] and [Ha], have this feature, but they require extra conditions such as special structure in the equations or invertibility of the derivative of the functional in a neighborhood about h^*.

Besides a constructive implicit function theorem, the other ingredients of the accurate strategies we have considered are two algorithms $\mathcal{A}$ and $\mathcal{B}$ that provide a suitable starting point h^* for the constructive implicit function theorem and that verify that the function satisfies the condition of the theorem. More precisely, we have to prove a lemma of the form:

Lemma 2.4. *There is a sequence of finite dimensional Banach spaces, $X^{(N)} \subset X_0$, and two algorithms, $\mathcal{A}$ and $\mathcal{B}$ such that:*

$i)$ *$\mathcal{A}$ yields an $h_N \in X^{(N)}$ in less than N operations. Moreover, if T has a zero in X_0 then:*

$\quad i.i)$ *$\lim h_N = h$ exists in X_0, and $T(h) = 0$.*

$\quad i.ii)$ *$\sup_{N,i} \mathcal{L}_i(h_N) < \infty$*

$ii)$ *For any h^* in some $X^{(N)}$, $\mathcal{B}$ yields a sequence of bounds:*

$$\mathcal{L}_1(h^*) \leq M_{1,\hat{N}}, \ldots, \mathcal{L}_n(h^*) \leq M_{n,\hat{N}} \quad \text{and} \quad \|T(h^*)\|_{X_1} \leq \mu_{\hat{N}}$$

The $M_{i,\hat{N}}$, $\mu_{\hat{N}}$ can be computed in less than $\hat{N}$ operations and:

$$\lim_{\hat{N}\to\infty} M_{i,\hat{N}} = \mathcal{L}_i(h^*) \quad 1 \leq i \leq n, \quad \text{and} \quad \lim_{\hat{N}\to\infty} \mu_{\hat{N}} = \|T(h^*)\|_{X_1}$$

The existence of these algorithms is, perhaps, not too surprising. If the Banach spaces we are considering are separable, any function can be approximated by functions in finite dimensional subspaces. That is, any function can be approximated by functions that can be described by a finite number of parameters. The set of these *finitely specified* functions can be systematically searched for candidates for solutions. If there was a solution, it is plausible that the solution could be systematically approached. The only non–trivial request is that the approximation is sufficiently well behaved not to make the functionals $\mathcal{L}_i$ blow up.

Since accurate strategies are algorithms, it is very natural to try and implement them on a computer. This brings forth some considerations we have neglected so far. Notably, some attention has to be paid to efficiency and we also have to pay attention to the fact that computers do not deal with all real numbers but with a finite set of them. Since small divisors introduce some amount of numerical instability, the effects of round–off could be very serious. In the next sections we take up these issues

3. Accurate strategies for Siegel and twist theorems

3.1. Preliminary considerations. The choice of the functional to be used is quite important. Even if two functionals have the same zeros, they could have very different properties. Out of the many functionals that are used in the mathematical literature to deal with these problems some are unsuitable for calculations and others can only be used to prove the K.A.M. theorem for sufficiently small values of the parameter. That is, finding solutions for them is equivalent to solving the problem, but only for some small enough values of the perturbation.

The choice of the sequence of spaces $X^{(N)}$ in algorithm $\mathcal{A}$ is also quite important as the $X^{(N)}$ correspond to the notion of discretization schemes in numerical analysis. For certain choices of $X^{(N)}$, it can happen that $\|T(h_N)\|$ decreases much more slowly than the $\mathcal{L}_i(h_N)$ increase in such a way that it would become impossible to use the implicit function theorem for any choice of h_N to verify the existence of a solution.

One instructive example of these difficulties is the following. The proof of Theorem 1.1 explained in [Ar] uses the functional:

$$(3.1) \qquad T[\Psi](z) \equiv \Psi^{-1} \circ f_\epsilon \circ \Psi(z) - az$$

The function Ψ is discretized using truncations of order N of the Taylor series at zero.

Since $\Psi(az) = f_\epsilon(\Psi(z))$ and a is an irrational rotation, it follows, in the case that f_ϵ is an entire function, that the maximal simply connected domain of definition of Ψ will be a circle. Nevertheless, the domain of definition of Ψ^{-1} will, in general, be a much more complicated domain (the cases where the maximal domain is a circle are classified in [Br].) Since the Taylor expansion of Ψ^{-1} converges only on a disk centered at zero and contained in this domain, $\|T[\Psi^N]\|$ will only converge to zero when we consider Ψ^N defined on circles so small that the image of this disk under $f_\epsilon \circ \Psi$ is contained in the domain of convergence of Ψ^{-1}.

This strategy was implemented in [LST],[LT] . The estimates they obtained, neglecting round–off errors, could only yield results for values of ϵ about 60% of the correct value. This is very close to the limit imposed by the previous argument. It is easy to produce examples in which this discrepancy becomes arbitrarily large.

Variants of this argument can be used to show that unless one uses carefully chosen ways of representing the $(\Psi^N)^{-1}$, the proof based on reducing the problem by a sequence of changes of variables and keeping track of Taylor coefficients or of estimates on disks do not lead to accurate strategies. This is somewhat disappointing since the proofs based on successive changes of variables are very convenient for analytical proofs. In [Ra], it is possible to find an accurate strategy based on the functional (3.1). The discretization used is interpolation on a complicated grid that has to be computed as the calculation proceeds.

Therefore, the choices of acceptable functionals is drastically reduced by the following considerations:

- The functional considered should only involve functions that can be approximated efficiently using the scheme of approximations chosen. Computing the functional should only involve operations that can be implemented effectively.

If we use Taylor expansions for the discretization and we want to understand a function in its maximal domain, the maximal domain should be a disk. If we use Taylor or Fourier series, we should not use inverse functions and the composition on the left should only involve very simple functions (e.g. entire.)

- The functions we are searching for should be functions of as few variables as possible.

The complexity of representing functions of several variables to a certain accuracy increases exponentially fast, with a large exponent, in the number of variables. Moreover, as the the perturbation parameter approaches a critical value, the solutions of the functional equation begin to have very large derivatives, so that they will be hard to approximate.

The choice of discretization schemes is, likewise, reduced by the following considerations.

- The hypothesis of K.A.M. theorems depend on the values of derivatives of the function of fairly high order. Hence the representation of functions considered should allow reliable evaluation of derivatives.

All the proofs of the K.A.M. theorem available so far require that the functions are several times differentiable and the size of derivatives of relatively high order plays a role. This restriction is probably unavoidable since there are a growing number of counterexamples with lower differentiability [He1]. This suggest that a discretization should specify a neighborhood on a space of functions which are highly differentiable. Since the neighborhoods for which the constructive implicit function theorem can be proven analytically are rather small, the discretization of functions we take should specify the high derivatives with high precision.

The most effective way of achieving precision for high derivatives we have found is to represent the function by Taylor polynomials (for functions defined in a complex disk) or Fourier series (for functions defined in a complex strip.) See [Ko] for a theoretical discussion on how to approximate functions when high precision is required on their derivatives.

An alternative that has been considered in [BZ] is to specify a trigonometric polynomial by storing the values it takes on a grid in a complex strip. If the numbers stored were infinitely precise, the representations would indeed be equivalent since interpolation allows an exact reconstruction of the polynomial. Nevertheless, it is well known that computer round–off leads to instabilities in the evaluation of derivatives of the interpolated functions. We suspect that the erratic behavior of the algorithm observed by [BZ] for many parameter values can be traced to the instability of interpolation for the reconstruction of derivatives.

• The calculations we are going to perform are intrinsically unstable.

Notice that the goal of the problem is to explore functions in domains as large as possible, getting as close to the domain of convergence as possible. (Informally, if the calculation was stable, we would "fix it" by studying a bigger domain.)

Notice also that since our goal is to compute estimates of $T[h^*]$, which are presumably very small, we are going to perform operations with large numbers to obtain eventually a very small number. In other words, there are a large number of cancellations. It is known that in these cases, round–off could be particularly harmful.

3.2. Strategy for the Siegel center theorem. For the Siegel center theorem, we choose the functional

$$T[\Psi](z) = f_\epsilon \circ \Psi(z) - \Psi(az).$$

Notice, that, as remarked before, Ψ has maximal domain a disk. We will see in moment that if $T[\Psi] = 0$ it follows that Ψ is univalent in its maximal domain of definition, so that $T[\Psi] = 0$ is completely equivalent to the functional equation in Theorem 1.1 iii).

Observe also that, if $T[\Psi] = 0$, then Ψ' does not take the value zero inside its maximal domain of definition.

A theorem of the form of Theorem 2.3 can be obtained by slightly modifying the proof of [Ze3] (full details can be found in [Ra].) The theorem proved in [Ra] takes X_0 to be the Banach space of analytic functions in a disk of radius r and X_1 the Banach space of analytic functions in a disk of radius $re^{-\delta}$ equipped with the norms $\|\Psi\|_{X_0} \equiv \sum_{k=0}^{\infty} |\Psi_k| r^k$ and $\|\Psi\|_{X_1} \equiv \sum_{k=0}^{\infty} |\Psi_k| r^k e^{-k\delta}$. We have found it convenient to use also the two norms $\||\Psi\||_{X_0} \equiv \sum_{k=0}^{\infty} |\Psi_k| k r^k$ and $\||\Psi\||_{X_1} \equiv \sum_{k=0}^{\infty} |\Psi_k| k r^k e^{-k\delta}$ in intermediate steps of the proof. It is intuitively clear that the more information we keep about the function Ψ^*, the sharper the final results will be. Therefore, the functionals $\mathcal{L}_i(\Psi^*)$ used in the proof are: $\|\Psi^{*\prime}\|_{X_0}$, $\|1/\Psi^{*\prime}\|_{X_0}$, $\||\Psi^{*\prime}\||_{X_0}$ and $\||1/\Psi^{*\prime}\||_{X_0}$.

With these choices, both r and δ are parameters which we have to adjust depending on the family f_ϵ. The problem of getting the best estimates for the Siegel center theorem becomes the problem of finding the best $re^{-\delta}$ for which its is possible to prove the theorem. A strategy will be accurate if it can prove the theorem for $re^{-\delta}$ arbitrarily close to the radius of the maximal domain of definition of Ψ.

The function g of the theorem is rather complicated to express analytically, since it involves using numbers which are arbitrary except for having to satisfy conditions involving other numbers already calculated. Nevertheless, calculating bounds for the function is reasonably straightforward to program since we can check whether the conditions in the arbitrary choices are met. The parameter δ also enters into g. The condition on the behavior of g as μ tends to zero is also readily verified for all possible δ. The explicit expressions for the conditions can be found in [Ra].

Producing approximations to the solution in the domain of definition is done by computing the Taylor expansions of Ψ. This can readily be done by matching equal powers. (See e.g. [Si] for details.) Computing arbitrarily good approximations to the auxiliary functionals can be done by using the usual rules of the algebra of polynomials. Notice that, since Ψ' does not take the value zero in the disk where it is defined, $1/\Psi'$ will be an analytic function there and the Taylor series obtained by synthetic division converges. It is possible to estimate the remainder of the truncated Taylor series for $1/\Psi^{*'}$ in the analytic norms we are considering. We again refer to [Ra].

The final strategy consists in computing Ψ and the auxiliary functionals with increasing degrees and exploring finer and finer grids in r and δ.

If we are considering a radius slightly smaller than the true one, the auxiliary functionals and their computed bounds will remain bounded whereas the remainder will go to zero. The theorem will be therefore verified with some finite amount of work.

3.3. Strategy for the twist mapping theorem. For the particular case of the standard mapping (1.4), (1.2) reduces to:

$$(3.2) \qquad T[\ell(x)] \equiv \ell(x+\omega) - 2\ell(x) + \ell(x-\omega) + \frac{\epsilon}{2\pi}\sin(2\pi(x+\ell(x))) = 0$$

where $\ell(x) = \ell(x+1)$ and $x + \ell(x)$ is to be a diffeomorphism of the circle.

This equation admits a constructive implicit function theorem of the form we discussed above. It is possible to prove the theorem by modifying slightly [Rü] or [SZ].

Remark. Even if for more general twist mappings the formalism of [Rü] or [SZ] leads to different functional equations, for the standard mapping they both agree. The iterative methods they use and the bounds they obtain are different.

Remark. It is also possible to prove a constructive implicit function theorem following the proof in [Bo]. The later appeals to the version of the Nash–Moser implicit function theorem in [Ha]. Since this reduction involves heavy use of inverses, which are difficult to estimate effectively, it seems that the results obtained would be numerically worse than the version in [Ze1], which is the one we have followed.

Remark. Either one of the three different formalisms mentioned above can be extended to higher dimensions. One important feature is that, to produce tori of dimension n in a $2n$ dimensional system, one only has to consider functions of n variables rather than the functions of $2n$ variables required by the formalisms based on transformations into normal forms (e. g. those of [Ze2]).

The implicit function theorem we have used is based on the method of [SZ]. The spaces we use are spaces of periodic functions analytic in a strip around the real axis.

$$X_0 = \{f(x)|\|f\|_0 \equiv \sum_k |\hat{f}_k|e^{|k|\rho_0} < \infty, \int f(x)dx = 0\}$$

$$X_1 = \{f(x)|\|f\|_1 \equiv \sum_k |\hat{f}_k|e^{|k|\rho_1} < \infty, \int f(x)dx = 0\}$$

where $0 < \rho_1 < \rho_0$. Again, we have also found it useful to introduce other norms $\||f\||_0 = \sum_k |\hat{f}_k|\|k|e^{|k|\rho_0}$ and $\||f\||_1 = \sum_k |\hat{f}_k|\|k|e^{|k|\rho_1}$ We then prove the theorem using the following auxiliary functionals $\mathcal{L}_i$: $\||\ell\||_0$, $\|\ell\|_0$, $\|1/(\ell'+1)\|_0$, $\|1/(\ell'+1)\|_1$, $\|1/(\ell'+1)(\ell' \circ T_\omega + 1)\|_0$ and $\|1/(\ell'+1)(\ell' \circ T_\omega + 1)\|_1$.

We refer to [Ra] for full details of the proof and the explicit expressions of the conditions.

Remark. Notice the condition $\int \ell = 0$ entering in the definition of the spaces X_0, X_1. This is a normalization to guarantee that the solution of (3.2) is unique. It can always be arranged since for any α, $T[\ell] \equiv 0 \iff T[\ell \circ T_\alpha - \alpha] \equiv 0$. Where $T_\alpha(x) = x + \alpha \bmod 1$.

If ℓ^N is a trigonometric polynomial, it is possible to compute $T[\ell^N]$ and $\mathcal{L}_i$ very efficiently. The Fourier expansion of $\sin(\ell^N)$ can be computed as follows: Write $\ell^N(x) = P(z) + \bar{P}(\bar{z})$ with $z = \exp(2\pi i x)$, $\bar{z}$ its complex conjugate, P is a polynomial and $\bar{P}$ is the polynomial whose coefficients are the complex conjugate of those of P. The sin and cos of a polynomial can be computed, as suggested in [Kn], by using the formulas $(\sin \circ P)' = \cos \circ P P'$, $(\cos \circ P)' = -\sin \circ P P'$. If we match coefficients of corresponding powers, we derive a recurrence relation for the coefficients of $\cos \circ P$, $\sin \circ P$. It is also possible to find bounds for the truncation error when the calculation stops at a finite order. Once we have computed the sin of the polynomials, it is very easy to use the addition formula of the sin to compute $\sin(2\pi(x + \ell^N(x)))$.

The computation of $1/(\ell'(x) + 1)$ is more tricky. In principle, evaluating the polynomial at sufficiently many points, computing the inverse at each point and, then, performing a discrete Fourier transform will produce the inverse up to any desired degree of accuracy. It is even possible to estimate the remainder and its derivatives. This turns out to be very unstable numerically. We have found it more efficient to use a constructive implicit function theorem between Banach spaces. If we find another trigonometric polynomial P such that $(\ell'+1)P = 1 + R$ and we can show that R is small, then, the inverse will be $P + P\sum_{i=1} R^i$. If we have estimates on R, we can estimate how much does P differ from the true inverse.

An algorithm to produce the approximate solution on which to verify the conditions of the constructive implicit function method, can be based on the usual *continuation methods*. It is possible to implement a Newton method for a truncation of the functional along the lines discussed above. When the solution for a parameter value is computed to enough accuracy, we increase the parameter and the order of the truncation and use a Newton method starting on the solution for the old parameter. If the increment of the parameter is small enough, the Newton method will converge and we will have a very accurate solution for a bigger parameter value. The process can be started on the parameter value corresponding to the integrable case, whose solutions are exactly known.

The final strategy consists on simultaneously computing approximate solutions using a numerical continuation method and trying to verify the hypothesis of Theorem 2.3 for increasingly finer grids of ρ_0 and ρ_1. It is not difficult to show that this strategy is accurate.

This strategy is accurate, that is, it can produce all circles of a particular Diophantine rotation number for which

i) The circles are analytic.

ii) The motion on the circles is analytically conjugate to a rotation.

iii) We can find a continuous family of solutions connecting them to the solutions of the unperturbed problem. (As any continuation method, the algorithm is powerless to find solutions that can be continued only to an isolated sets of parameters).

If we had maps more complicated than the standard map, in order to derive the analogue of (3.2), we would require some global conditions which depend on the formalism we use (either that of [Rü], [SZ], [Bo].) For example, [Bo] requires

iv) The circle is the graph of a function from the circle to the reals.

In principle the algorithm could fail to find circles that do not satisfy either of *i*) – *iv*) above. Nevertheless, it is possible to show by other methods that the conditions *i*) – *iv*) are implied by some weaker ones. For circles with rotation ω, the golden mean invariant under (1.4) we have:

I) [SZ] prove a theorem that implies that all the circles which are $C^{7+\epsilon}$ are analytic.

II) [He2] shows that on all analytic circles, the motion is analytically conjugate to a rotation (more modern proofs are [Yo], [KO].)

III) If one finds an approximate circle by any method other than continuation, it is possible to use it in the strategy instead. For the so called " *reversible* " maps, which appear naturally in mechanics, there are methods to compute approximate circles which do not use continuation [DeV], [Gr].

IV) [Ma2] shows that all invariant circles are graphs of Lipschitz curves (A more modern proof is [Fa].) Renormalization group analysis [McK], gives evidence that, for many systems, the circles remain as differentiable for all the values of the parameter up to a critical value, for which they become much less differentiable and they disappear for bigger values.

In higher dimensions, *II*) and *IV*) are not available, so the situation is more confusing. Some partial results are available [He4], [Ka]. See also the contribution of M. Muldoon to this volume.

Remark. Since it is possible to compute the terms in the perturbation expansion in powers of ϵ, it seems to be reasonable to use the sum of a finite number of them to produce a good starting value. Unfortunately, there is, at the moment, no way to show that this strategy is accurate because it could, in principle, happen that there were complex singularities in the expansion off the real axis but very close to the origin. Moreover, it seems to happen that summing the series leads to severe cancellations that make the procedure much more prone to numerical errors than the self–correcting Newton method which can refine the guess till the numerical remainder is of thre same order of magnitude as the round–off. Such an strategy has been considered in [CC] using also an implicit function theorem based on [SZ]. Their implementation requires the use of interval arithmetic to manipulate the perturbative expansion in powers of ϵ. Since interval arithmetic cannot profit from cancellations, the results they obtain are pessimistic. Nevertheless, the computation of the power series expansions could have intrinsic interest, since there are some surprising regularities [BC].

4. Converse bounds

4.1. Siegel theorem.   For the Siegel Center theorem, it is possible to use univalent function theory to prove not only converse K.A.M. bounds, but also to prove estimates on the speed of convergence of the algorithms we discussed before to the right answer. Here it is much more convenient to formulate the problem as trying to find the maximal domain of definition of Ψ for a given function rather than the problem of finding the best scaling.

Lemma 4.1. If the function f_1 is entire, the function Ψ satisfying iii) of Theorem 1.1 is univalent in its maximal domain of definition.

Proof. Assume that there were two points z_1, z_2 in the domain of definition of Ψ such that $\Psi(z_1) = \Psi(z_2)$. Using the functional equation satisfied by Ψ we can conclude that for all $n \in \mathbf{N}$, $\Psi(a^n z_1) = \Psi(a^n z_2)$.

If $z_1 = z_2 = 0$, there is nothing to prove, so we can assume that $z_1 \neq 0$. For $z \in \{a^n z_1\}_{n=0}^{\infty}$, $\Psi(z) = \Psi(z(z_2/z_1))$. This set is infinite and has accumulation points, hence the above functional equation has to hold in the maximal domain of Ψ. If $z_1 \neq z_2$ this leads to a contradiction with $\Psi'(0) = 1$.

Univalent functions satisfy many a–priori bounds. Perhaps the most famous are the Bieberbach–De Branges bounds for the coefficients. If $\Psi'(0) = 1$ and Ψ is univalent in a disk of radius ϵ, then:

(4.1). $$\epsilon^{n-1}|\Psi_n| \leq n \quad n \geq 1$$

Since the Taylor coefficients of Ψ can be computed by algebraic operations from those of f_1,

$$\epsilon_N = \min_{2 \leq N} \left(\frac{n}{|\Psi_n|} \right)^{\frac{1}{n-1}}$$

is a decreasing sequence of computable numbers. By (4.1), each one of them is an upper bound to the Siegel radius. By the Hadamard formula for the radius of convergence, these bounds converge to the Siegel radius. Therefore, the strategy is accurate.

Remark. It is quite possible to use the stronger Milin–DeBranges inequality in similar manner.

When f is a polynomial it is possible to use other methods. It is well known that we can find $0 < R < \infty$ in such a way that $|f(z)| > 1.1|z|$ whenever $|z| > R$. So that, whenever $|z| > R$, $f^n(z)$ goes to infinity. On the other hand, if z belonged to the range of Ψ, we would have $f^n(z) = f^n(\Psi(w)) = \Psi(a^n w)$ so that $f^n(z)$ would come arbitrarily close to z. We conclude that Area(Range$(\Psi)) \leq \pi R^2$.

The area of the range of a univalent function defined on a disk of radius ϵ^* can be related to the coefficients of the Taylor expansion by the well known "*Area formula*":

$$\text{Area}(\text{Range}(\Psi)) = \sum_{n=1}^{\infty} |\Psi_n|^2 \epsilon^{*2n}$$

It is clear that any ϵ_N verifying $\sum_{n=1}^{N} |\Psi_n|^2 \epsilon_N{}^{2n} > R^2$ is an upper bound for the Siegel radius. Since the coefficients Ψ_N can be computed, it is also possible to compute, in a finite number of operations, ϵ_N. As N becomes large, the ϵ_N get arbitrarily close to optimal.

Again using the Hadamard formula, it is easy to check that the bounds obtained in this way converge to the Siegel radius.

Univalent functions also satisfy a–priori bounds on how big they and their derivatives or the reciprocal of their derivatives be on domains slightly smaller than the maximal domain of definition. Using the detailed form of the smallness conditions in the constructive implicit function theorem, the Koebe distortion theorem as Bieberbach–DeBranges inequalities it is possible to prove:

Theorem 4.2. *Assume that f in Theorem 1.1 is a polynomial. For any $0 < p < 1$ there exists an N_0 that depends on p and f such that if $N > N_0$ and ϵ^* is the Siegel radius, then h^* the Nth order Taylor polynomial of Ψ verifies the conditions of the constructive implicit function theorem on a disk of radius $\epsilon^* e^{-N^{-p}}$.*

Informally, by using the accurate strategy described above, working with N terms we can guarantee that the lower bounds are correct up to a factor $e^{-N^{-p}}$.

We refer to [Ra] for the proof.

4.2. Twist mapping. In [Ma] it is shown that a class of mappings involving (1.4), for $|\epsilon| > 4/3$ there is no invariant circle. Since the class contains elements that saturate the bound, it is clear that any improvement has to use the specific form of (1.4) and, hence, involve some calculation. In [MP], it is shown by extremely elegant computer assisted methods that for the standard map, when $|\epsilon| > 63/64$ there are no non–trivial invariant circles.

We would also like to call attention to [Au] which, provides with a method that could easily be made rigorous. [OS] introduces and implements a criterion of non existence for invariant circles with a fixed rotation number.

5. Computer Implementation

5.1. Interval arithmetic. Since a strategy involves an algorithm it is natural to use a computer for an implementation. Unfortunately, most of today's computers are equipped only with an approximate form of arithmetic operating on a subset of the rational numbers, which makes the results of a numerical computation inadmissible as elements of a proof. A way of overcoming this difficulty without venturing very far from the traditional numerical analysis lore is the use of interval arithmetic.

The basic idea , introduced by R. Moore in the 60's is to introduce a basic data structure – the interval – which represents all the real numbers contained between two representables. It is possible to write routines that given any pair of intervals, produce another interval guaranteed to contain the result of an arithmetic operation performed between elements of the input intervals. If this were impossible, because of overflow , because we divide by an interval containing zero, or any other reason, they raise an exception. If a program runs without raising an exception, we may be assured that the bounds reported for an algebraic expression are rigorous. We refer to [Mo], [KM] for a more detailed discussion of interval arithmetic.

5.2. Implementation of Interval Arithmetic. Our programs have been written on a VAX 11-750 whose technical description appears in [Dec]. Of all the different classes of numbers that the machine is capable of representing, the two that interest us are "machine integers:"

$$\mathcal{I} = \left\{ I \mid I \in \mathbf{Z}, \ -2^{31} \leq I < 2^{31} \right\}$$

and the "*representable numbers* " which simulate the reals. Out of the many floating point systems supported by the VAX, we have used the D–float, which corresponds to the type `double` in many C compilers:

$$\mathcal{R} = \left\{ (s, e, m) \mid s = +1, -1; \ 0 \leq e \leq 2^8 - 1; \ 0 \leq m \leq 2^{56} - 1, \ e, m \in \mathbf{Z} \right\}.$$

For $e \neq 0$ any $(s, e, m) \in \mathcal{R}$ represents:

$$s \cdot 2^{(e-128)} \cdot \left(\frac{m + 2^{56}}{2^{57}} \right)$$

If $e = 0, s = +1$ then (s, e, m) represents 0 (regardless of m), and $e = 0, s = -1$ is unused.

There are several properties of the machine operations on these numbers which we have used. All of them can be found in the DEC manual and we checked them in some cases. We have also checked by looking at the assembler listing generated that the compiler generates calls to the VAX floating point instructions rather than calling library routines.

i) Comparisons of numbers in $\mathcal{I}$ or $\mathcal{R}$ are done by comparing bits, so the results of a comparison are exact.

ii) Arithmetic operations on machine integers are performed correctly modulo 2^{32}. An overflow is reported by the machine as a "exception", but the computation proceeds.

iv) Conversion of an integer to a real is done exactly. Since the mantissa of a real is longer than the integer and the internal representations are almost the same, it is hard to imagine that this is done otherwise. Nevertheless, we checked some borderline cases.

v) Arithmetic operations on machine reals produce results in $\mathcal{R}$, which are correct to "$\frac{1}{2}$ the least significant bit."

vi) Floating point overflow is reported by the hardware as an exception. We have relied on the fact that the code generated by a UNIX compiler, by default, stops execution when an overflow occurs.

The key to implementing interval arithmetic is to write two functions:

$$\text{up}: \mathcal{R} \to \mathcal{R} \quad \text{and} \quad \text{dn}: \mathcal{R} \to \mathcal{R}$$

which respectively increment (decrement) the least significant bit of the machine real to which they are applied (reporting overflows if they occur.) Then, if $x, y \in \mathcal{R}$ and $+_c$ is the computer's addition operation and $+$ denoteds the true mathematical operation, we have:

$$\text{dn}(x +_c y) \leq x + y \leq \text{up}(x +_c y),$$

and similarly for the other arithmetic operations.

For efficiency, we wrote up dn in assembler, but the rest of the arithmetic routines to manipulate intervals are written in C. We also found it useful to include routines to manipulate bounds which can be interpreted as intervals whose endpoints are at infinity. We also included routines to manipulate subsets of complex numbers either *rectangles*, products of intervals, or *balls* specified by a center and an upper bound on the radius. Operations with rectangles run faster, but there are instances where using balls produces sharper bounds. We have also routines that compute some elementary functions on the types described above.

5.3. Arithmetic on Banach algebras of analytic functions. It is possible to extend the basic ideas of interval analysis to function space. We work with neighborhoods specified by a finite number of representable numbers and write routines that implement computer operations that bound the mathematical operations. The resulting neighborhood is required to include all the possible outcomes when the operands range over the input neighborhoods. If interval arithmetic takes care automatically of round–off error, the interval arithmetic in function space, sometimes called "*ultra–arithmetic*" will take care of truncation error.

There are several proposals in the literature [Mo], [KM] each one, designed with some particular problem in mind.

The set up we have found best adapted for our purposes is truncation in power series introduced by Lanford. (See [La2], [EKW], [EW] .) We have found convenient to add some more information to the basic data structure.

If $f_0 \ldots f_n$ are intervals (or rectangles or balls) and g_0, g_1, h_0, h_1 are bounds, we define, with the notation used in **3.2**:

$$
\begin{aligned}
\mathcal{N}(f_1, ..., f_n; g_0, g_1, h_0, h_1) = \{K(z) \mid {} & K = K_p + K_g + K_h, \\
& K_p(0) = K_g(0) = K_h(0) = 0. \\
& (K_p)_m \in f_m \text{ for } 1 \le m \le n, \ (K_p)_m = 0, \ m > n \\
& \|K_g\|_0 \le g_0, \||K_g\||_0 \le g_1, \\
& (K_h)_m = 0 \text{ for } 0 \le m \le n, \ \|K_h\|_0 \le h_0, \||K_h\||_0 \le h_1\}
\end{aligned}
$$

K_g, K_h are the error terms.

Analogous definitions can be introduced for periodic functions using Fourier coefficients and the norms introduced in **3.3** . In that case, however, we have not found it useful to keep the error divided into high order and general. For some purposes, it is useful to use neighborhoods in which only $\| \ \|_0$ is used in the estimate of the errors.

The arithmetic operations between these neighborhoods are easy to implement remembering that, when the coefficient of order zero vanishes $\|f \cdot g\|_0 \le \|f\|_0 \|g\|_0$ and $\||f \cdot g\||_0 \le \|f\|_0 \||g\||_0 + \||f\||_0 \|g\|_0$.

It is also possible to implement other operations such as composition, integration, etc. as well as some transcendental functions.

For problems in celestial mechanics, this approach has shortcomings. Typically, the most interesting problems in celestial mechanics involve many variables. Using the representation discussed above, the number or terms required would exceed the capacity of the largest computers. Nevertheless, for many problems, only a very small number of Fourier coefficients is relevant. A sytematic package to deal with sparse Fourier and Taylor series of many variables with rational coefficients has been developed by K. Meyer and D. Schmidt. They have used it to settle stability problems for isolated parameter values where standard K.A.M. theory fails and it becomes necessary to study subsequent terms. [MS].

5.4. Implementation of arithmetic in Banach algebras. We have written routines in C that implement all the operations described above.

The routines return pointers to the places where the results are stored. This allows us to use successive calls without making explicit assignments to intermediate variables. The alternative or returning structures was discarded due to the fact that it is slow and that with many compilers does not work. The degree of the polynomials was controlled by a global variable, so that some intermediate delicate calculations could be carried to a higher degree.

One advantage of this modular organization is that the programs can be reused and its is possible to write check routines that enhance reliability. Given the similarity between Fourier series and complex polynomials, some of the code could be shared by the programs for the Siegel center and the twist mapping problems. L. Seco could use substantial part of the package in a SUN 3 for an project independent of ours and make some enhancements [Se] (see also his contribution to this volume.) Another advantage of this modularity is that it makes possible to isolate the weakest spots and fine–tune them.

Once one modifies the rounding routines, the rest of the program is machine independent except for questions of efficiency. We have complied them and run test routines successfully on MS–DOS machines and SUN's. We expect they are not difficult to move to other machines.

For the details of the implementation, including listings as well as some other optimizations in the Siegel theorem, we refer to [Ra].

5.5. Numerical analysis considerations. It is well known that interval analysis is very bad at detecting cancellations. Straightforward translation into interval arithmetic will produce the result that, if $x \in (-1, 1)$, then $x - x \in (-2 - \epsilon, 2 + \epsilon)$, which, is certainly correct but far from optimal.

For problems such as K.A.M theory in which cancellations play an important role, mathematically equivalent expressions could lead to different, but overlapping, results.

If the estimates are far too conservative, it is possible sometimes to get an understanding of what is happening by looking at the width of intervals at selected places.

Notice that the results of conventional numerical analysis are notoriously imprecise if some intermediate result involves calculating the difference of two very similar numbers. The problem is that there is no record that this thing has happened. Interval arithmetic, can be used to spot places where destructive looses of precision occur. With hindsight, some of them could have been guessed from the beginning, but others are more difficult to understand. Even if we were not interested in getting rigorous proofs, interval arithmetic can be an invaluable debugging tool to write more robust programs.

Notice also that interval arithmetic can produce results in regimes where the conventional tests of numerical analysis prescribe stopping. In our case, for example, we can be confident of our results for values of the parameter bigger than those for which reruns in single and double precision of reasonable implementations produce different results.

5.6. Results. We have run our programs for a few hours on a VAX 11/750 and obtained:

Theorem 5.1. *For f_ϵ as in (1.3) the results of Theorem 1.1 hold if $|\epsilon| \leq .306$ and are false if $|\epsilon| \geq .342$.*

Remark. The converse K.A.M. bounds above are obtained using the area formula. Using the Bieberbach–De Branges bounds, we obtain only .360.

Theorem 5.2. *The family of diffeomorphisms F_ϵ as in (1.4) has a topologically non–trivial analytic invariant circle with golden mean rotation number for $\epsilon = .91$.*

Remark. We have verified the theorem for many other values of the perturbation parameter between 0 and 0.91. Due to limitations in time, we have not yet covered the whole interval, but we have not found any indication that the theorem fails for an intermediate value. Numerically, it seems that it is possible to interpolate smoothly between all the computed circles. The difficulty of the computation and all the measures of roughness of the computed circle increase monotonically with ϵ. See also [CC] for another computer assisted proof that covers the interval $[0, 0.65]$ in parameter space.

We also recall the result of [MP]:

Theorem 5.3. For F_ϵ as (1.4) the results of Theorem 1.2 are false if $|\epsilon| \geq 63/64$.

In both examples we have proved K.A.M. bounds that are 90% of converse K.A.M. bounds.

REFERENCES

[Ar] V. I. Arnol'd: *"Geometric methods in the theory of ordinary differential equations"*, Springer Verlag, NY (1983).

[Au] S. Aubry: The twist map, the extended Frenkel–Kontorova model and the devil's staircase. Physica **7D**, 240–258 (1983).

[BC] A. Berretti, L. Chierchia: Univ. Roma preprint.

[BZ] D. Braess, E. Zehnder: On the numerical treatment of a small divisor problem. Numer. Math. 269–292 (1982).

[Bo] J.–B. Bost: Tores invariantes des systémes dynamiques hamiltoniens. Seminaire Bourbaki Asterisque **133–134**, 113–158 (1986).

[Br] Brolin: Invariant sets under iteration of rational functions. Arkiv for Math. 6, 103–144 (1965).

[CC] A. Celletti, L. Chierchia: Construction of analytic K.A.M. surfaces and effective stability bounds. Comm. Math. Phys. **118**, 119–161 (1988).

[Dec] Digital Equipment Corporation: *"VAX architecture handbook"*, Digital Equipment Corp (1981).

[DeV] DeVogelere: On the structure of symmetric periodic solutions of conservative systems, with applications. In *"Contributions to the theory of nonlinear oscillations IV"*. Princeton Univ. Press, Princeton 53–84 (1958).

[Di] J. Dieudonné: *"Foundations of modern analysis"*, Academic Press, N.Y. (1960).

[EKW] J-P. Eckmann, H. Koch and P. Wittwer: A computer assisted proof of universality in area preserving maps. Mem. of the A.M.S. **289**, (1984).

[EW] J-P. Eckmann, P. Wittwer: *"Computer methods and Borel summability applied to Feigenbaum's equation"*, Springer Verlag, NY (1985).

[Fa] A. Fathi Une interpretation plus topologique de la démonstration du théorème de Birkhoff: appendix to [He1] 39–47 (1983).

[Gr] J. Greene: A method for determining the stochastic transition. Jour. Math. Phys. **20**, 1183–1201 (1979).

[Ha] R. S. Hamilton: The inverse function theorem of Nash and Moser. Bull. A.M.S. **7**, 65–222 (1982).

[He1] Michael R. Herman: Sur les courbes invariantes par les difféomorphismes de l'anneau vol 2. Astérique **144**, (1986).

[He2] Sur la conjugaison différentiable des difféomorphismes du cercle a des rotations. Pub. Mat. du I.H.E.S. **49**, (1978).

[He3] Simple proofs of local conjugacy theorems for diffeomorphisms of the circle with almost every rotation number. M.S.R.I. preprint (1984).

[He4] Existence et non–existence de tores invariantes par des diffeomorphismes symplectiques. preprint (1988).

[KM] E. W. Kaucher, W. Miranker: *"Self-Validating numerics for function space problems"*, Academic Press, Orlando (1984).

[KO] Y. Katznelson, D. Ornstein: the differentiability of the conjugation of certain diffeomorphisms of the circle. preprint (1987).

[Ka] A. Katok: Minimal orbits for small perturbations of completely integrable hamiltonian systems. preprint (1988).

[Kn] D. E. Knuth: " *The art of computer programming vol II: Seminumerical algorithms* , 2nd *edition*", Addison Wesley (1980).

[Ko] A. N. Kolmogorov: Various approaches to an estimate of the difficulty of an approximate definition and calculation of functions. In "*Proc. Int. Cong. Math., Stockholm*". 369–376

[LST] C.A. Liverani, G. Servizi, and G. Turchetti: Some K.A.M. estimates for C.L. Siegel's center theorem. Lett. Nuovo Cimento **39**, (1974).

[LT] C.A. Liverani, G. Turchetti: Improved K.A.M. estimates for the Siegel radius. Jour. Stat. Phys. **45**, 1071–1086 (1986).

[La] O. E. Lanford III: Computer assisted proofs in analysis. In "*Proc. International Congress of Mathematicians, Berkeley 1986*". A.M.S. Providence 1385–1394 (1988).

[La2] O.E. Lanford III: Computer assisted proofs in analysis. Physica **124A**, 465-470 (1984).

[MP] R. MacKay, I.C. Percival: Converse K.A.M.: theory and practice. Comm. Math. Phys. **98**, 469-512 (1985).

[MS] K. Meyer, D. Schmidt: The stability of the Lagrange triangular point and a theorem of Arnol'd. Jour. Diff. Eq. **62**, 222-236 (1986).

[Ma] J. Mather: Non-existence of invariant circles. Ergo. Th. & Dynam. Sys. **4**, 301-309 (1984).

[Mo1] J. Moser: Is the solar system stable?. Math. Intelligencer **1**, 65-71 (1978).

[Mo2] J. Moser: On invariant curves of area preserving mappings of an annulus. Nachr. Akad. Wiss. Göttingen Math Phys **K1**, 1-20 (1962).

[Mo3] R.E. Moore: "*Methods and applications of interval analysis*", S.I.A.M. Philadelphia (1979).

[OS] A. Olvera, C. Simó: An obstruction method for the destruction of invariant curves. Physica **26D**, 181–192 (1987).

[Po] H. Poincaré: "*Les methodes nouvelles de la méchanique céleste*", Gauthier Villars, Paris (1891-1899).

[Rü] H. Rüssmann: On a new proof of Moser's twist mapping theorem. Cel. Mech. **14**, 19–31 (1976).

[Ra] D. Rana: "*Proof of accurate upper and lower bounds to stability domains in small denominator problems*", Thesis, Princeton Univ. (1987).

[SZ] D. Salamon , E. Zehnder: K.A.M. theory in configuration space. Comment. Math. Helv. **64**, 84–132 (1989).

[Se] L. Seco: Lower bounds for the ground state energy of atoms. Thesis, Princeton Univ. (1989).

[Si] C. L. Siegel: Iteration of analytic functions. Ann. of Math. **43**, 607–612 (1942).

[St] J. Stark: An exhaustive criterion for the non–existence of invariant circles for area–preserving twist maps. Comm. Math. Phys. **117**, 177–189 (1988).

[Yo] J.-C. Yoccoz: conjugaison différentiable des difféomorphismes du cercle dont le nombre de rotation vérifie une condition diophantiene. Ann. scient. Ec. Norm. Sup. **17**, 333-359 (1984).

[Ze1] E. Zehnder: Generalized implicit function theorems with applications to some small divisor problems I. Comm. Pure and Appl. Math. **28** , 91–140 (1975).

[Ze2] E. Zehnder: Generalized implicit function theorems with applications to some small divisor problems II. Comm. Pure and Appl. Math. **29** , 49–111 (1976).

[Ze3] E. Zehnder: A Simple proof of a generalization of a theorem by C.L. Siegel. Springer Verlag, N.Y., Lec. Notes in Math. **597**, (1970).

A SOFTWARE TOOL FOR ANALYSIS IN FUNCTION SPACES

J.-P. ECKMANN*, A. MALASPINAS*, AND S. OLIFFSON KAMPHORST*

Abstract. A tool is presented which allows for an efficient description of Banach spaces of analytic functions. Based on this description, an extended Pascal and its associated compiler understanding the usual operations on such spaces is generated. Furthermore, based on the mathematical context, the relevant subroutines needed to implement the function calls are produced.

1. Introduction. A fairly large number of operations in Banach spaces, and more particularly in spaces of analytic functions, are constructive. This means that together with an existence theorem, they also provide a bound on the corresponding object. Typical such results are the contraction mapping principle and its variants, or the fact that restriction of a set of analytic functions to a compact subdomain is a compact operator. In the first case, effective bounds on the fixed point of the contracting maps are obtained, and in the second case, the Cauchy formula provides a bound on the restricted function in terms of the original one.

The effort to find those bounds which are useful and sufficiently strong to solve a given problem may vary considerably from one problem to the next, and the number of elementary operations to verify a bound may be very large. On the other hand, it has been demonstrated that, with necessary precautions concerning rounding issues, computers can be used to perform this task in the context of functional analysis as it is sketched above. With the aid of computers, problems of computational complexity beyond human capacity can be attacked, and the method has been used with success in several non-trivial examples [EKW, KW, EW1, L].

One of the difficulties of the approach seems to be that the computer programs which perform these bounds are often hard to read, and hence to verify, because there is no adequate way to transcribe mathematical facts about Banach spaces into the current programming languages. In fact, similar difficulties occur in other fields of research:

- The efficient description of compilers and operating systems and their semantics [BJ].

- The relation between mathematical facts and their implementation as programs [Kn].

The present work can be viewed as a further attempt to gain experience with this class of problems. It is our conviction that in the difficult field of teaching "ideas" to computers, progress is made by solving increasingly general classes of problems, such as the ones mentioned above. Understanding what they mean, and what their essence is, should fruitfully complement those parts of "knowledge" contained in data bases.

---

*Department of Theoretical Physics, University of Geneva, 1211 Geneva 4, Switzerland

Many proposals and implementations for allowing scientific notation in programming have been made. The examples of Ada and Pascal-SC come closest to our project. What is new here is that large quantities of contextually implied subroutine code are automatically generated from the *mathematical* description. The actions which we generate are sometimes nonlinear functions of the structures involved (see below).

This paper deals with the fundamentals and a detailed description of a software tool for analysis in function spaces. In a companion paper [EW2], an extensive description of an example is given. The example gives a complete proof of the so-called "Feigenbaum conjectures" for maps of the interval and it uses analytic functions in two variables.

2. Overview. To facilitate the programming of computer assisted proofs in analysis, we have developed a tool which consists of three interlocking components:

1) A small *programming language* called *Mini* in which one can describe efficiently the vector spaces or, more particularly, the spaces of polynomials, one wishes to consider.

2) A *compiler generator* which, using the descriptions given under 1), defines an extension of Pascal† called *Lang*, in such a way that the common vector space operations ($+$, $-$, assignment, scalar multiplication, norm) can be programmed in usual mathematical notation. The grammar of Lang is described in Yacc (the standard compiler-compiler of Unix). A simple compilation leads to a preprocessor which converts Lang to Pascal.

3) A *subroutine generator*, which generates those subroutine calls which are needed to implement the function calls the Lang compiler generates when it translates the extended Pascal to Pascal.

The system exists in two versions: One version is based on Pascal, and other is based on the C programming language. Both versions run under Unix BSD4.2. However, our description below will deal only with the interface based on a somewhat extended Pascal, which provides a more checkable version. This "extension" of Pascal does not support the "with ... do" construction. The translation of the grammar of Mini to the C-version (which looks more like C rather than Pascal) is easy.

3. An Example. The aim of Sects. 3–7 is to explain the mathematical framework on which the construction of Mini is based. This framework is less general than similar other proposals made in the literature, but more directly applicable to concrete problems. The intended use of Mini is in giving bounds on Banach space operations. We begin with an example dealing with a space of analytic functions on the unit disk, which we denote by A_0. We represent a function f in A_0 as

$$f(z) = \sum_{i=0}^{\infty} f_i z^i \ .$$

---

† A version extending C is also available

We denote by A the subset of A_0 consisting of functions with real f_i and with finite "l_1-norm"

$$\|f\| \equiv \sum_{i=0}^{\infty} |f_i| \ .$$

A ball B (of degree n) in A is a subset of A given by

$$B(I_0,\ldots,I_n,h)$$
(1)
$$= \{f \in A | f_i \in I_k \text{ for } i = 0,\ldots,n \ , \ \sum_{m=n+1}^{\infty} |f_m| < h\} \ .$$

Here, the I_j are real intervals, and h is a nonnegative real number. We claim that the usual operations on analytic functions carry over to balls of the above type, and that these operations are in fact constructive. This means that they can be performed by computers. As a first example, we consider *addition*. Note that if $f \in B(I_0,\ldots,I_n,h)$ and $g \in B(I_0',\ldots,I_n',h')$ then

(2)
$$f + g \in B(I_0 + I_0',\ldots,I_n + I_n',h + h'),$$

where $I + I'$ is the addition of intervals:

$$I + I' = \{x \in R | x = x_1 + x_1', x_1 \in I, x_1' \in I'\}.$$

It is clear from the above bound that the addition of balls can be implemented on a computer provided rounding issues are handled carefully. *The purpose of Mini is to make this kind of computations easy to formulate and to extend Pascal in such a way that ordinary mathematical notation can be applied to calculations with balls.* We show now how to implement the structure of B in Mini. In addition to the usual type "real" of Pascal, there are three predefined types in Mini, called "s", "u", and "l". The type s (segment) is a data structure describing the two endpoints of a real interval, and all operations on objects of type s are implemented with rigorous bounds, so that , e.g. the addition of two intervals I, I' leads to a third interval I'' (represented again with 2 "real" numbers as endpoints) such that

$$I + I' \subset I'' \ .$$

The type "u" represents (non-negative) real numbers and is used for upper bounds, such as h above. The type "l" is used for the lower bounds.

The exact implicit definitions of "s", "u', and "l" are (in Pascal):

```
TYPE u = RECORD
    value :  real ;
END;


TYPE 1 = RECORD
```

```
   value :  real ;
END;

TYPE s = RECORD
   lower :  1 ;
   upper :  u ;
END ;
```

In Mini, the description of **B** as given by Eq. (1) (for $n = 5$) takes the following form:

Example 1.
```
TYPE p = POLYNOMIAL OVER s
CONST n=5 ;
VAR i :  integer ;
BEGIN
FOR i := 0 TO n DO
   coef :  s
END ;

TYPE b = VECTOR OVER s
BEGIN
   poly :  p ;
   BOUND high :  u ;
END ;
```

The word "bound" signals that high is used to store an absolute value, see Sect. 8. If Example 1 is given as an input to Mini, the following things happen.

1) Two new types are implicitly declared, namely

```
   CONST loop $1 = 5 ;

   TYPE p = RECORD
   coef :  ARRAY[0 .   . loop $1] OF s ;
   END ;

   TYPE b = RECORD
     poly :  p;
     high:  u;
   END ;
```

2) Subroutines are generated which implement a large set of operations (see Table 2 below). In particular, the addition of two objects of type b according to Eq. (2) will be defined. More precisely, a subroutine bSUM will be generated, which implements the sum of two balls as described above. Similarly, bsLMULT will implement "scalar multiplication" by an interval, i.e., if $\mathbf{G}(I_0,\ldots,I_n,h)$ and I are given then we want to find a ball $\mathbf{G}'$ which contains all $y \cdot f$ for $y \in I$ and $f \in \mathbf{G}(I_0,\ldots,I_n,h)$. Clearly, a reasonable choice of $\mathbf{B}'$ is

$$(3) \qquad \mathbf{B}' \equiv \mathbf{B}(I^*I_0,\ldots,I^*I_n,|I|h),$$

where

$$I^*I' \equiv \{y = xx'|x \in I, x' \in I'\},$$

and

$$|I| \equiv \max\{y = |x| \mid x \in I\}.$$

3) A Yacc description of the Lang grammar is generated which defines an extension of Pascal by types l, u, s, p, and b and in which the mathematical symbols take their conventional meaning. Thus, $b1 + b2$ means the addition of two balls in the sense of Eq. (2) and $s1 * b2$ implements the bound of Eq. (3).

Note: The first letters of each variable *must* coincide with the type name (except for reals and integers). Also a variable whose first letters coincides with a type name will be taken by the Lang grammar to be of that type.

4. Bounds in Banach Spaces. In Mini, constructions of the type of Example 1 are supported in great generality. The underlying mathematical structure is as follows:

Let H be a Banach space. Let $E_0,\ldots,E_n$ be unit vectors in H and let $H_1,\ldots,H_m$ be subspaces of H, spanning together H. We call this a decomposition of H. The decomposition of $x \in H$ into these subspaces need not be unique, and the H_i may be one-dimensional, multidimensional or infinite dimensional, but not empty. In the case of the balls $\mathbf{B}$ of Sect. 3, we have $m = 1, E_i = z^i, i = 0,\ldots,n$, and H_1 is the set of functions in $\mathbf{A}$ which are of order z^{n+1}. We give $n+1$ real intervals $s_i, i = 0,\ldots,n$ and m non-negative numbers $u_i, i = 1,\ldots,m$ and we define

$$H(s,u) = \{x \in H | x = \sum_{i=0}^{n} \sigma_i E_i + \sum_{i=1}^{m} y_i \text{ with}$$

$$\sigma_i \in s_i, y_i \in H_i, \quad \text{and} \quad \|y_i\| \leq u_i\},$$

(Similar definitions are possible for complex vector spaces.) Clearly, if

$$x \in H(s,u) \quad \text{and} \quad x' \in H(s',u'),$$

then

$$(4) \qquad x + x' \in H(s + s', u + u'),$$

while for $\lambda \in \mathbf{R}$, we have

(5) $$\lambda x \in H(\lambda \cdot s, |\lambda|u).$$

Here, $s + s'$ is the componentwise addition of intervals, i.e. $(s + s')_i = s_i + s'_i$, while $\cdot$ is the componentwise multiplication by a scalar.

We also have a bound on the norm, for $x \in H(s, u)$:

(6) $$\|x\| \le \sum_{i=0}^{n} |s_i| + \sum_{i=1}^{m} u_i \, ,$$

where

$$|s| = \max\{|z|, \ z \in s\}.$$

The Eqs. (4)–(6) lead naturally to constructive algorithms, which can be automatically generated from the description of H (and which carry the generic names SUM, LMULT, and ABS). The Lang grammar generated by Mini will in turn recognize $+, *$, and $|\cdot|$ as the corresponding operation symbols.

Some further operations are defined: The difference of $x \in H(s, u)$ and $x' \in H(s', u')$ is bounded by

(7) $$x - x' \in H(s - s', u + u'),$$

which leads to a subroutine DIFF, and a unary minus (NEG). Finally, it is useful to have a notion of zero, and some output routines (See again Table 2.)

5. Polynomials. We recall the example of Sect. 3, dealing with analytic functions on the unit disk. Clearly, the definition of the polynomial part of the function lends itself to an automatic generation of several operations which can be performed in the (truncated) polynomial algebra. In Mini, a polynomial is *one* indexed array of coefficients. Thus, the definition of polynomials is more restricted than that of vectors. On the other hand, Mini will generate more subroutines for polynomials and extend the Lang grammar correspondingly. Apart from those operations defined for every vector type, the following operations are automatically generated for polynomials:

-The (truncated) product, $h(z) = f(z) \cdot g(z)$,

-Evaluation, $y = f(z_0)$,

-Dilation and translation of arguments, $g(z) = f(az + b)$,

-Derivatives of every order, $g(z) = \partial_z^n f(z)$,

-The (truncated) inverse, $g(z) = 1/f(z)$.

Of course, all these operations are generated with rigorous error bounds on the (interval) coefficients. Polynomials in several variables are allowed. The Lang grammar will understand that if p is a polynomial type, then *all variables starting with the letter* p *will be of type* p. On the other hand, with this convention the grammar will imply that operations like

p1 * p2
p1+p2
a * p1

take their standard mathematical meaning, namely (truncated) product, sum, and multiplication by a scalar of the corresponding field. Furthermore, p1(#=a) means evaluation of the polynomial at a, and p1(#:a,b) means p1(a· z+b), where z is the formal variable of p1.

Finally, in the Lang grammar generated by Mini, |p1| means the sum of the absolute values of the coefficients of p1.

6. Products. When one considers analytic functions, as in the example of Sect. 3, it is desirable to be able to give a bound on their product. A constructive bound is in fact possible as we now show. Assume $\mathbf{B} = \mathbf{B}(I_0,\ldots,I_n,h)$ and $\mathbf{B}' = \mathbf{B}(I'_0,\ldots,I'_n,h')$ are given, and we are to find a ball $\mathbf{B}''$ containing all $f\ f'$ with $f \in \mathbf{B}, f' \in \mathbf{B}'$. A possible choice is $\mathbf{B}'' = \mathbf{B}(J_0,\ldots,J_n,k)$, with

(a)
$$J_l = \sum_{p=0}^{l} I_p I'_{l-p}, \quad \text{for } l = 0,\ldots,n,$$

(b)
$$k = \sum_{p+q>n} |I_p|\,|I'_q|$$

(8)(c)
$$+ \sum_{p=0}^{n} |I_p| h'$$

(d)
$$+ \sum_{p=0}^{n} h|I'_p|$$

(e)
$$+ h\ h'\ .$$

The bound (8) may seem canonical, but in practice, (see e.g. [EW1]), there occur cases where other choices may be preferable, or where no canonical choice is possible (cf. also [M]). We therefore have designed Mini so that it does not generate a bound automatically. But Mini provides a syntactic construct to aid in the generation of the product of two objects of type vector such as b. In Example 2 below, we implement the natural choice of Eq. (8), corresponding to the definition of balls given above (Example 1). In Mini, this is described as follows (we add the letters (a)–(e) to indicate the correspondence with Eq. (8)):

Example 2.

```
DEFINE b*b -> b
CONST n=5;
BEGIN
poly * poly -> poly;                          (a)
IF (i$ 1+i $2 > n) THEN
    poly * poly -> high;                       (b)
poly * high -> high;                           (c)
high * poly -> high;                           (d)
high * high -> high;                           (e)
END;
```

The idea is that all implicit summations can be omitted from the description of the bounds. The variables i$1, i$2 are derived from the variable name for the declaration of the polynomial part, in our case, the variable i of Example 1. Here, i$1 and i$2 refer to the left and right factors, respectively. Of course, the Lang grammar will understand b1*b2 as a bound of the kind just described.

Finally, if p is a polynomial and b any object over the same number field as p, for which multiplication is defined, then the command

Example 3.

```
DEFINE p o b ;
```

generates the composition

$$p \circ b = \sum_{i=0}^{n} p_i b^i .$$

In other words, $p \circ b(z) = p(b(z))$. Also, the notation p1(#=b2) will be understood as p1 o b2 in the extension of Pascal generated by Mini.

A deeper consequence of the definition of the product is that the *inverse* can be automatically defined.

7. Inverses. We have seen how products can be defined using Mini. Using this product, we want to define the "inverse" of a vector in H, where H is as in Sect. 4. We *assume* that $E_0 \in H$ spans the one dimensional subspace corresponding to the identity for the product $H \times H \to H$. (In the case of a polynomial with higher order terms as in Examples 1–2, E_0 corresponds to the constant function 1.) We then *define* the inverse of $(1+\nu)$, as $\sum_{k=0}^{\infty}(-\nu)^k$, where $(-\nu)^k$ is the k-fold product of $-\nu$. Here, we assume that the E_0-component of ν vanishes. (In the case of bounds on functions, this definition is reasonable.) We now discuss how we bound the infinite sum. If H is of the form $E_0, \ldots, E_n$, $H_1, \ldots, H_m$, as above, and if $E_0, \ldots, E_n$ are the coefficients of a polynomial (in one or several variables) of maximal degree M, then we consider

$$\sum_{k=0}^{\infty}(-\nu)^k = \sum_{k=0}^{M}(-\nu)^k + \sum_{k=0}^{\infty}(-\nu)^k(-\nu)^{M+1}.$$

The first term is a finite sum and can be evaluated by using the sum and product operators. Similarly $(-\nu)^{M+1}$ is easily evaluated. Note now that $(-\nu)^{M+1}$ *lies entirely in the subspace spanned by* $H_1, \ldots, H_m$, since the lowest order occurring in ν is 1. (This argument applies to any Euclidean ring with inverse. Mini will generate an incorrect inverse in other cases, or when the product does not accumulate "errors" only in $H_1, \ldots, H_m$.) We now view the left multiplication by $(-\nu)$ as a *left action* on $H_1, \ldots, H_m$. This action is defined by the definition of the product in H and defines a matrix operator A on the corresponding bounds. It is a simple matter to invert $(1 + A)$ as a matrix. If the declaration of the product defines a finite dimensional algebra in which the inverse is really given by the sum $\Sigma(-\nu)^k$, then the above construction will give sharp bounds to within rounding errors.

We illustrate the construction of the inverse in detail for the Example 1–2. We assume for simplicity that the degree (const n) is only 1. Thus we want to invert

$$(1, \tau, \alpha) \rightarrow 1 + \tau x + h(x) , \quad \text{with} \quad \|h\| \leq \alpha ,$$

(τ is an interval, x is the formal variable, $h(x) = \sum_{j=2}^{\infty} h_j x^j$). Applying the general ideas explained above, we get

$[1 + \tau x + h(x)]^{-1}$

$$= 1 - \tau x - h(x) + \sum_{k=0}^{\infty}(-\tau x - h(x))^k[\tau^2 x^2 + 2\tau x h(x) + h^2(x)].$$

Note now that the term in square brackets contains only terms of order higher than 1 in x, and is bounded (in the l_1-norm) by

$$|\tau|^2 + 2|\tau|\alpha + \alpha^2 ,$$

where $|\tau| = \sup\{|y|, \ y \in \tau\}$. Similarly, the 1×1 matrix A is induced by the map

$$h_1(x) \rightarrow -\tau x h_1(x) - h(x)h_1(x).$$

We get

$$\| - \tau x h_1(x) - h(x)h_1(x)\| \leq (|\tau| + \alpha)\|h_1\| .$$

The matrix A acts on the bound, i.e. if $\|h_1\| = \alpha'$, then we get

$$A\alpha' = (|\tau| + \alpha)\alpha' .$$

Thus, if $|\tau| + \alpha < 1$, then

$$(1 + \tau x + h(x))^{-1} = 1 - \tau x + \tilde{h}(x),$$

with

$$\|\tilde{h}(x)\| \leq \alpha + (1 - (|\tau| + \alpha))^{-1}(|\tau|^2 + 2|\tau|\alpha + \alpha^2).$$

Again, Mini generates these operations as INV and the Lang grammar is augmented to accept the construction 1/b1 as the inverse and b1/b2 as the quotient.

8. The Grammar of Mini. We have now described the main features of Mini. The grammar of Mini, given in detail in Table 1, allows for polynomials in several variables, for indices in the higher order terms, as well as for simple conditions in the choice of index sets. We use BNF notation where {...} denotes 0 or more occurrences of the item, and (...) denotes zero or one occurrence. Upper case words and symbols in quotes '...' denote terminals, while the lower case words are non-terminals. Finally | denotes alternatives.

Table 1. The grammar of Mini, somewhat simplified.

```
/*TYPES*/
/ * * * * * * */
new                    :  new_part {new_part}

new_part               :  TYPE identifier '=' VECTOR OVER field
                               (declarations_part) block ' ; '
                       |TYPE identifier '=' POLYNOMIAL OVER field
                               (declarations_part) short_block ' ; '
                       | DEFINE identifier '*' identifier '->' identifier
                               (declarations_part) action_list ' ; '
                       | DEFINE identifier 'o' identifier ' ; '

field                  :  identifier

block                  :  BEGIN statement {' ; ' statement}  END

short_block            :  BEGIN for_clause {for_clause}
                               name ':' type_name {' ; '} END

action_list               :BEGIN action  {' ; ' action}  END

statement                 :  {for_clause} simple_component

simple_component          :  name {' , ' name} ' : '  type_name
                          | BOUND name {' , ' name} ' : '  type_name


/*DECLARATION - PART*/
/* * * * * * * * * * * * * * * * */

declarations_part      :  {declaration}

declaration            :  constant_definition_part
                       | variable_decl_part

constant_definition_part:  CONST constant_definition
```

```
                              {' ; ' constant_definition} ' ; '

constant_definition    :   identifier '=' constant

constant               :unsigned_integer (as in Pascal)

variable_decl_part     :  VAR variable_declaration
                             {' ; ' variable_declaration} ' ; '

variable_declaration   :identifier {' ; ' identifier} ' : ' INTEGER


/*LOOP CONSTRUCTS WITH EMBEDDED IF - THEN*/
/* * * * * * * * * * * * * * * * * * * * * * * * * * * * */

for_clause      :  FOR control_variable ' := '
                       initial_value TO final_value DO
                   | IF expression THEN

control_variable :  identifier

initial_value   :  expression

final_value     :  expression

expression      :  simple_expression
                   | simple_expression
                       relational_op simple_expression

simple_expression :  term
                   | ' - ' term
                   | ' + ' term
                   | simple_expression adding_operator term

adding_operator  :   ' +'
                   | ' - '
                   | OR

term             :  factor
                   | term multiplying_operator factor

relational_op    :  '<' | '>' | '>=' | '>=' | '<>' | '='

multiplying_operator ' * ' | ' / ' | AND | DIV | MOD

factor           :  variable
```

```
                     | unsigned_integer (as in Pascal)
                     | ' ( ' expression ' ) '
                     | NOT factor

variable          :  identifier


/*PRODUCTS WITH BOUNDS*/
/* * * * * * * * * * * * * * * * */

action            :  {for_clause} simple_action

simple_action     :  (prefactor ' * ') indexed_identifier ' * '

prefactor         :  ' " ' any text ' " '


/ *GENERIC STUFF*/
/* * * * * * * * * * * */

name              :  identifier

type_name         :  identifier
indexed_identifier :  identifier (' [ ' expression_list ' ] ')
expression_list   :  expression {' , ' expression}

identifier        :  A-Za-z followed by zero or more A-Za-z0-9
```

We next describe the semantics of Mini. The language Mini has four main structural elements, namely the description of *vectors, polynomials*, the definition of *bounds* in products, and *composition*.

We start by describing *vectors*. Intuitively, the description of a new type of "vector space" is a "direct sum" of vector spaces which have already been defined. The vector spaces (vectors) are constructs as in Sect. 4 and are over a number field which can be one of the types real or s. If the user wants to supply another field, he has to choose a name for it, e.g. "c" for the complex numbers and must provide a set of basic subroutines. Some of these subroutines are cZERO, cONE, ucABS, cDIFF, cNEG, cPROD, ciCONST, ucLMULT. The name of these subroutines must be prefixed by the name of the field (i.e. 'c' in our example). The action is obvious, except for cNEG (the unary minus), ciCONST (conversion of an integer to the new type), and ucLMULT (product of $|c|$ with u). A list is given in Table 3. We suggest the following procedure to create a new number field: First define it as a vector space, e.g. the complex numbers are two-component vectors over the "reals" (or "s" in case of rigorous bounds). This vector space can then be used in the "OVER" construction. Mini will signal that this vector space is used as a field, and the user must check that the necessary operations are (correctly) implemented.

159

Vector spaces are recursively constructed from smaller spaces. The basic vector spaces from which bigger spaces can be constructed are the reals, and intervals (called segments). Segments are a predeclared type, called s. furthermore, there is a sort of affine space, of predeclared type u which serves to allow components in a vector which are upper bounds. These types have been described in Sect. 3. The components of a vector space have names (as for components of a Pascal record), and must be of an already defined type. See the Example 1 given above.

A second, more general possibility is the definition of indexed components. They can be defined with a combination of for statements and if statements. It is *illegal* to use an indexed component name more than once in a declaration. Suppose we want to define components called f_i of type s numbered 1 to 10, with the exception of 3. Then we would define

Example 4.

```
TYPE  q = VECTOR OVER s
VAR  i : integer ;
BEGIN
    FOR  i := 1 TO 10 DO
    IF i<> 3 THEN
    f :  s ;
END;
```

The variable i *must* be declared as an integer (in Pascal fashion) before the "begin". It is desirable to be able to define indexed objects whose upper bound depends on a parameter. In *Mini*, a constant definition is allowed. The user may want to define the constant definitions in an "include file" which however must be accessible to all files generated by Mini. In particular, Mini generates a separate program whose sole purpose is to compute effective array dimensions, because arrays are "packed" by Mini. It must of course know these constants.

A *polynomial* is a special case of a vector and must have exactly one component. In the definition of a polynomial, in principle any combination of for statements is allowed, but the implementation will fail unless *all* degrees below the maximal order are present (and no negative degrees occur). So

Example 5.

```
TYPE  p = POLYNOMIAL OVER real
VAR  i , j : integer ;
BEGIN
    FOR  i := 0 TO 5 DO
     FOR  j := 0 TO 2*i + 3*i*i DO
      coef  : :  real ;
END;
```

```
is allowed but

TYPE p = POLYNOMIAL OVER real
VAR i :  integer ;
BEGIN
   For i :  = 0 TO 5 DO
    IF i<> 2 THEN
      coef  : real ;
END;
```

will lead to disaster (there would be no problem if we only defined a vector and not a polynomial).

To describe the bounds in *products* of vectors, the notation of the Example 2 together with the grammar of Table 1 should make the intentions clear. To be more precise, we give the detailed semantics for the construction $a^*b \to c$, inside a definition of a product $t_1 * t_2 \to t_3$.

We consider first the case when the three types t_1, t_2, t_3 are equal (to a type t), as in Example 2. Then, a, b, c must be names of components of t. If a, b, c are all names of polynomials, then $a^*b \to c$ defines the truncated product. Wherever one of the variables a, b is a polynomial, summation over the indices is implied. In this case, if the summation indices are needed explicitly (as in prefactors, see below) they are obtained from the original names by appending "$1" for a and "$2" for b. The range of summation can be restricted by a sequence of conditional statements as in Example 2.

If, in the definition of c as a component of t, one has given the attribute BOUND to c, as in Example 1, then the command $a^*b \to c$ is meant to accumulate the absolute values of the products. In this case, c can be over the same number field as a and b, *or* of type u (i.e. an upper bound). If c was not given the attribute BOUND then c must be of the same type as a and b (in fact, if c is of type u, it is wise to give it the attribute BOUND).

If the three types t_1, t_2, t_3 are not equal, the rules are in principle the same, however, meaningless combinations of subtypes will not be detected by Mini.

Mini distinguishes four cases in the construction of products:

1) The truncated polynomial product.
2) The polynomial product leading to "higher order" terms.
3) Polynomial times non-polynomial.
4) Non-polynomial times polynomial.

In case 1), Mini generates all necessary terms. In case 2), if the original names of the loop variables were x, y, ... then one refers to the corresponding variables in the left factor as x$1, y$1,... and in the right factor as x$2, y$2, ... The $ distinguishes the name to make it unique. (We suggest that '$' should not be part of a variable declared by the user.) So, as we have seen,

IF (i$1 + i$2 > n) THEN
  poly * poly -> high ;

in Example 2 means
$high:=high$

$$+ \sum_{i\$1+i\$2>n} |bleft.poly.coef[i\$1]^*bright.poly.coef[i\$2]| \, .$$

In cases 3) and 4), there is no appended "$", since only one of the factors uses the
original indices. In other words, to refer to indices in terms of the form "poly * high
-> high", of Example 1, use the index i of the original definition of poly. Thus one
might write

IF (i > n) THEN
  poly * high -> high;

We also allow for an extension of the bounds, so that *weight factors* are possible.
The preceding example can be augmented to

IF (i$1 + i$2 >n) THEN
  "ui const(i$1*i$1+i$2* i$2)" * poly * poly -> high;

which then means
$high:=high$

$$+ \sum_{i\$1+i\$2>n} |uiconst(i\$1^*i\$1 + i\$2^*i\$2)^*$$
$$bleft.poly.coef[i\$1]^*bright.poly.coef[i\$2]| \, .$$

Note that the user must convert the result to the type of the right hand term (in
our case "u").

REMARKS.

1) Mini does not automatically symmetrize the definitions, so that the user is
   responsible for this.

2) Note that summation and indices are *only* implied for polynomials. For
   other indexed compounds, such as an array of bounds, a Pascal-like notation
   must be used. To simplify notation, h[i] denotes the element defined by the
   value i. So if we define

      FOR i : = 7 TO 19 DO
      IF i <> 8 THEN

```
h: u;
```

then h[9] identifies the element with index i=9 although effective storage will be in an array of consecutive indices starting at 0. This correspondence is hidden from the user.

We next address the issue of the *identity*. If the user defines a product $t^*t \rightarrow t$, then Mini will attempt to define an identity, called tONE, by defining it as the polynomial which is constant $= 1$, with all other terms equal to zero. This fails obviously in more complicated cases, where an identity might still be uniqely defined. In this case, the user must replace tONE by a routine of his own.

In case the identity is defined, Mini will generate a subroutine tINV which computes a bound on the (left) inverse for the multiplication. This inverse is obtained by solving the corresponding linear problem. If no bounds occurred, then this is done in normal arithmetic, otherwise interval arithmetic is automatically applied. See Sect. 7 for details.

9. Using the Tools. The user prepares a file containing descriptions of new types written in Mini. He then executes

mini <filename .

A multitude of files will be generated and in particular a file called *info* which describes the set of newly generated subroutines. In the case of input from Example 1 together with 2 this will be:

Table 2. Information File

```
INFORMATION ON LOOPS FOR STRUCTURE p
```

1.

We represent the component(s)

    coef

of the structure p

by array(s) [0 . . loop $1].

The procedure init $1 (in initloops.p) initializes these arrays.

The procedure init $ $ calls all init $n .

The program calcconst.p calculates the constant

    loop$1

INFORMATION ON SUBROUTINES

The call pSHOW(p1) prints p1.

The call bSHOW(b1) prints b1.

Setting p1 to zero is implemented as p1 :=pZERO

Setting b1 to zero is implemented as b1 :=bZERO

p1 :=p2+p3 is implemented as p1 :=pSUM(p2 , p3)

p1 :=p2-p3 is implemented as p1 :=pDIFF(p2 ,p3)

b1 :=b2+b3 is implemented as b1 :=bSUM(b2 , b3)

b1 : =b2 - b3 is implemented as b1 :=bDIFF (b2 , b3)

p1 :=s2* p3 is implemented as p1 :=psLMULT(s2 , p3)

b1 :=s2* b3 is implemented as b1 :=bsLMULT(s2 , b3)

p1 := -p2 is implemented as p1 :=pNEG(p2)

b1 := -b2 is implemented as b1 :=bNEG(b2)

u1 :=|p2| is implemented as u1 :=upABS(p2)

u1 := |b2| is implemented as u1 :=ubABS(b2)

s1 :=p2(#=x3) is implemented as s1 :=spVALUE(P2 , x3)

 p2 is a polynomial evaluated at xi ,

 the argument(s) are of type s .

p1 := p2(#:x3 , y3) is implemented as p1 :=pDILATE(p2, x3 , y3)

 p2 is a polynomial evaluated at xi*(i'th variable) +yi .

 The xi and yi are of type s .

p1 :=p2* p3 is implemented as p1 :=pPROD(p2 , p3)

 This is the truncated product of polynomials.

p1 := 1/p2, the inverse of p2, is implemented as p1 :=pINV(p2)

 This is the truncated inverse of polynomials .

p1 :=p2/p3 is implemented as p1 :=pQUOT(p2 , p3)

The derivate of p1 of order ni is implemented as

 pDERIVE(p1 , ni).

b1 := b2* b3 is implemented as b1 :=bPROD(b2 , b3)

b1 := b2/b3 is implemented as b1 :=bQUOT(b2 , b3)

b1 := 1/b2, the inverse of b2, is implemented as b1 :=bINV(b2)

 The product with bounds is given by

 the definition b * b -> b.

If type comes from polynomials then identity is defined as bONE

Composition b1 :=p2(#=bi) is implemented as

 b1 :=bpCOMP(p2 , bi)

    Mini not only generates subroutines, and the corresponding information, but
also an extension to the grammar of Pascal. This grammar is called *Lang* and

depends on the input to Mini. It is written in Yacc (a common compiler-compiler of Unix).

The command

    make lang

generates a compiler for the "Lang" extension of Pascal. This compiler, (or rather, preprocessor), translates a program from Lang into Pascal, generating function calls for all the standard operations. So in *Lang*, after input from Examples 1–2 to mini, one may write a file "main.lang", such as

```
PROGRAM test (input, output);
PROCEDURE actions;
 VAR b1, b2, b3 :  b;
 BEGIN
  b2 :=bONE:
  b3 :=bZERO;
  b1 := b2 * b3;
  bSHOW(b1);
END;

BEGIN
init $ $;
actions ;
END.
```

One then performs

    make main p

which invokes the Lang preprocessor which acts on "main.lang". It will produce a file which looks about like

```
PROGRAM test (input, output);
PROCEDURE actions;
 VAR b1 , b2, b3 :  b;
 BEGIN
  b2 :=bONE;
  b3 :=bZERO;
  b1 := bprod(b2, b3);
  bSHOW(bONE);
```

```
END
BEGIN
init $ $;
actions;
END .
```

Note that Mini did generate bPROD and bSHOW from the Example 2 (and from Example 1). Note also that the file info contained the information about the existence of bPROD, *and* the legal notation in *Lang*, i.e. b1 :=b2*b3;.

The command

```
make program
```

will put all the pieces together and generate an executable Pascal program.

We specify some particulars of Lang, using Example 1–2. There are now 5 new types, namely 1, u, s, p, and p. In Lang, *every variable name starting with* b *will designate a variable of type* b. Similarly for p, s, l, and u. This simplifies the legibility of programs and is necessary by the way we have implemented the analyzer for Lang. The user should exert extreme caution to avoid the following difficulties.

a) Upper and lower case are distinguished.

b) Having a type f and a type ff could lead to disaster for the lexical analyzer.

c) The letters s, l, u are already reserved as initial letters for variable names.

d) The above rules serve only as an aid to the Lang analyzer. They do not make the usual Pascal declarations unnecessary.

The issue of whether the arithmetic operations are rigorously rounded on the computer is not a question of how Mini operates, but only of how the lowest level routines (see Table 3) assure rounding in the good directions.

Table 3. Low level routines over a number field f.

```
fNEG              -f1
fPROD             f1*f2
f fLMULT          f1*f2
f SUM             f1+f2
fDIFF             f1-f2
f INV             1/f1
fQUOT             f1/f2
u fLMULT          |f1|*u2
f iCONST          conversion of an integer to type f
```

In particular, the routines for the number field "real" do not provide for rigorous rounding, but the routines for "s", "u", and "I" do. This depends on the machine implementation of arithmetic, and in the software which is available from the authors, this is done for the IEEE standard.

Acknowledgements. We thank V. Baladi, D. Buchs, and P. Wittwer for very helpful discussions. This work has been supported by the Fonds National Suisse (JPE, AM) and by CAPES, Brazil (SOK).

REFERENCES

[BJ]   D. BJORNER, C.B. JONES editors, *The Vienna Development Method: The Meta-Language*, Lecture Notes in Computer Science 61, Springer-Verlag, Berlin (1978).

[EKW]   J.-P. ECKMANN, H. KOCH, P. WITTWER, *A computer-assisted proof of universality for area-preserving maps*, Memoirs Am. Math. Soc. 47, 289 (1984).

[EW1]   J.-P. ECKMANN, P. WITTWER, *Computer methods and Borel summability applied to Feigenbaum's equations*, Lecture Notes in Physics 227, Springer-Verlag, Berlin (1985).

[EW2]   J.-P. ECKMANN, P. WITTWER, *A Complete Proof of the Feigenbaum Conjectures*, J. Stat. Phys., in print.

[Kn]   D. KNUTH, *Literate Programming*, Computer Journal 27, 97 (1984).

[KW]   H. KOCH, P. WITTWER, *A Non-Gaussian Renormalization Group Fixed Point for Hierarchical Scalar Lattice Field Theories*, Comm. Math. Phys. 106, 495 (1986).

[L]   O.E. LANFORD, *A computer-assisted proof of the Feigenbaum conjectures*, Bull. AMS, New Series 6, 127 (1984).

[M]   W.L. MIRANKER, *Ultra-Arithmetic: The Digital Computer Set in Function Space*, A New Approach to Scientific Computation, Academic Press (1983).

APPENDIX. SOME ADDITIONAL FEATURES OF "LANG'.

The language Lang is created from the definitions given in Mini. In addition, it has a set of properties which may be useful in writing computer assisted proofs.

1. The types "s" , "u", "I" are predefined: s are intervals, u are upper bounds and l are lower bounds.

2. The common arithmetic operations are defined for the 3 types above, but the rules for using "I" and "u" are very restrictive. The operations u+u, l+l, u*u, l*l lead to a result of type u, l, u, l respectively. (The last condition is somewhat dangerous and applies only to nonnegative lower bounds.) The quantity -u is a lower bound, and similarly 1/u, -l, 1/l are of type l, u, u respectively.

3. A special set of operations yielding s-factors is defined:

   ⟨s⟩ an interval whose two endpoints coincide and are very close to the center of the interval s.

   [u] an interval whose two endpoints coincide and are equal to u.

   [i] here, i is a Pascal integer expression. Then [i] is an interval whose two endpoints coincide and are equal to the expression. (By the grammar, confusion with indexed arrays of Pascal is not possible).

   + -u here, u is a u-term and + -u denotes the interval [-u,u], (of type s).

   1/s The inverse of s. This is defined as $\{x^{-1}|x \in s\}$. (the notation 1/ must be rigorously followed.)

4. The functions upper(s) and lower(s) extract the upper and lower end of an interval s, yielding type u and l. Note: lower and upper are reserved words.

5. |s| yields a u equal to max $\{|x|, x \in s\}$.

6. u**i, s** i denote the power, as in Fortran. i must be an integer expression which is nonnegative. The meaning of s**i is

$$s^{**}i = \{x^i|x \in s\},$$

not

$$s^{**}i = \{x_1 x_2 \cdots x_i \mid x_j \in s, j = 1, \ldots, i\}.$$

7. Upper and lower case : Lang distinguishes upper and lower case letters. All Pascal tokens are accepted either in upper case or in lower case. The special types, such as s,u,l are lower case. If the user defines a new type in Mini, then Lang will only match exact case context. So if you define a type c, then cos will be a token of type c in Lang but Cos will be a standard identifier name (and so will COS, but not cOS).

EQUATION SOLVING BY SYMBOLIC COMPUTATION

ANTHONY C. HEARN*

In proving theorems in analysis, we are often concerned about establishing the equivalence of two algebraic expressions. It is therefore natural to use computer programs that do symbolic manipulations for this purpose. There are now many such programs available, but since I happen to be the principal author of one of these, namely REDUCE [1], the examples I use will be biased towards this system. However, the ideas can be expressed equally well in other algebraic manipulation systems.

Given the nature of existing algebra systems, it is appropriate to concentrate on proofs of theorems that can be expressed as algebraic equations. In other words, we shall consider sets of equations like

$$(1) \qquad\qquad f_i(x_1, x_2, ..., x_n) = 0.$$

Since a large number of diverse problems can be expressed in this form, we can consider many proofs in analysis by a study of such equations.

There are many ways in which the equations in (1) can be considered. The first of these involves considering (1) as an identity. In other words, is it possible to prove that this set of equations is satisfied for all values of $x_1, ..., x_n$? For example, in program verification, one often needs to prove that simple expressions like

$$(2) \qquad\qquad (2x + 3x - 4x)^2 - x^2$$

are zero for all x. The "proof" of such a result is trivial for most algebra systems. Such a proof consists of reducing the expression in (2) to a *canonical form*, at which point the theorem can be established, since the canonical form in such a case is zero. Early verification systems used theorem provers to establish such results, but these were slow and cumbersome. In the early 1970's, verification system designers switched to using algebra systems for such tasks, thus making their systems more efficient [2].

The design of canonical simplifiers has been a key problem in computer algebra for many of years. The depressing news, as discussed in [3] is that the canonical reduction of a general algebraic expression is an undecidable problem. Fortunately, most expressions of practical interest do have a canonical form. In particular, if we limit ourselves to polynomials, there is obviously no difficulty.

The specific problem we are considering here is a little simpler than I have described. It is sufficient to reduce the expression to *normal form* to prove (or disprove) the required result. Normal form reduction does not guarantee a unique representation as a canonical form does, but it does recognize equivalence to zero,

*The RAND Corporation, Santa Monica, CA 90406-2138

which is all we need. Furthermore, if we are only interested in disproving a result, it is sufficient to show that the expression on the left-hand-side of (1) is non-zero for a subset of the variables involved. In the extreme case, replacing all variables by numerical values reduces the problem to a simple calculation. If the result of this calculation is clearly not zero, no further computation is necessary. However, I don't think that any existing algebra system does such numerical checking automatically. A user must therefore make such checks explicitly, although it would be relatively easy to include automatic checking in any existing system.

A more common problem for algebra systems is to determine for what explicit values of the variables the set of equations is satisfied. Ideally, one would like closed-form algebraic expressions as the result. However, it is well known that this is difficult in general. Consider just a single polynomial equation in one variable. If the degree is greater than four, there is no general closed-form solution involving only algebraic numbers and radicals. In addition, a closed form solution to a cubic or quartic is so complicated in general that it is of little use. In such cases, it is necessary to use numerical or combined numerical-symbolic methods to obtain a tractable solution. One might therefore be tempted to conclude that there is little to be gained in investing much effort in finding exact solutions of equations. However, if one is concerned with scientific problem solving rather than abstract mathematics, things are not so bleak. There are many occasions when exact solutions *can* be found. In the world of physics, for example, where physical laws are often expressed as differential or integral equations, we often see problems with closed form solutions. The "solution" may only be an approximation to the exact result, but that often provides much more insight than a table of numbers found by numerical methods. Consequently, there continues to be a sustained research effort concerned with finding closed form solutions to sets of equations. Sophisticated techniques for finding closed form solutions to integrals and ordinary and partial differential equations have been developed during the past decade or so, motivated often by a physical problem for which a closed form solution is thought to exist.

Since many interesting problems in science and engineering can be expressed in terms of polynomial equations, there has been considerable research for several decades in finding solutions to such equations. One might therefore be tempted to believe that there was little left to do in this area. However, the symbolic case still provides many opportunities for new discoveries. In particular, there are two relatively new innovative ideas that I believe will have a long lasting effect. One already has a big following. The other has had very little publicity, but I am convinced will become more important in the years ahead.

The first of these deals with non-linear equations. In the past, we have often used linear approximations to systems better expressed in terms of non-linear equations because of the sheer difficulty in dealing with the latter types of equations. In 1965, Bruno Buchberger wrote a dissertation [4] presenting a method for effectively dealing with multivariate polynomial equations of arbitrary degree. This method, based on what he called *Gröbner bases* after his supervisor, received little attention from the mathematical community at that time. In the mid-seventies, the computer algebra community discovered this work, and research in this area has really accelerated during the past few years. It has now become the basis for powerful techniques for working with such polynomial equations and there are over two hundred papers describing work in this area. What makes it important is that it not only has elegant theoretical underpinnings, but is also of great use in practical problem solving. Another attractive thing about this technique is that it is easy to understand. When used to solve sets of non-linear equations, the method is a generalization of the elimination methods used for solving linear equations. In the latter case, elimination reduces such a set to triangular form. One can then read off the value of one variable from the last row of the triangle, and by back substitution determine values for all other variables. When there are a finite number of solutions to the set of non-linear equations the Gröbner method reduces the set to a form in which the last equation contains just one variable, but is no longer linear as in the linear equation case. This can then be solved either exactly, if possible, or with standard root finding techniques. By back substitution, one can then construct the complete solution set for the problem. A more extensive account of this technique can be found in Bernhard Kutzler's paper in these proceedings [5]. To give you an idea of its power for scientific problem solving, some calculations were recently completed by a group in West Berlin, using a Gröbner basis program written in REDUCE and run on a Cray X-MP [6]. These calculations involved finding analytic solutions for chemical reaction systems which had previously only been studied numerically. One of the most impressive involves 50 polynomials in 37 variables [7]. The analytic result showed that some of the solutions found numerically and thought to represent stable chemical behavior were in fact unstable.

The second important idea I wish to discuss relates to solving sets of linear equations. The formalism traditionally used for this problem is that of matrix manipulation. The manipulation of matrices has been a stimulus to computer development since the beginnings of digital computation. The standard numerical algorithms manipulate rows of matrices sequentially according to consecutive indexing just as computers access memories sequentially according to consecutive addresses. In other words, each representation reflects the other.

This duality works well for numerical calculations, and a considerable literature of techniques has been developed during the past fifty years. On the other hand, many classical matrix algorithms use partitioning for their effectiveness, thus presenting relationships between blocks of data rather than indexed vectors. For example, if we want the product of two matrices, it is possible to partition each matrix into blocks, and, assuming that the blocks are compatible, compute the product as follows:

$$
\begin{bmatrix} A_1 & B_1 \\ C_1 & D_1 \end{bmatrix} \times \begin{bmatrix} A_2 & B_2 \\ C_2 & D_2 \end{bmatrix} = \begin{bmatrix} A_1A_2 + B_1C_2 & A_1B_2 + B_1D_2 \\ C_1A_2 + D_1C_2 & C_1B_2 + D_1D_2 \end{bmatrix}.
$$

In the linear storage model, the blocks A_1, A_2 and so on are not available directly, and it is often very costly to assemble them from the elements of the whole matrix. This is unfortunate, since algorithms based on block partitioning abound in matrix theory. In fact, adaptations of some of these have given rise to asymptotically fast algorithms [8]. On the other hand, most algebraic computation systems, especially those such as REDUCE that use Lisp as the embedding language, use a heap model of memory which is much more suited to such block representations. However, existing systems mirror the iterative approach used by numerical systems. In 1984, David Wise [9] proposed a representation for symbolic matrices which had such a partitioned representation. Like Buchberger's early experience, Wise's work has not yet influenced many people. However, I am convinced that it is an important new technique for symbolic matrix manipulation. The representation he used is called a *quadtree* representation.

The basic idea is to represent any d-dimensional array as a 2^d-ary tree. Thus a vector would be represented by a binary tree, and a matrix by a 4-ary tree or quadtree. We consider only matrices from now on. As a special case, a unit or zero matrix is represented as a *scalar*, and only one copy of the element is stored. Otherwise it is a quadruple of four equally-ordered submatrices. To make the recursion work smoothly, a matrix of order n is embedded in a $2^{\lceil \log n \rceil} \times 2^{\lceil \log n \rceil}$ matrix, justified to the lower right corner, with padding to the north and west. Apart from the principal diagonal, this padding can be 0. On the principal diagonal, it can be 0 or 1 depending on whether inverses or determinants are required. Either choice prescribes a *normal form* for such matrices.

A suitable Lisp representation for such an object is

$$(T\langle \text{ integer } \rangle)$$

for a scalar, and

$$(\text{NIL } \langle \text{qmat} \rangle \langle \text{qmat} \rangle \langle \text{qmat} \rangle \langle \text{qmat} \rangle)$$

for a quadruple, where the recursively defined <qmat> entries represent the northwest, northeast, southwest and southeast quadrants respectively. To complete the representation, we need to add at the top level information about its order. So the complete standard form for a matrix appends an integer defining the order to this structure. Thus the matrix

$$\left[\begin{array}{cc} A & B \\ C & D \end{array} \right]$$

would be represented as

$$(2 \text{ NIL } (T \ A) \ (T \ B) \ (T \ C) \ (T \ D))$$

whereas the matrix

$$\left[\begin{array}{cccc} A11 & A12 & A13 & A14 \\ A21 & A22 & A23 & A24 \\ 0 & 0 & A33 & A34 \\ 0 & 0 & A43 & A44 \end{array} \right]$$

would have the representation

$$(4 \text{ NIL } (\text{NIL } (T \ A11) \ (T \ A12) \ (T \ A21) \ (T \ A22))$$
$$(\text{NIL } (T \ A13) \ (T \ A14) \ (T \ A23) \ (T \ A24))$$
$$(T \ 0)$$
$$(\text{NIL } (T \ A33) \ (T \ A34) \ (T \ A43) \ (T \ A44))).$$

Operations such as matrix addition and multiplication are now straightforward to define.

Just because it has an elegant representation does not by itself qualify a technique for my list of innovative, lasting contributions. So what else is important about the quadtree method? There are two things I can mention. The first is the performance of the quadtree representation compared with the conventional vector-oriented representation found in REDUCE. It is well known that the performance of matrix algorithms can depend critically on the structure of the given matrices. Extreme cases such as dense or diagonal matrices can exhibit quite different behavior depending on the algorithms used. Tridiagonal or triangular matrices, falling somewhere between the two extremes, can have a different performance again. Just as is true in the numerical case, there has been a lot of research into finding good representations and algorithms for adding and multiplying symbolic matrices in these various classes. Abdali and Wise [10] in a series of experiments show that the quadtree representation is efficient over this whole spectrum of matrix forms. In fact, except for small order dense matrices, the quadtree representation is always better than the standard REDUCE representation, and for large order tridiagonal and diagonal matrices is astoundingly better. In other words, the quadtree representation is equally effective regardless of the structure of the matrix. There is therefore little incentive to design special purpose representations for special cases. This makes the maintenance of a complete symbolic matrix package much easier, since there is much less code to support. In addition, an improvement for one class of matrix will usually apply to other classes as well.

Another important feature of this quadtree representation is its effectiveness in a multiprocessing environment. The partitioning used in building quadtree matrices breaks a calculation into independent sub-calculations, each of which can itself be further partitioned. One could therefore distribute these sub-calculations, each with their own pieces of the partitioned matrices, across processors. Conventional matrix calculations usually require whole matrices to be distributed in order to work in a multiprocessing environment.

I should also note in passing that the Gröbner basis method also has the potential for parallel operation. For example, recent developments [11] take advantage of factors found in terms arising during the computation of the Gröbner base. If such factors are found, it is possible in some circumstances to find further terms in the base by an independent inspection of such factors. In other words, one could distribute the computation of the basis from each such factor to other processors.

The case for the quadtree representation is however not yet as solid as that for the Gröbner method. Matrix inverses and determinants are currently cumbersome in this representation. One way to calculate these is to represent a partitioned matrix in the form:

$$\left[\begin{array}{cc} A & B \\ C & D \end{array} \right] = \left[\begin{array}{cc} A & 0 \\ C & I \end{array} \right] \times \left[\begin{array}{cc} I & A^{-1}B \\ 0 & \Delta \end{array} \right]$$

where $\Delta = D - CA^{-1}B$. From this representation, it is easy to show that

$$\left[\begin{array}{cc} A & B \\ C & D \end{array} \right]^{-1} = \left[\begin{array}{cc} A^{-1} + A^{-1}B\Delta^{-1}CA^{-1} & -A^{-1}B\Delta^{-1} \\ -\Delta^{-1}CA^{-1} & \Delta^{-1} \end{array} \right]$$

and that

$$det \left[\begin{array}{cc} A & B \\ C & D \end{array} \right] = det(A)det(\Delta).$$

To compute the inverse of a nonsingular matrix, both A and Δ must be nonsingular. If they are singular, the matrix must be rearranged to avoid this. Furthermore, the computation of the above determinant requires the computation of the inverse of A, so both determinant and inverse require considerable computation. The computation of inverses by this method is less efficient for dense and tridiagonal matrices than the standard REDUCE procedure, which uses the Bareiss two-step elimination method [12]. The computation of determinants by minor expansion is also more efficient in general than the above method, since there are fewer operations involved.

If the quadtree method is to have long lasting impact, there has to be a better method for finding determinants and inverses. However, at an equivalent point in its development, the Gröbner method was also far less efficient than today's implementations. The point is that once people concentrate on such techniques, improvements are bound to follow. I am confident that this will be equally true for the quadtree method as it has been for the Gröbner method.

REFERENCES

[1] ANTHONY C. HEARN, REDUCE user's manual, Version 3.3. Report CP 78, The RAND Corporation, (July 1987).

[2] R. LONDON AND D. R. MUSSER, *The application of a symbolic mathematical system to program verification*, Proc. ACM 74 (1974), 265–273.

[3] B. BUCHBERGER AND R. LOOS, *Algebraic simplification*, in Computer algebra: symbolic and algebraic computation, ed. B. Buchberger, G.E.Collins and R. Loos, 11-44, Springer-Verlag, Wien (2nd edition) (1983).

[4] B. BUCHBERGER, *An Algorithm for Finding a Basis for the Residue Class Ring of a Zero-dimensional Polynomial Ideal (in German)*, PhD thesis, Math. Inst, Univ. of Innsbruck, Austria (1965).

[5] B. KUTZLER, *Deciding a class of Euclidean geometry theorems with Buchberger's algorithm*, these proceedings.

[6] H. MELENK, H. M. MÖLLER, AND W. NEUN, *Symbolic solution of large stationary chemical kinetics problems*, Impact of Computing in Science and Engineering, 1(2): (June 1989), 138–167.

[7] L. A. FARROW AND D. EDELSON, *The steady-state approximation: fact or fiction?*, Int. Journ. Chemical Kinetics 6:(1974), 787–800.

[8] V. STRASSEN, *Gaussian elimination is not optimal*, Numer. Math., 13(4): (August 1969), 354–356.

[9] D. S. WISE, *Representing matrices as quadtrees for parallel processors (extended abstract)*, SIGSAM Bulletin 18(3): (August 1984), 24–25.

[10] S. K. ABDALI AND D. S. WISE, *Experiments with quadtree representation of matrices*, In Proc. of ISSAC '88, 358, Springer-Verlag (1988) 96–108.

174

3

[11] H. MELENK, H. M. MÖLLER, AND W. NEUN, *On Gröbner bases computation on a super-computer using REDUCE*, Preprint SC 88-2, Konrad-Zuse-Zentrum für Informationstechnik Berlin, (January 1988),.

[12] E. H. BAREISS, *Sylvester's identity and multistep integer-preserving gaussian elimination.*, Math. Comp.22: (July 1968), 565–578.

DECIDING A CLASS OF EUCLIDEAN GEOMETRY THEOREMS WITH BUCHBERGER'S ALGORITHM

BERNHARD KUTZLER*

Abstract. Buchberger's method of Gröbner bases can be used to decide a certain class of theorems in elementary Euclidean geometry. Moreover, the method can also be used to find subsidiary conditions that are necessary in order to transform an "almost valid" formulation of a geometry theorem into a valid one. The introduction surveys all alternative approaches to automated geometry theorem proving, giving references to the corresponding literature. After explaining how to obtain correct algebraic translations of geometry theorems, Buchberger's method of Gröbner bases is shortly reviewed. Then the application of Buchberger's algorithm to geometry theorem proving is explained in all details. Finally, a computing time statistics on 20 plane Euclidean geometry theorems of growing complexity is given.

Key words. automated geometry theorem proving, Buchberger's algorithm, Gröbner bases.

1. Introduction. Geometry was one of the first sciences of mankind and one of the origins of mathematics. It was a purely empirical science in its beginnings (Babylonia, Egypt) and was turned into a deductive science by the Greeks. While interest has waned until recently, the field has received an enormous boost in the last few years, when highly sophisticated application areas like robotics, computer aided design, molecular conformation, and others received wide recognition. Today, *geometric reasoning*, the field concerned with algorithmic problem solving for geometric objects, is one of the most promising areas of computer science.

One aspect of this topic, namely automated geometry theorem proving, has gained a lot of attention during the past few years. The first geometry theorem prover, was the "Geometry Machine" of (GELERNTER 1963). He built a natural deduction theorem prover capable of proving geometry theorems formulated in a suitable geometric theory. One of its best achievements was a proof that the base angles of an isosceles triangle are equal. There followed many similar attempts, among them the utilization of the logic programming language PROLOG by (COELHO AND PEREIRA 1979,1986). However, the successes of provers based on this *logical approach to automated geometry theorem proving* were only moderate. A rather complete survey of the literature on such provers is given in an appendix of (KUTZLER 1988).

An alternative to treating geometry in form of a logical theory has been proposed by R. Descartes in 1637. He established analytic coordinate geometry as a suitable algebraic approach, which turned out to be a very powerful means. E.T. Bell, in (BELL 1937, p. 21), remarked: "Though the idea behind it all is childishly simple,

---

*RISC (Research Institute of Symbolic Computation), Johannes Kepler University, A-4040 Linz, Austria. The work was supported by the "Österreichischer Fond zur Förderung der wissenschaftlichen Forschung" (project no. P6763).

yet the method of analytic geometry is so powerful that very ordinary boys of seventeen can use it to prove results which would have baffled the greatest of the Greek geometers — Euclid, Archimedes, and Apollonius."

One of the first major results obtained with the *algebraic approach to automated geometry theorem proving* was Tarski's decision procedure for the elementary part of Euclidean geometry, see (TARSKI 1948). But his method is beyond any applicability since it is of exponential worst case complexity even for formulae containing only a single quantified variable and no free variables. No experimental results for any geometry theorems are known. (COLLINS 1975) devised the method of cylindrical algebraic decomposition as the key part of a new algorithm equivalent to Tarski's procedure. Although Collins' method is of much better complexity, experiments in (KUTZLER 1988) document its impracticality for non-trivial examples.

A considerable milestone was the method developed in (WU 1978,1984) and its refinements in (CHOU 1985). It certainly stimulated most of the research on this topic including our work on applying Buchberger's algorithm. But recent investigations in (KUTZLER 1988,1989) demonstrated a severe flaw in Wu's method, namely the inadequacy of his way of translating geometry theorems into algebraic problems. From this new point of view, almost all results obtained with Wu's algorithm can not be regarded as proofs but only as "near-proofs" in a sense fully explained in the above reference.

2. Algebraization of geometry theorems. Let some geometric conjecture be given. The central goal is to find a (mechanically generated) proof in case there exists one. Throughout this paper we use the following example:

(Ex) Let ABC be a triangle and M be the midpoint of the centroidal line through C. Then the line through A and M subdivides side BC in the ratio 2:1.

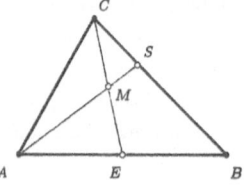

First of all, one has to *formulate* the theorem in a suitable logical theory. For the purpose of this paper we choose the well known theory of *Plane Euclidean Geometry* (= $\mathcal{PEG}$). (However, the ideas can be applied to many other geometries as well.) Since the languages of usual axiomatizations of $\mathcal{PEG}$ by Hilbert, Tarski, and others are not very convenient to use, we have established a suitable specification language by extending $\mathcal{PEG}$ by definitions[1] with various useful predicates like *collinear, online,* etc. (see *(Kutzler 1988,1989)*). The formulation of *(Ex)* in this extended theory is:

---

[1]An extension by definitions does not change the expressiveness of the theory.

($\forall$ point A, B, C, E, M, S)
($triangle(A, B, C) \wedge midpoint(E, A, B) \wedge midpoint(M, E, C) \wedge$
$intersection(S, B, C, A, M) \Rightarrow double(B, S, S, C))$

At this point the provers based on the logical approaches start their work of finding a proof. The basic idea of the algebraic approach is to transform this $\mathcal{PEG}$-formula into a corresponding algebraic problem (called the *algebraization*) and to solve this algebraic problem by computer algebra methods.[2] By such an algebraization we mean a translation of the formula into an algebraic model of $\mathcal{PEG}$. We choose *Descartes' analytic coordinate geometry* ($= \mathcal{AG}$) and obtain the following algebraic translation of *(Ex)*:

($\forall x_B, x_C, y_C, x_E, x_M, y_M, x_S, y_S \in \mathbf{R}$)
($x_b y_C \neq 0 \wedge x_B - 2x_E = 0 \wedge x_E + x_C - 2x_M = 0 \wedge y_C - 2y_M = 0 \wedge$
$x_M^2 + y_M^2 \neq 0 \wedge x_B^2 - 2x_B x_C + x_C^2 + y_C^2 \neq 0 \wedge x_C y_M - x_B y_M + x_M y_C \neq 0 \wedge$
$y_C x_S - y_C x_B + y_S x_B - y_S x_C = 0 \wedge x_S y_M - x_M y_S = 0 \Rightarrow$
$x_B^2 - 4x_C^2 + 8x_S x_C - 2x_S x_B - 3x_S^2 - 4y_C^2 + 8y_C y_S - 3y_S^2 = 0$)

Here, (x_α, y_α) are the Cartesian coordinates of point α. We also have set x_A, y_A, y_B to zero, because a special choice of coordinate axes (here: origin in A, x-axis through B) does not effect the validity of the theorem. The algebraic translation of the atomic formulae like $triangle(A, B, C)$, etc. is straightforward using a "$\mathcal{PEG}$–$\mathcal{AG}$ dictionary". The design of such a dictionary requires some initial effort. For our specification language this has been done in all detail in (KUTZLER 1988).

The full second order theory $\mathcal{PEG}$ is undecidable. Its elementary part (denoted by $\mathcal{PEG}_{elem}$) is decidable. But even the best decision procedure that is available today, namely Collins' method, is impracticable. The only existing implementation of Collins' algorithm in the SAC-2 computer algebra system does not allow to solve examples with more than five variables within three hours of CPU-time.

3. Buchberger's Gröbner Bases method. In 1965, B. Buchberger developed the method of Gröbner bases. The basic idea of this method is to solve a problem for an aribitrary set of multivariate polynomials F by transforming F into a standard form G (the *Gröbner basis* form) and solving the problem for G. For many classes of problems, certain useful properties of Gröbner bases allow a fast solution for G. The algorithm for transforming F into G has become known as the *Gröbner bases algorithm* or as *Buchberger's algorithm*.

The original application was for finding a basis for the residue class ring of a zero-dimensional ideal (BUCHBERGER 1965).[3] Later he described how to use Gröbner bases for solving systems of algebraic equations (BUCHBERGER 1970). There followed many more results by Buchberger and others. Today, the Gröbner

---

[2] The soundness of this algebraic approach is based on the fact that $\mathcal{PEG}$ is a categorical theory, i.e. all models of $\mathcal{PEG}$ are isomorphic.

[3] The method is named after Prof. W. Gröbner, the thesis advisor of B. Buchberger, who stimulated the research on the subject.

bases method provides solutions to a wide range of problems that can be formulated using finite sets of multivariate polynomials. In this section, we describe the main ideas of the method and sketch the algorithm in its basic version. A thorough introduction to the method of Gröbner bases is (BUCHBERGER 1985), where also a complete reference to the literature is given. Details on applications of the method to various geometric problems are given in (BUCHBERGER 1987).

Let K be a field and $K[y_1, \ldots, y_m]$ be the ring of m-variate polynomials over K. We use f, g, h as typed variables for polynomials in $K[y_1, \ldots, y_m]$, F, G for finite subsets of $K[y_1, \ldots, y_m]$, t, u for power products of the form $y_1^{p_1} \cdots y_m^{p_m}$, a, b for elements in K, and i, j for natural numbers. Let some total, admissible[4] ordering $<$ on the power products be given.[5] It is straightforward to extend the ordering to arbitrary polynomials. For a fixed ordering $<$, $coeff(g, t)$ denotes the coefficient of t in g, $lpp(g)$ denotes the leading power product of g w.r.t. $<$, and $lc(g)$ denotes the coefficient of $lpp(g)$.

A non-zero polynomial f gives rise to the following reduction relation: $g \to_f h$ *(g reduces to h modulo f)* iff there exists $u, b \neq 0$ such that $b \cdot u \cdot lpp(f)$ is identical with some monomial of g and $h = g - b \cdot u \cdot f$. Such a polynomial reduction step $g \to_f h$ can be viewed as a generalized division, deleting a monomial in g. h is strictly smaller than g w.r.t. $<$.

Additional notations are: $g \to_F h$ *(g reduces to h modulo F)* iff there exists $f \in F$ such that $g \to_f h$; $\underline{g}_F$ *(g is irreducible modulo F)* iff there exists no h such that $g \to_F h$; $\to_f^+$ $(\to_F^+)$ denotes the transitive closure of $\to_f$ $(\to_F)$; $\leftrightarrow_f^*$ $(\leftrightarrow_F^*)$ denotes the symmetric, reflexive and transitive closure of $\to_f$ $(\to_F)$; *h is a normal form of g modulo F* iff $g \to_F^* h$ and $\underline{h}_F$. In the sequel, let NF denote an algorithm for computing, for given F, g a normal form of g modulo F. Such an algorithm is called a *normal form algorithm*. The *ideal generated by* F is the set $Ideal_{K[y_1, \ldots, y_m]}(F) := \{\sum_{f \in F} h_f \cdot f \mid h_f \in K[y_1, \ldots, y_m]\}$. If the domain is clear from the context, $Ideal(F)$ is written for short. The *zeros of* F is the set $Zero(F) := \{(a_1, \ldots, a_m)\} \in \mathbb{C}^n \mid f_{y_1, \ldots, y_m}[a_1, \ldots, a_m] = 0$ for all $f \in F\}$.

EXAMPLE 3.1. Choose $K = \mathbb{Q}$, $n = 3$, the lexical ordering, $g = y_2^2 + y_1 y_2 y_3 - y_1^2$, and $f_1 = y_1 y_3 - y_2$, $f_2 = y_1 y_2 - y_1$, $F = \{f_1, f_2\}$. Then, g can be reduced to $h_1 = 2y_2^2 - y_1^2$ modulo f_1 using $u = y_2$ and $b = 1$. The second monomial of g is deleted by this reduction. h_1 *is a normal form of g modulo F*, because h_1 cannot be reduced further by f_1 or f_2. On the other hand, g can also be reduced to $h_2 = y_1 y_3 + y_2^2 - y_1^2$ modulo f_2 using $u = y_3$ and $b = 1$. h_2 further reduces to $h_3 = y_2^2 + y_2 - y_1^2$ modulo f_1 and $\underline{h_3}_F$. Therefore, h_3 *is a normal form of g modulo F*.

As can be seen from this example, in general, the normal form of a polyno-

---

[4]An ordering $<$ is called *admissible* iff $1 = y_1^0 \cdots y_m^0$ is minimal under $<$ and multiplication by a power product preserves the ordering.

[5]Examples for such orderings are the total degree ordering $(1 < y_1 < y_2 < y_1^2 < y_1 y_2 < y_2^2 < y_2^3 < \ldots$ in the bivariate case) and the lexical ordering $(1 < y_1 < y_1^2 < \ldots < y_2 < y_1 y_2 < y_1^2 y_2 < \ldots < y_2^2 < y_1 y_2^2 < \ldots$ in the bivariate case).

mial g modulo a basis F is not unique, since there may exist several "reduction paths" leading to different normal forms. Bases for which this cannot happen are emphasized in Buchberger's theory:

DEFINITION 3.1. (BUCHBERGER 1965) G is called a *Gröbner basis* iff each g has a unique normal form modulo G.

This *canonical simplification* property of Gröbner basis is the key to a large number of applications. For making the *Gröbner bases method* (i.e. solving a problem for a set F by transforming F into Gröbner basis form G and solving the problem for G) constructive one needs an algorithm for constructing, for a given set F, a set G, such that $Ideal(F) = Ideal(G)$ and G is a Gröbner basis. Such an algorithm has been given in (BUCHBERGER 1965). Before this algorithm is presented, its two main algorithmic ideas, namely "completion" and "critical pairs" are sketched:

Whenever a polynomial g gives rise to two different normal forms h_1, h_2 modulo an arbitrary basis F, it suffices to add the difference $h_1 - h_2$ to F in order to enforce that these two reduction paths (modulo $F \cup \{h_1 - h_2\}$) yield the same normal form. It is easy to check that the ideal of the basis has not been changed by this *completion*. This completion procedure is not effective, because infinitely many polynomials g would have to be tested. Buchberger found that it suffices to consider only finitely many such polynomials, namely all $S-polynomials$ of pairs of elements of F. These pairs of elements of the basis are called *critical pairs*. The $S-polynomials$ are the "minimal" polynomials where distinct reductions can occur.

DEFINITION 3.2. (BUCHBERGER 1965) $S-polynomial(f_1, f_2) := u_1 \cdot f_1 - \frac{lc(f_1)}{lc(f_2)} \cdot u_2 \cdot f_2$, where u_1, u_2 are such that $u_1 \cdot f_1$ and $u_2 \cdot f_2$ have the same (smallest possible) leading power product (= least common multiple of $lpp(f_1)$ and $lpp(f_2)$).

By the construction, an $S-polynomial$ of two elements of F lies in the ideal generated by F, hence its normal form must be zero. For the completion step, therefore, it suffices to compute one normal form and to add it to the basis in case it does not vanish. This yields the following algorithm. (Here, NF is any normal form algorithm.)

ALGORITHM 3.1. (BUCHBERGER 1965)
$\quad G := F$
$\quad B := \{(f_1, f_2) \mid f_1, f_2 \in G, f_1 \neq f_2\}$
$\quad \underline{while}\ B \neq \emptyset\ \underline{do}$
$\quad\quad (f_1, f_2) := \text{a pair in } B$
$\quad\quad B := B - \{\{f_1, f_2\}\}$
$\quad\quad h := NF(G, S-polynomial(f_1, f_2))$
$\quad\quad \underline{if}\ h \neq 0\ \underline{then}(B := B \cup \{\{g, h\} \mid g \in G\}; G := G \cup \{h\})$

A Gröbner basis as defined above and as constructed by this algorithm is not necessarily unique. Different normal form algorithms may lead to different Gröbner bases. A Gröbner basis G is called *reduced* iff g is irreducible modulo $G - \{g\}$ and $lc(g) = 1$ for all $g \in G$. Reduced Gröbner bases are unique, for their construction the basic algorithm has to be adjusted accordingly. The above algorithm, though structurally very simple, is of extremely high complexity. In (BUCHBERGER 1979) an improved version of the algorithm is given that significantly improves the computing times. The main idea was to find criteria that allow to detect whether an $S - polynomial$ will reduce to zero without actually having to do the (sometimes very expensive) normal form computation. Two such criteria are contained in Buchberger's improved algorithm. In the sequel, let GB denote an algorithm for computing, for a given F a reduced Gröbner basis of F.

Below, some of the most important properties of Gröbner bases are summarized.

THEOREM 3.1. (BUCHBERGER 1965,1970)

- *(Ideal membership)*
  For all F, f: $f \in Ideal(F)$ iff $NF(GB(F), f) = 0$.

- *(Radical membership)*
  For all F, f: f vanishes on all common zeros of F iff $1 \in GB(F \cup \{z \cdot f - 1\})$, where z is a new indeterminate.

- *(Solvability of polynomial equations)*
  For all F: $Zero(F) = \emptyset$ iff $1 \in GB(F)$.

- *(Finite solvability of polynomial equations)*
  For all F: F has only finitely many solutions iff
  for all $1 \le i \le m$ there exists an $f \in GB(F)$ such that $lpp(f)$ is a power of y_i.

- *(Elimination ideals, solution of polynomial equations)* (TRINKS 1978)
  Let $<$ be the lexical ordering defined by $y_1 < y_2 < \ldots < y_m$.
  Then, for all F, $1 \le i \le n$: $GB(F) \cap K[y_1, \ldots, y_i]$ is a Gröbner basis for the "i-th elimination ideal" generated by F, i.e. for $Ideal(F) \cap K[y_1, \ldots, y_i]$.

- *(Ideal intersection)*
  Let $<$ be the lexical ordering defined by $y_1 < y_2 < \ldots < y_m < z$, where z is a new variable. Then, for all F, G: $GB(\{z \cdot f \mid f \in F\} \cup \{(z-1) \cdot g \mid g \in G\}) \cap K[y_1, \ldots, y_m]$ is a Gröbner basis for $Ideal(F) \cap Ideal(G)$.

A large amount of current research on Gröbner bases aims at further speeding up the algorithm. Today, the Gröbner bases method is a central technique in computer algebra with many applications in numerous fields. [6]

4. Deciding certain geometry theorems. Collins' algorithm yields a decision method for the real numbers and, therefore, a decision method for the full

---

[6]It is sometimes remarked that Gröbner bases have been found already in 1964 by Hironaka (and were called *Standard bases* there). But Hironaka gave only an indirect proof of the existence of such bases. The main merit definitely lies in making this notion constructive, as it was done by Buchberger.

elementary part of Euclidean geometry (and any other suitable geometry that is interpreted over the real numbers like, for inst ance, Minkowskian geometry.) In contrast to that, Buchberger's method works over the complex numbers. Its application, therefore, is limited to geometry theorems that do not involve order or, equivalently, whose formulation in $\mathcal{PEG}$ do not involve the basic geometric predicate *between*. Moreover, the algebraic translation has to be of a certain form as is requested by the following lemma proposed by (B. BUCHBERGER, personal communication). The proof of this lemma describes the method how to apply Buchberger's algorithm GB.

LEMMA 4.1. *Let Φ be an arbitrary Boolean combination of the polynomial equations $f_1 = 0, \ldots, f_n = 0$, $f_i \in \mathbb{Q}[y_1, \ldots y_m]$. Let $y = (y_1, \ldots, y_m)$ and $a = (a_1, \ldots, a_m)$. Algorithm GB can be used to decide $(\forall a \in \mathbb{C}^m)\Phi_y[a]$.*

PROOF: The following proof, actually, constitutes an effective decision procedure. For legibility we write $(\forall a \in \mathbb{C}^m)\Phi_y[a]$ shortly as
(1) $(\forall a)\Phi[a]$.
By applying a disjunctive normal form algorithm to Φ, (1) can be transformed into the equivalent formula
(2) $(\forall a)((f_{1,1}^*[a]\ell_{1,1}0 \wedge \ldots \wedge f_{1,r_1}^*[a]\ell_{1,r_1}0) \vee \ldots \vee (f_{k,1}^*[a]\ell_{k,1}0 \wedge \ldots \wedge f_{k,r_k}^*[a]\ell_{k,r_k}0))$,
where $f_{i,j}^* \in \{f_1, \ldots, f_n\}$ and $\ell_{i,j} \in \{=, \neq\}$. This, clearly, is equivalent to
(3) $\neg(\exists a)((f_{1,1}^*[a]\ell'_{1,1}0 \vee \ldots \vee f_{1,r_1}^*[a]\ell'_{1,r_1}0) \wedge \ldots \wedge (f_{k,1}^*[a]\ell'_{k,1}0 \vee \ldots \vee f_{k,r_k}^*[a]\ell'_{k,r_k}0))$,
where $\ell'_{i,j}$ is the opposite of $\ell_{i,j}$. Application of Rabinowitsch's trick[7] yields the equivalent formula
(4) $\neg(\exists a, b)((g_{1,1}[a, b] = 0 \vee \ldots \vee g_{1,r_1}[a, b] = 0) \wedge \ldots \wedge (g_{k,1}[a, b] = 0 \vee \ldots \vee g_{k,r_k}[a, b] = 0))$,
where $b := (b_{1,1}, \ldots, b_{k,r_k})$ and $g_{i,j} \in \mathbb{Q}[y_1, \ldots, y_m, z_{1,1}, \ldots z_{k,r_k}]$ ($z_{1,1}, \ldots, z_{k,r_k}$ new variables) such that $g_{i,j} := f_{i,j}^*$ in case $\ell'_{i,j}$ is the '='-symbol and $g_{i,j} := f_{i,j}^* \cdot z_{i,j} - 1$ otherwise. Finally, (4) is equivalent to
(5) $\neg(\exists a, b)(g_{1,1} \cdots g_{1,r_1}[a, b] = 0 \wedge \ldots \wedge g_{k,1} \cdots g_{k,r_k}[a, b] = 0)$,
which can be decided using Algorithm GB by applying *(Solvability of polynomial equations)* from Theorem 3.1. $\qquad \Omega$

For our geometry theorem proving application, Lemma 4.1 gives the following theorem:

THEOREM 4.1. *Buchberger's Gröbner bases method yields a decision procedure for those theorems τ of $\mathcal{PEG}_{elem}$ that fulfill the following two properties :*

(a) $\tau_{\mathcal{AG}}$ (i.e. the translation of τ into $\mathcal{AG}$) is of the form $(\forall y_1, \ldots, y_m \in \mathbb{R})\Phi$, where Φ is an arbitrary Boolean combination of polynomial equations in $y_1, \ldots, y_m$ with coefficients in $\mathbb{Q}$.

(b) $\tau_{\mathcal{AG}}$ is valid over the complex numbers, i.e. $(\forall y_1, \ldots, y_m \in \mathbb{C})\Phi$ holds.

Property *(a)* certainly is a strong restriction, since it excludes all theorems that

---
[7] $(\exists a)f(a) \neq 0 \Leftrightarrow (\exists a, b)f(a) \cdot b - 1 = 0$, b a new variable

involve order or the existence of certain objects. In fact, it is not necessarily the theorem itself that lies in this class or not, but the choice of the geometric specification language and the choice of the algebraic interpretations of the language's elements decide whether a theorem me ets *(a)* or not. We illustrate this observation with the following simple example: *"If A, B, C are collinear and B, C, D are collinear then A, C, D are collinear."* Suppose the theory includes just the incidence predicate *on*. Then the theorem has to be formulated as follows:

$$(\forall\ points\ A, B, C, D)((\exists\ line\ \ell_1)(on(A, \ell_1) \wedge on(B, \ell_1) \wedge on(C, \ell_1)) \wedge$$
$$(\exists\ line\ \ell_2)(on(A, \ell_2) \wedge on(B, \ell_2) \wedge on(D, \ell_2)) \Rightarrow$$
$$(\exists\ line\ \ell_3)(on(A, \ell_3) \wedge on(C, \ell_3) \wedge on(D, \ell_3)))$$

The algebraic translation of this formula is

$$(\forall x_A, y_A, x_B, y_B, x_C, y_C, x_D, y_D \in \mathbf{R})($$
$$(\exists a_1, b_1, c_1 \in \mathbf{R})((a_1 \neq 0 \vee b_1 \neq 0) \wedge$$
$$a_1 x_A + b_1 y_A + c_1 = 0 \wedge a_1 x_B + b_1 y_B + c_1 = 0 \wedge a_1 x_C + b_1 y_C + c_1 = 0) \wedge$$
$$(\exists a_2, b_2, c_2 \in \mathbf{R})((a_2 \neq 0 \vee b_2 \neq 0) \wedge$$
$$a_2 x_A + b_2 y_A + c_2 = 0 \wedge a_2 x_B + b_2 y_B + c_2 = 0 \wedge a_2 x_D + b_2 y_D + c_2 = 0) \Rightarrow$$
$$(\exists a_3, b_3, c_3 \in \mathbf{R})((a_3 \neq 0 \vee b_3 \neq 0) \wedge$$
$$a_3 x_A + b_3 y_A + c_3 = 0 \wedge a_3 x_C + b_3 y_C + c_3 = 0 \wedge a_3 x_D + b_3 y_D + c_3 = 0)),$$

which is not of the requested type *(a)*. Extending the theory by definition with a predicate *collinear* such that $collinear(A, B, C) \Leftrightarrow (\exists\ line\ \ell)(on(A, \ell) \wedge on(B, \ell) \wedge on(C, \ell))$ allows a much shorter formulation:

$$(\forall\ points\ A, B, C, D)$$
$$(collinear(A, B, C) \wedge collinear(A, B, D) \Rightarrow collinear(A, C, D)).$$

Using the interpretation $collinear_{\mathcal{A}\mathcal{G}}((x_A, y_A), (x_B, y_B), (x_C, y_C)) :\Leftrightarrow (y_B - y_A)x_C + (x_A - y_B)y_C + x_B y_A - x_A y_B = 0$ yields the following algebraic translation meeting *(a)*:

$$(\forall x_A, y_A, x_B, y_B, x_C, y_C, x_D, y_D \in \mathbf{R})$$
$$((y_B - y_A)x_C + (x_A - y_B)y_C + x_B y_A - x_A y_B = 0 \wedge$$
$$(y_B - y_A)x_D + (x_A - y_B)y_D + x_B y_A - x_A y_B = 0 \Rightarrow$$
$$(y_C - y_A)x_D + (x_A - y_C)y_D + x_C y_A - x_A y_C = 0)$$

Our specification language from (KUTZLER 1988,1989) has been designed under this aspect. Answering the question whether there exists an extension by definitions of, let say $\mathcal{P}\mathcal{E}\mathcal{G}$, together with suitable algebraic interpretations, such that a given geometry theorem is of type *(a)* certainly is an interesting future research goal.

As our experiments and the experiments of Wu and Chou showed, most Euclidean geometry theorems are also valid over the complex numbers. Property *(b)*, therefore, is not very restrictive. However, there is no way other than using an algorithm like Collins' method to decide whether a concrete theorem meets *(b)* or not. The significance of Theorem 4.1 for proving Euclidean geometry theorems, therefore, is as follows: In case the algebraic translation of a geometry theorem fulfills property *(a)*, it can be attacked by Buchberger's method: One has to determine the set of products $\{g_{1,1}\cdots g_{1,r_1},\ldots,g_{k,1}\cdots g_{k,r_k}\}$ mentioned in the above proof and apply to it algorithm *GB*. If the result is $\{1\}$, the geometry theorem is proved. Otherwise no conclusion can be drawn, since the theorem is wrong over **C** but might still be true over **R**.

Buchberger's algorithm is available in most computer algebra systems. However, the implementations vary greatly in efficiency. We experimented with R. Gebauer's implementation in the SCRATCHPAD II computer algebra system. On an IBM 4341 our test theorem *(Ex)* was proved in 9.09 seconds. More experiments are reported in Section 7.

5. Almost valid formulations.

As became clear from the preceding sections, the algebraic approach to automated geometry theorem proving requires two preparatory steps: (α) the formulation of the geometric situation as a theorem in a suitable logical theory and (β) the translation of this theorem into algebraic form. (β) has been completely investigated in (KUTZLER 1988,1989) and can, in particular, be fully mechanized. (α) is the step from the informal description to a formal description, hence, no formal argument about its correctness is possible. Although the user is responsible for (α), a rich specification language certainly facilitates it, cf. the simple example from the last section.

But it still can happen that a formulation slightly differs from the "intended" geometry theorem and, therefore, no proof is found. The following example is of that kind: *"On the two sides AC and BC of triangle ABC, two squares ACDE and CBGF are drawn. M is the midpoint of side AB. Then the length of DF is twice the length of CM."* A formulation in $\mathcal{PEG}$ is

$(\forall\ points\ A,B,C,D,F,M)(triangle(A,B,C) \wedge congruent(A,C,C,D) \wedge$
$rightangle(A,C,D) \wedge congruent(B,C,C,F) \wedge rightangle(B,C,F) \wedge$
$midpoint(M,A,B) \Rightarrow double(F,D,M,C))$

In this formulation the orientation of the two squares $ACDE$ and $CBGF$ is not determined, hence, it allows the following four cases, in only two of which the theorem holds (the leftmost and the rightmost). The intended case is the leftmost.

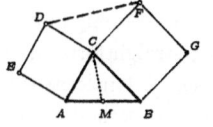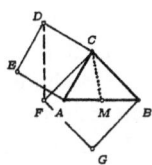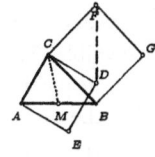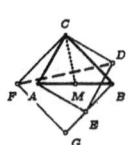

For the algebraic translation of this example there exists a polynomial d that "distinguishes" the valid from the invalid cases. We call the formulation of a theorem for which this is the case *almost valid*. Finding such a polynomial certainly is worthwhile, since it can be used to transform the almost valid formulation into worthwhile, since it can be used to transform the almost valid formulation into a valid one. However, its geometric value can be judged only if its retranslation into geometric form can be done. But for this task no general solution exists so far. Existing implementations are restricted to heuristical methods only. Therefore, proving a geometry theorem subject to an algebraic subsidiary conditions without doing further investigations on this condition can only be regarded as a "near-proof" in the sense that the theorem is likely to be valid in this or a slighlty modified version.

For finding such a polynomial one restricts interest to geometry theorems, whose algebraic translation is of the form:

$$(\forall a \in \mathbf{R}^m)(h_1(a) = 0 \land \ldots \land h_n(a) = 0 \land h_1'(a) \neq 0 \land \ldots \land h_k'(a) \neq 0 \Rightarrow c(a) = 0),$$

where h_i, h_j' are the polynomials corresponding to the hypotheses of the theorem and c is the polynomial corresponding to its conjecture. The algebraic problem of automated geometry theorem proving, then, becomes

PROBLEM 5.1. Given polynomials $h_1, \ldots, h_n, h_1', \ldots, h_k', c \in \mathbf{Q}[y_1, \ldots, y_m]$. Decide whether $(\forall a \in \mathbf{R}^m)(h_1(a) = 0 \land \ldots \land h_n(a) = 0 \land h_1'(a) \neq 0 \land \ldots \land h_k'(a) \neq 0 \Rightarrow c(a) = 0)$ is valid. In case it is, the corresponding geometry theorem is proved. Otherwise, find a polynomial $d \in \mathbf{Q}[y_1, \ldots, y_m]$ such that $(\forall a \in \mathbf{R}^m)(h_1(a) = 0 \land \ldots \land h_n(a) = 0 \land h_1'(a) \neq 0 \land \ldots \land h_k'(a) \neq 0 \land d(a) \neq 0 \Rightarrow c(a) = 0)$ and $\neg(\forall a \in \mathbf{R}^m)(h_1(a) = 0 \land \ldots \land h_n(a) = 0 \land h_1'(a) \neq 0 \land \ldots \land h_k'(a) \neq 0 \Rightarrow d(a) = 0)$, or report that no such polynomial exists. In case a polynomial d is found, its retranslation into geometric form gives the missing hypothesis.

In his method, Wu used only the "finding part" of this problem, most probably because his careless translations almost always had hypotheses missing (even if the formulation was correct). We refer to this as *Wu's finding problem*. Employing J.F. Ritt's method of Characteristic sets, Wu gave an algorithmic solution for the finding problem considered over an algebraically closed field. Wu's prover cannot solve the above Problem 5.1. It generates (superfluous) subsidiary conditions also for correct theorems.

6. Finding subsidiary conditions. Kapur described a method how to use Buchberger's algorithm for solving Wu's finding problem over an algebraically closed field. The remarkable fact about his procedure is that it suffices to test only finitely many polynomials as possible candidates for d. It is, in fact, straightforward to see that Kapur's prover, in particular, solves the above Problem 5.1 in case $\mathbf{R}$ is replaced by $\mathbf{C}$. Hence, for the case of Euclidean geometry, again, the method cannot be used to prove a geometry theorem false. In the sequel let $H = \{h_1, \ldots, h_n, z_1 \cdot h_1' - 1, \ldots, z_k \cdot h_k' - 1\}$, $z_1, \ldots, y_k$ new variables. We use y as an abbreviation for $y_1, \ldots, y_m$.

ALGORITHM 6.1. (KAPUR 1986)
  $(in:\ H, c, y)$

$G := GB(H \cup \{z \cdot c - 1\})$ (using the lexical ordering such that $y < z$)
$\underline{if}\ 1 \in G\ \underline{then}\ \underline{return}$ 'theorem proved'
$\{g_1, \ldots, g_t\} := G \cap \mathbf{Q}[y]$
$\underline{do}\ \underline{for}\ 1 \leq i \leq t$
$\quad G_i := GB(H \cup \{z \cdot g_i - 1\})$
$\quad \underline{if}\ 1 \notin G_i\ \underline{then}\ \underline{return}$ 'theorem proved under the
$\qquad\qquad\qquad\qquad\qquad\qquad$ nondegeneracy condition' $\{g_i\}$
$\quad \underline{enddo}$
$\underline{return}$ 'theorem not proved'

This prover was implemented in SCRATCHPAD II by (KUSCHE, KUTZLER AND MAYR 1987). On an IBM 4341 our test theorem (Ex) was proved in 60.78 seconds. The example from the last section was proved to be valid subject to the (algebraic!) condition $y_A x_F - y_B x_D \neq 0$ in 363.56 seconds. Kapur's prover requires more computing time than the prover from Section 4 but, on the other hand, is more powerful since it can generate subsidiary conditions for almost valid formulations.

Both the algorithm inherent in Lemma 4.1 and Kapur's prover are refutational theorem provers. Our prover given below is constructive in the sense that it provides useful information about the geometric object considered. The algorithm requires as additional input a distinction of the variables y into independent variables $u = y_{i_1}, \ldots, y_{i_s}$ (i.e. those variables corresponding to points that can be chosen arbitrarily) and dependent variables $x = y_1, \ldots, y_{i_1-1}, y_{i_1+1}, \ldots, y_{i_s-1}, y_{i_s+1}, \ldots, y_m$ (i.e. those variables corresponding to points that are constructed subject to conditions).

ALGORITHM 6.2. (KUTZLER AND STIFTER 1986)
$\quad (in:\ H, c, u, x)$
$\quad G := GB(H)$ (using a lexical ordering such that $u < x$)
$\quad \underline{if}\ G \cap \mathbf{Q}[u] \neq \emptyset\ \underline{then}\ \underline{return}$ 'wrong choice of indep. variables'
$\quad r := NF(c, G)$
$\quad \underline{if}\ r = 0\ \underline{then}\ \underline{return}$ 'theorem proved'
$\quad (pr, D) := ITPSRED(r, G, u, x)$
$\quad \underline{if}\ pr = 0\ \underline{then}\ \underline{return}$ 'theorem proved under the nondegeneracy
$\qquad\qquad\qquad\qquad\qquad\qquad$ conditions' D
$\qquad \underline{else};\underline{return}$ 'theorem not proved'

$ITPSRED\ (in:\ G, r, u, x;\ out:\ r, D)$
$D := \emptyset$
$\underline{do}\ \underline{while}\ (\exists g \in G, p) r \stackrel{\curvearrowright}{}_g p$
$\quad choose\ g \in G, p\ such\ that\ r \stackrel{\curvearrowright}{}_g p$
$\quad r := NF(p, G)$
$\quad D := D \cup \{lc_{\mathbf{Q}(u)[x]}(g)\}$
$\quad \underline{enddo}$

Here, $ITPSRED$ is a modified version of a normal form algorithm, using a new notion of reduction, called *pseudoreduction*, which we introduced in 1986 and which is defined as follows. (Below, $lc_{\mathbf{Q}(u)[x]}(g)$ denotes the leading coefficient of g regarded as a polynomial in $\mathbf{Q}(u)[x]$.)

DEFINITION 6.1. *r u-pseudoreduces to p modulo g* iff $lc_{\mathbf{Q}(u)[x]}(g) \cdot r \rightarrow_g^+ p$ and $\underline{p}_g$. For abbreviation we write $r \stackrel{u}{\leadsto}_g p$.

This prover is more in the spirit of the "Gröbner bases method": In a first step the hypotheses polynomials are transformed into Gröbner basis form G, in a second step the conjecture polynomial is processed with respect to G. This prover solves a slightly different algebraic problem and is, in fact, not so powerful as Kapur's prover (extensive experiments showed that it can prove only 90–95% of the examples proved by Kapur's method). On the other hand it is often faster, the preprocessing of the hypothesis allows to check various conjectures for the same hypotheses with a minimal additional effort and, finally, the Gröbner basis form of the hypotheses gives useful information about the geometric object itself, since it "solves" the hypotheses polynomials for the dependent variables. The test theorem *(Ex)* gives the following basis (polynomials containing the new variables introduced by Rabinowitsch's trick are not displayed), from which it is easy to compute the coordinates of the constructed points E, M, S as soon as the triangle (i.e. the coordinates of B, C) are fixed.

$$y_S - \tfrac{2}{3}y_C$$
$$x_S - \tfrac{2}{3}x_C - \tfrac{1}{3}x_B$$
$$y_M - \tfrac{1}{2}y_C$$
$$x_M - \tfrac{1}{2}x_C - \tfrac{1}{4}x_B$$
$$x_E - \tfrac{1}{2}x_B$$

Again, the prover was implemented in SCRATCHPAD II by (KUSCHE, KUTZLER AND MAYR 1987). The computation of the Gröbner basis took 30.61 seconds on an IBM 4341, the (pseudo)reduction of the conjecture took 0.34 seconds. The example was proved without any subsidiary conditions.

We also gave the following algorithm as an alternative to Algorithm 6.2. While in the above prover all computations were done over the rational numbers, this prover does all computations over the rational function field obtained by adjoining the independent variables u to $\mathbf{Q}$. This allows to use the ordinary normal form computation for step 2. However, the determination of the necessary subsidiary polynomial d, if any, is quite complicated and involves complete tracing of Buchberger's algorithm.

ALGORITHM 6.3 (KUTZLER AND STIFTER 1986)
$(in: H, c, u, x)$
$G := GB(H)$ (computed in $\mathbf{Q}(u)[x]$)
if $1 \in G$ *then return* 'wrong choice of indep. variables.'
$r := NF(c, G)$ (computed in $\mathbf{Q}(u)[x]$)
$D :=$ all denominators appearing in $c = \sum_{h \in H} f_h \cdot h$
if $r = 0$ *then return* 'theorem proved under the nondegeneracy
conditions' D
else return 'theorem not proved'

The determination of D is not contained in our implementation (KUSCHE, KUT-ZLER AND MAYR 1987). The test theorem *(Ex)* took 9.58 seconds for the Gröbner basis and 0.74 seconds for the normal form computation. The Gröbner basis includes the same polynomials as above.

7. Experiments. For our experiments we have selected twenty representative theorems from plane Euclidean geometry, all taken from the existing literature on automated geometry theorem proving. All examples are referenced only by their name, full descriptions (including the formulation in $\mathcal{PEG}$ and the translation into $\mathcal{AG}$) can be found in (KUTZLER 1988). Some of the examples have been used in two variants.

The table summarizes the computing times for the four provers discussed in this paper, namely the prover based on the Theorem 4.1, the prover of Kapur (Algorithm 6.1), and the provers of Kutzler/Stifter (Algorithms 6.2 and 6.3). For the latter three methods the times for two main steps are given separately. The symbol '∞' in dicates that the computation was aborted after approximately 3 hours CPU time, the symbol '†' indicates that the computation could not be completed within 4MB memory. For the answers we use the following symbols:

'✓' — theorem proved;

'★' — theorem proved subject to an algebraic subsidiary condition;

'∅' — theorem not proved (prover stopped but no proof was found);

'?' — theorem proved or proved subject to a condition;

Example	var	times in seconds using SCRATCHPAD II on an IBM 4341							
		Thm. 4.1		*Alg. 6.1 (Kapur)*		*Alg. 6.2 (Ku/Sti)*		*Alg. 6.3 (Ku/Sti)*	
congr. of halves	10	11.48	✓	13.13+	✓	7.93+	0.55 ✓	8.88+	1.83 ?
_"_	8	1.45	✓	1.73+	✓	0.80+	0.48 ✓	2.64+	1.85 ?
Thales' Thm	6	1.15	✓	1.64+	✓	1.33+	0.04 ✓	3.96+	0.08 ?
Thales' Inverse	6	2.29	✓	3.22+	✓	3.71+	0.02 ✓	6.01+	0.02 ?
△ circumctr	6	0.81	✓	1.39+	✓	1.06+	0.11 ✓	2.81+	0.41 ?
△ orthoctr	6	0.73	✓	1.20+	✓	0.95+	0.08 ✓	2.71+	0.27 ?
△ centroid	11	3.44	✓	5.89+	✓	4.21+	0.18 ✓	6.34+	0.65 ?
_"_	9	1.80	✓	1.80+	✓	0.96+	0.42 ✓	3.00+	1.59 ?
△ centr line	8	4.64	✓	5.08+	✓	3.20+	0.40 ✓	6.02+	1.23 ?
Euler line	13	20.52	✓	31.81+	✓	30.86+	0.59 ✓	9.06+	1.84 ?
centr line midpnt	11	7.55	✓	9.71+	✓	6.47+	0.28 ✓	10.70+	0.76 ?
Ninepointcircle	12	94.14	✓	944.65+	✓	1556.36+	0.27 ✓	15.15+	2.30 ?
_"_	9	32.70	✓	51.70+	✓	49.49+	1.14 ✓	7.70+	8.69 ?
isosceles △	11	257.59	✓	1140.11+	✓	∞		32.64+	1.92 ?
squares on △	9	3.52	∅	15.66+	13.13 ★	7.23+	0.42 ∅	4.20+	1.30 ∅
_"_	13	3.57	✓	3.81+	✓	2.38+	0.77 ✓	3.90+	1.50 ?
circle secants	11	4.94	✓	9.76+	✓	9.90+	0.88 ✓	11.13+	6.85 ?
Simson's Thm	10	6678.26	∅	∞		5264.62+	193.55 ★	7.63+	16.33 ?
compl quadrangle	20	14.59	✓	∞		∞		13.60+	6.73 ?
Pascal's Thm	25	†		∞		∞		∞	
Desargue's Thm	21	∞		∞		∞		41.72+	36.61 ?
Desargue's Inverse	25	∞		∞		∞		3948.21+	3.25 ?
_"_	20	7736.87	✓	∞		∞		3686.39+	0.60 ?
5-Star	16	∞		∞		∞		4133.45+	408.08 ?
MacLane 8-3	10	6768.85	∅	∞		∞		159.65+	0.13 ∅

8. Conclusion. Buchberger's method of Gröbner Bases certainly is a powerful technique for proving geometry theorems. Its practical applicability is much better than for the provers based on the logical approach or for Collins' method. However, the price for this is the provers' incapability of detecting wrong theorems.

In more general, algebraic methods are becoming more and more important for solving non-linear problems in geometry. A comprehensive survey on first results is (BUCHBERGER, COLLINS AND KUTZLER 1988).

The main shortcoming of most existing attempts to use algebraic methods for solving geometry problems is that useful *geometric* information, in many cases, cannot be employed by these methods. Often, geometry theorems that are readily proved by an experienced geometer by "looking at it from the right side" are not mastered by the machine provers within reasonable time, because the algebraic problem does not reflect this fact. A worthwhile goal for future research, therefore, is to find ways of transforming *geometric reasoning techniques* to the algebra level.

REFERENCES

B. BUCHBERGER 1970, *Ein algorithmisches Kriterium für die Lösbarkeit eines algebraischen Gleichungssystems (An algorithmic criterion for the solvability of algebraic systems of equations)*, Aequationes Mathematicae, vol. 4, pp. 374–383.

B. BUCHBERGER 1979, *A criterion for detecting unnecessary reductions in the construction of Gröbner bases*, Proc. EUROSAM'79, Marseille, June 1979, ed. W. Ng, Lecture Notes in Computer Science, vol. 72, pp. 3–21, Springer.

B. BUCHBERGER 1985, *Gröbner bases: An algorithmic method in polynomial ideal theory*, N.K. Bose (ed.): "Multidimensional Systems Theory", pp. 184–232, D. Reidel Publ. Comp., Dordrecht.

B. BUCHBERGER 1987, *Applications of Gröbner bases in non-linear computational geometry*, Proc. Workshop on Scientific Software, IMA, Minneapolis, March 1987, e. J.R. Rice, pp. 59–88, Springer Verlag; also: Proc. Int. Symp. Trends in Computer Algebra, Bad Neuenahr, FRG, May 19–21, 1987, ed. R. Janßen, Lecture Notes in Computer Science, pp. 52–80, Springer Verlag Berlin.

B. BUCHBERGER, G.E. COLLINS AND B. KUTZLER 1988, *Algebraic methods for geometric reasoning*, Annual Review of Computer Science, vol. 3, pp. 85–119.

S.C. CHOU 1985, *Proving and discovering geometry theorems using Wu's algorithm*, PhD Thesis, U Texas at Austin, USA, 78 p.

H. COELHO AND L.M. PEREIRA 1979, *GEOM: A PROLOG geometry theorem prover*, Laboratório Nacional de Engenharia Civil Memória no. 525, Ministerio de Habitacao e Obras Publicas, Portugal.

H. COELHO AND L.M. PEREIRA 1986, *Automated reasoning in geometry theorem proving with PROLOG*, J. Automated Reasoning, vol. 2, pp. 329–390.

G.E. COLLINS 1975, *Quantifier elimination for the elementary theory of real closed fields by cylindrical algebraic decomposition*, Proc. 2nd GI Conf. Automata Theory and Formal Languages, ed. H. Brakhage, Lecture Notes in Computer Science, vol. 33, pp. 134–183, Springer Berlin.

H. GELERNTER 1963, *Realization of a geometry theorem proving machine*, E.A. Feigenbaum, J.

Feldman (eds.): "Computers and Thought", pp. 134–152, McGraw Hill.

D. KAPUR 1986, *Geometry theorem proving using Hilbert's Nullstellensatz*, Proc. Symp. Symbolic and Algebraic Computation (SYMSAC'86), Waterloo, Canada, July 21–23, 1986, ed. B.W. Char, pp. 202–208, ACM Press.

K. KUSCHE, B. KUTZLER AND H. MAYR 1987, *Implementation of a geometry theorem proving package in SCRATCHPAD II*, Proc. Int. Symp. Symbolic and Algebraic Computation (EURO-CAL'87), Leipzig, GDR, June 2–5, 1987, ed. J. Davenport, Lecture Notes in Computer Science, Springer Verlag (to appear).

B. KUTZLER AND S. STIFTER 1986, *Automated geometry theorem proving using Buchberger's algorithm*, Proc. Symp. Symbolic and Algebraic Computation (SYMSAC'86), Waterloo, Canada, July 21–23, 1986, ed. B.W. Char, pp. 209–214, ACM Press.

B. KUTZLER 1988, *Algebraic approaches to automated geometry theorem proving*, PhD Thesis, U Linz, Austria, 161 p.

B. KUTZLER 1989, *Careful algebraic translations of geometry theorems*, Proc. Int. Symp. Symbolic and Algebraic Computation (ISSAC'89), July 17–19, 1989, Portland, Oregon, USA (to appear).

A. TARSKI 1948, *A decision method for elementary algebra and geometry*, RAND Corp., Santa Monica; also: University of California Press, Los Angeles (2nd edition 1951).

W. TRINKS 1978. *Über B. Buchberger's Verfahren, Systeme algebraischer Gleichungen zu lösen (On B. Buchberger's method for solving systems of algebraic equations)*, J. Number Theory, vol. 10, pp. 475–488.

W.T. WU 1978, *On the decision problem and the mechanization of theorem proving in elementary geometry*, Scientia Sinica, vol. 21, pp. 159–172; also: Contemporary Mathematics, vol. 29 (1984), pp. 213–324.

W.T. WU 1984, *Basic principles of mechanical theorem proving in elementary geometries*, J. Systems Sciences and Mathematical Sciences, vol. 4, pp. 207–235.

LIE TRANSFORM TUTORIAL - II*

KENNETH R. MEYER†

I. Introduction. This survey paper is an extension of Meyer (1990) since it contains complete proofs of the main theorems and some generalizations of Lie transform theory. However, the first part of this paper deals with the applications of Lie transforms to various perturbation problems leaving the technical proofs to the later sections.

Over the years many different techniques have been developed for handing various perturbation problems. Some are suited for a few special problems while others are quite general, but almost all were developed before the computer age. To our knowledge only one general technique was developed specifically to be used in conjunction with a computer algebra system, namely the method of Lie transforms. It is truly an algorithm in the sense of modern computer science: a clearly defined iterative procedure.

The method was first given in Deprit (1969) for Hamiltonian systems of differential equations, then generalized to arbitrary systems of differential equations by Kamel (1970) and Henrard (1970). The predecessor of this method was a limited set of formulas given in Hori (1966). All these papers appeared in astronomy journals which are far from the usual journals of perturbation analysis. Through the seventies only a few papers on this subject appeared outside the astronomy literature. Recently, several books have presented the method but only in the limited context in which it was initially developed.

In this paper we would like to indicate the great generality of the method by illustrating how it can be used to solve perturbation problems that are typically solved by other methods, often special ad hoc methods. In most cases we have chosen the simplest standard examples. There are many topics of current research that are not considered here since this is to be a tutorial, not a summary of new results.

Below we will indicate how the method of Lie transforms can be used to: calculate the function given by the implicit function theorem; calculate the coordinates given in the splitting lemma of catastrophe theory; calculate the center and stable manifolds of a critical point; calculate a limit cycle or an invariant torus; calculate the Poincare normal form for a center; do classical averaging to arbitrary order; calculate Floquet exponents; calculate the Darboux coordinates of symplectic geometry. All these seemingly distinct calculations can be done with one simple algorithm – the method of Lie transforms.

Most of the first part of the paper consists of examples of problems that can be solved by Lie transforms, without spending too much time on the derivation or the

---

*This research was supported by a grant from ACMP/DARPA administered by NIST.

†Departments of Mathematics and Computer Science, University of Cincinnati, Cincinnati, Ohio 45221

theory. One main theorem summarizes the power of the method and it is given in Section II. The proof of this general theorem is postponed until Section VIII. The middle sections are all examples.

Section VIII is written independently of most of the paper so if you are interested in the proof itself you can skip the examples and go directly from Section II to Section VIII. On a first reading this might be the best approach.

II. The Main Theorem of the Theory. In the traditional setting of perturbation theory one is given a differential equation depending on a small parameter ε. When $\varepsilon = 0$ the differential equation is simple and well understood, say for example a harmonic oscillator. The problem is to understand the solutions of the equations when ε is non-zero but small. To gain generality think of any smooth tensor field defined on some open set $D \subset \mathbf{R}^n$ depending on a small parameter. The tensor field might be a function; a contravariant vector field, i.e. an ordinary differential equation; a covariant vector field, i.e. a differential form; a Riemannian metric; a symplectic structure; or any of the other classical tensors of differential geometry. The important thing about these objects is that there is a Lie derivative defined for them.

Let F be a smooth tensor field defined on an open set $D \in \mathbf{R}^n$, that is for each point $x \in D$ there is assigned a unique tensor, $F(x)$, of a fixed type say p-covariant and q-contravariant. Let W be a smooth autonomous ordinary differential equation defined on D, i.e. a contravariant vector field on D, and let $\phi(\tau, \xi)$ be the solution of the equation which satisfies $\phi(0, \xi) = \xi$. The Lie derivative, $\mathcal{L}_W F$, is simply the directional derivative of F in the direction of W and is a tensor field of the same type as F itself. The general definition is given in any non-elementary book on differential geometry and in Section VIII. For now we shall simply give examples.

Differential geometry has used many different notations which still persist today making a general presentation difficult. For example the object W given above might be called an autonomous differential equation on D and so W is thought of as a smooth function from D into $\mathbf{R}^n$ and is denoted by

$$(1) \qquad \frac{dx}{d\tau} = W(x).$$

Then W is considered as a column vector with components $W^1, \ldots, W^n$. In classical tensor terminology W is 1-contravariant and we write W^i where i is a free index ranging from 1 to n – here the superscript tells you it is contravariant. More recent notation is

$$(2) \qquad \sum_{i=1}^{n} W^i(x)\frac{\partial}{\partial x^i} = W^1(x)\frac{\partial}{\partial x^1} + \cdots + W^n(x)\frac{\partial}{\partial x^n}.$$

In any case let $\phi(\tau, \xi)$ be the solution satisfying the initial condition $\phi(0, \xi) = \xi$. The simplest tensor field is a smooth function $f : D \to \mathbf{R}^1$, i.e. to each point of D you assign a scalar. The Lie derivative of f along W, $\mathcal{L}_W f$, is a smooth function from D to $\mathbf{R}^1$ also and is defined by

$$(3) \qquad \mathcal{L}_W f(x) = \frac{\partial}{\partial \tau} f(\phi(\tau, \xi)) \Big|_{\tau=0} = \nabla f(x) \cdot W(x),$$

the dot product of the gradient of f and W.

The next simplest tensor field is a vector field, either covariant or contravariant. First let χ be a contravariant vector field or differential equation on D. Using differential equation notation for χ we write

$$(4) \qquad \chi : \ \dot{x} = F(x)$$

where $\dot{} = \dfrac{d}{dt}$. The column vector F is a representation of the contravariant vector field χ in the x coordinates. Do not confuse t and τ they are different parameters for different vector fields. Changing variables in (4) from x to ξ by $x = \phi(\tau, \xi)$ where τ is simply a parameter gives

$$(5) \qquad \dot{\xi} = \left(\frac{\partial \phi}{\partial \xi}(\tau, \xi) \right)^{-1} F(\phi(\tau, \xi)) = G(\tau, \xi).$$

G is the representation of χ in the new coordinate system ξ. The Lie derivative, $\mathcal{L}_W \chi$, is defined by

$$(6) \qquad \mathcal{L}_W \chi(x) = \frac{\partial}{\partial \tau} \, G(\tau, \xi) \bigg|_{\tau = 0} = \frac{\partial F}{\partial x} W(x) - \frac{\partial W}{\partial x}(x) F(x).$$

Note that x and ξ are the same when $\tau = 0$. $\mathcal{L}_W \chi$ is a smooth contravariant vector field on D. We usually abuse the notation and confuse the vector field χ with its representation F in a coordinate system by writing $\mathcal{L}_W F$ for (6).

Let η be a 1-covariant vector field on D, i.e. a differential form, so

$$(7) \qquad \eta = \sum_{i=1}^{n} h_i(x) dx^i.$$

Think of h as the column vector $(h_1, \ldots, h_n)^T$ and change variables from x to ξ by $x = \phi(\tau, \xi)$ to get

$$(8) \qquad \eta = \sum_{i=1}^{n} k_i(\xi) d\xi^i$$

where k is a column vector related to h by

$$(9) \qquad k(\tau, \xi) = \frac{\partial \phi}{\partial \xi}(\tau, \xi)^T \, h(\phi(\tau, \xi))$$

The vector k is the components of the differential form η in the new coordinates ξ. The Lie derivative of η in the direction of W, $\mathcal{L}_W \eta$, is a one form whose component vector is given by

$$(10) \qquad \mathcal{L}_W \eta(x) = \frac{\partial}{\partial \tau} k(\tau, \xi) \bigg|_{\tau = 0} = \frac{\partial h}{\partial x}(x)^T W + \frac{\partial W}{\partial x}(x)^T h(x).$$

The Lie derivative of other tensor fields in the direction W are defined in the same way and the reader can find a complete discussion in a book on differential geometry.

Let $\mathcal{F}_{pq} = \mathcal{F}_{pq}(D)$ denote the vector space of all smooth p-covariant and q-contravariant tensor fields D. A symmetric notation for $\mathcal{L}_W K$ is $[K, W]$, the Lie bracket of K and W. For fixed W the map $\mathcal{L}_W = [\cdot, W]$ is a linear operator from $\mathcal{F}_{pq}$ into itself. The set, $\mathcal{V} = \mathcal{V}(D) = \mathcal{F}_{01}(D)$, of all smooth contravariant vector fields on D is a vector space and $[K, \cdot]$, for fixed K, is a linear from $\mathcal{V}$ into $\mathcal{F}_{pq}$. Thus $[\cdot, \cdot] : \mathcal{F}_{pq} \times \mathcal{V} \to \mathcal{F}_{pq}$ is bilinear.

Suppose that the perturbation problem is given as a tensor field Z_* on D which has a formal expansion in a small parameter ε. In many cases ε is simply a scale parameter. Consider

$$(11) \qquad Z_* = Z_*(x, \varepsilon) = \sum_{j=0}^{\infty} \left(\frac{\varepsilon^j}{j!} \right) Z_j^0(x)$$

where each Z_j^0 is a tensor field of fixed type. Specifically assume that

$$(12) \qquad Z_j^0 \in \mathcal{P}_j \subset \mathcal{F}_{pq}, \quad \text{for } j = 0, 1, 2, \ldots$$

where $\mathcal{P}_j$ is a linear subspace of $\mathcal{F}_{pq}$. In order to simplify the problem the method of normal forms seeks a near identity change of variables of the form $x = \xi + O(\varepsilon)$ such that the tensor field Z_* in the new coordinates is simpler.

The traditional approach is simple: assume a general series for the change of variables, substitute it in the series for Z_*, collect terms, and try to choose the coefficients in the change of variables series so that the tensor Z_* in the new coordinates is as simple as possible. For simple problems that will suffice, however there are several disadvantages to this approach. The bookkeeping of the terms of the series can become a major problem especially if the problem has some special structure or symmetry. For example if Z_* is a Hamiltonian vector field one would want the vector field in the new coordinates to be Hamiltonian also. Or if Z_* is invariant under some symmetry group one would want this to be true in the new coordinates also. Figuring out what the form of the nth term in new series can be quite difficult using the straight plug and chug method. Also, this procedure is not easily coded in a symbolic computer language.

Hori (1966) was interested in perturbation theory for Hamiltonian vector fields and suggested that the near identity transformation be given as the solution of an autonomous ordinary differential equation. Unfortunately, not all near identity transformations are solutions of autonomous equations and so Hori was not able to develop a general theory. Deprit (1969) took Hori's idea one step further by using non-autonomous equations. He was able to give a simple set of recursive formulas that overcomes the objections given above. Hori and Deprit worked with Hamiltonian systems, but soon afterwards Kamil (1970) and Henrard (1970) considered the general case.

Thus to simplify the perturbation problem given by Z_* in (11) we seek a near identity change of coordinates of the form

$$(13) \qquad x = x(\xi, \varepsilon) = \xi + \cdots$$

where $x(\xi, \varepsilon)$ is constructed as a formal solution of the system of equations and initial conditions

$$(14) \qquad \frac{dx}{d\varepsilon} = W(x, \varepsilon) = \sum_{j=0}^{\infty} \left(\frac{\varepsilon^j}{j!} \right) W_{j+1}(x), \quad x(0) = \xi.$$

It can easily be shown that for any change of coordinates of the form (13) there is a unique differential equation of the form (14) for which it is the solution function. The W above is a smooth vector field on D for each ε, so we take

$$(15) \qquad W_j \in \mathcal{R}_i \subset \mathcal{V}, \quad \text{for all } i = 0, 1, 2, \ldots$$

where $\mathcal{R}_i$ is a linear subspace of $\mathcal{V}$, the space of smooth vector fields on D. The problem defined by Z_* may have some special symmetry, like a reflective symmetry, or a special structure, like being Hamiltonian, and this is reflected in the assumption that we have identified the subspace $\mathcal{P}_i$ to which the Z_i belong. To preserve this symmetry or structure it may be necessary to restrict the change of variables by requiring the W_i to lie in the subspaces $\mathcal{R}_i$.

In the new coordinates ξ the tensor $Z_*(x, \varepsilon)$ becomes

$$(16) \qquad Z^* = Z^*(\xi, \varepsilon) = \sum_{j=0}^{\infty} \left(\frac{\varepsilon^j}{j!} \right) Z_0^j(\xi).$$

We say (13) or (14) transforms (11) into (16). Also we shall say the tensor Z^* in (16) is in normal form and hence simplified by definition if we have identified subspaces $\mathcal{J}_i$, $i = 1, 2 \ldots$ such that

$$(17) \qquad Z_0^i \in \mathcal{J}_i \subset \mathcal{P}_i, \quad \text{for } i = 1, 2, 3, \ldots$$

The fundamental theorem of the theory is:

THEOREM 1. *Assume i)* $[\mathcal{P}_i, \mathcal{R}_j] \subset \mathcal{P}_{i+j}$ $i, j = 1, 2, 3, \ldots$ *and ii) for any* $i = 1, 2, 3, \ldots$ *and for any* $A \in \mathcal{P}_i$ *there exists* $B \in \mathcal{J}_i$ *and* $C \in \mathcal{R}_i$ *such that*

$$(18) \qquad B = A + [Z_0^0, C].$$

Then one can compute a formal expansion for W as given in (14) with $W_i \in \mathcal{R}_i$ for all i which transforms (11) to (16) where $Z_0^i \in \mathcal{J}_i$ for all i.

The proof of this theorem in almost this level of generality can be found in Meyer and Schmidth (1977) and is given in sightly more generality in Section VIII, see Theorem 5. The proof is completely constructive in the sense that an effective algorithm is given to find the expansion of W and Z^* term by term. In practice Z_0 is given and so one takes the subspaces $\mathcal{P}_i$ as small as possible. The spaces $\mathcal{J}_i$ and $\mathcal{R}_i$ come from an analysis of the equation in (18).

III. Function Applications. In this section we will show some applications of the method of Lie transforms when the problem involves simply functions as opposed to vector fields.

The implicit function theorem. One of the fundamental theorems of analysis is the implicit function theorem. We will show how to compute the implicitly defined function using Lie transforms.

Consider a function (or formal power series) $f(u, x)$ defined in neighborhood of the origin in $\mathbf{R}^m \times \mathbf{R}^n$ into $\mathbf{R}^n$ such that $f(0, 0) = 0$ and $\frac{\partial f}{\partial x}(0, 0) = D$ is nonsingular. Then the implicit function theorem asserts that there is an analytic function (or formal power series) $\psi(u)$ defined in a neighborhood of the origin in $\mathbf{R}^m$ into $\mathbf{R}^n$ such that $\psi(0) = 0$ and $f(u, \psi(u)) \equiv 0$. Introduce a small parameter ε by scaling $u \to \varepsilon^2 u$, $x \to \varepsilon x$ and $f \to \varepsilon^{-1} f$, that is define F_* by

$$(1) \qquad F_*(u, x, \varepsilon) = \varepsilon^{-1} f(\varepsilon^2 u, \varepsilon x) = \sum_{i=0}^{\infty} \left(\frac{\varepsilon^i}{i!} \right) F_i^0(u, x)$$

and $F_0^0(u, x) = Dx$. Let x be the variable and treat u simply as a parameter in the problem. The functions $F_i^0(u, x)$ are vectors of polynomials in u and x and so let $\mathcal{P}_i$ be the vector space of such vectors of polynomials in u and x.

By Theorem 1 we must be able to solve (18) where A is any polynomial. In this case the Lie bracket is $[F_0^0, C] = DC$. Clearly we can solve $[F_0^0, C] + A = B$ by taking $B = 0$ and $C = -D^{-1}A$. Thus if we define $\mathcal{I}_i = \{0\}$ and $\mathcal{R}_i = \mathcal{P}_i$, then for any $A \in \mathcal{P}_i$, we can solve (18) for $B \in \mathcal{I}_i = \{0\}$ and $C \in \mathcal{R}_i = \mathcal{P}_i$. Thus one can compute a transformation such that $F(u, \xi, \varepsilon) = D\xi$. But $F^*(u, \xi, \varepsilon) = F_*(u, \phi(u, \xi, \varepsilon), \varepsilon) = \varepsilon^{-1} f(\varepsilon^2 u, \varepsilon\phi(u, \xi, \varepsilon))$. So $\phi(u, o, 1) = \psi(u)$ satisfies $f(u, \psi(u)) \equiv 0$. This shows that the implicit function can be computed by Lie transforms. In general the method of Lie transforms only produces a formal series, but in this case the implicit function theorem assures that formal series converges when the series for f does. Note that the parameter ε was only used to order the terms in the series since it was set to 1 in the end.

The splitting lemma. The splitting lemma is an important tool in the analysis of critical points of a function and catastrophe theory (see Poston and Stewart (1978)). Let $V(x)$ be a real value analytic function defined in a neighborhood of the origin in $\mathbf{R}^n$ and $x \in \mathbf{R}^n$. Assume that the origin is a critical point for V and for simplicity assume that $V(0) = 0$. Assume that the rank of the Hessian, $\frac{\partial^2 V}{\partial x^2}(0)$, is s, $0 \le s \le n$. Then splitting lemma says that there is a change of coordinates $x = \phi(y)$ such that in the new coordinates

$$(2) \qquad V(y) = (\pm y_1^2 \pm \cdots \pm y_s^2)/2 + v(y_{s+1}, \cdots, y_n).$$

Scale by $x \to \varepsilon x$, and $V \to \varepsilon^{-2} V$ or define

$$(3) \qquad U_*(x, \varepsilon) = \varepsilon^{-2} V(\varepsilon x) = \sum_{i=0}^{\infty} \left(\frac{\varepsilon^i}{i!} \right) U_i^0(x).$$

Here the $U_i^0(x, \mu)$ are polynomials in x of degree $i + 2$, so let $\mathcal{P}_i$ be the vector space of such polynomials. $U_0^0(x)$ is a quadratic form in x and so by making a linear change of variables if necessary we may assume that

$$(4) \qquad U_0^0(x) = (\pm x_1^2 \pm x_2^2 \pm \cdots \pm x_s^2)/2.$$

To solve (II.18) let

$$(5) \qquad C = cx_1^{k_1} \cdots x_n^{k_n}$$

be a monomials of degree $i + 2$ and where $c = (c_1, \ldots, c_n)^T$ is an n-vector. then

$$(6) \qquad [U_0^0, C] = \pm c_1 x_1^{k_1+1} x_2^{k_2} \cdots x_n^{k_n} \pm \cdots \pm c_s x_1^{k_1} x_2^{k_2} \cdots x_s^{k_s+1} \cdots x_n^{k_n}$$

so the kernel of $[U_0^0, C]$ consists of all homogeneous polynomials of degree $i + 2$ in $x_s, \ldots, x_n$ and the range of $[U_0^0, C]$ consists of the span of all monomials which contain one of $x_1, \ldots, x_s$ to a positive power or equivalently those polynomials which are zero when $x_1 = \cdots = x_s = 0$. Thus we can solve (II.18) by taking $\mathcal{P}_i$ as the space of all scalar homogeneous polynomials of degree $i + 2$, $\mathcal{I}_i$ the subspace of $\mathcal{P}_i$ consisting of all scalar homogeneous polynomials of degree $i + 2$ in $x_s, \ldots, x_n$ alone, and $\mathcal{R}_i$ the space of all n-vectors of homogeneous polynomials of degree $i + 1$ in $x_1, \ldots, x_n$.

Thus the method of Lie transforms will construct a change of coordinates so that in the new coordinate

$$(7) \qquad U^*(y, \varepsilon) = \sum_{i=0}^{\infty} \left(\frac{\varepsilon^i}{i!} \right) U_0^i(y)$$

where for $i \geq 1$ the $U_0^i(y)$ depend only on $y_s, \ldots, y_n$. Setting $\varepsilon = 1$ gives the form given by the splitting lemma in (2).

In Meyer and Schmidt (1987) the problem for finding bifurcations of relative equilibria in the N-body problem was reduced to finding the bifurcation of critical points of the potential constrained to a constant moment of inertia manifold. The constraint equation was solved by the method of Lie transforms to compute the implicitly defined function. Then by applying the splitting lemma algorithm we obtained the bifurcation equations in a form that could be analyzed by hand.

IV. Autonomous Differential Equations. In this section we will show how the Theorem 1 can be used to study autonomous differential equations. There are many more applications than the ones given here.

The classical normal form. Consider the equation

$$(1) \qquad \dot{x} = Lx + f(x)$$

where $x \in \mathbf{R}^n$, L is an $n \times n$ constant matrix, f is an analytic function defined in a neighborhood of the origin in $\mathbf{R}^n$ whose series expansion starts with second degree terms. Scale the equations by $x \to \varepsilon x$ and divide the equation by ε so that (1) becomes

$$(2) \qquad \dot{x} = \sum_{i=0}^{\infty} \left(\frac{\varepsilon^i}{i!} \right) F_j^0(x),$$

where $F_0^0(x) = Lx$ and F_i^0 is an n-vector of homogeneous polynomials of degree $i+1$ so let $\mathcal{P}_i$ be the space of all such polynomials.

Assume that L is diagonal so $L = \mathrm{diag}(\lambda_1, \ldots, \lambda_n)$. In order to solve (II.18) let

$$(3) \qquad \begin{aligned} &A = ax^k, \ B = bx^k, \ C = cx^k \\ &k = (k_1, \ldots, k_n), \ x = (x_1, \ldots, x_n), \quad x^k = x_1^{k_1} \cdots x_n^{k_n} \end{aligned}$$

and substitute into (II.18) to get

$$(4) \qquad bx^k = ax^k + (L - (\Sigma k_s \lambda_s)I)cx^k.$$

The coefficient matrix, $L - (\Sigma K_s \lambda_s)I$, of cx^k is diagonal with entries $\lambda_j - \Sigma k_s \lambda_s$. So to solve (II.18) take

$$(5) \qquad \begin{aligned} &c_j = \frac{-a_j}{\lambda_j - \Sigma k_s \lambda_s}, \ b_j = 0 \quad \text{when} \ \lambda_j - \Sigma k_s \lambda_s \neq 0 \\ &c_j = 0, \ b_j = a_j \quad \text{when} \ \lambda_j - \Sigma k_s \lambda_s = 0 \end{aligned}$$

Let $e_j = (0, \ldots, 0, 1, 0, \ldots, 0)^T$ be the standard basis for $\mathbf{R}^n$. From the above we define

$$(6) \qquad \begin{aligned} &\mathcal{I}_i = \mathrm{span}\{e_j x^k : \lambda_j - \Sigma k_s \lambda_s = 0, \ \Sigma k_s = i+1\} \\ &\mathcal{R}_i = \mathrm{span}\{e_j x^k : \lambda_j - \Sigma k_s \lambda_s \neq 0, \ \Sigma k_s = i+1\} \end{aligned}$$

so the condition in ii) of the Theorem 1 is satisfied. So (27) can be formally transformed to

$$(7) \qquad \dot{y} = \sum_{i=0}^{\infty} \left(\frac{\varepsilon^i}{i!} \right) F_0^i(y),$$

where $F_0^i \in \mathcal{I}_i$ for all $i \geq 1$. Setting $\varepsilon = 1$ brings the equations to the form

$$(8) \qquad \dot{y} = Ly + g(y)$$

where the terms in g lie in some $\mathcal{I}_i$. It is easy to check that a term $h(y)$ is in some $\mathcal{I}_i$ if and only if $h(e^{Lt}y) = e^{Lt}h(y)$ for all y and t. Thus g in (7) satisfies

$$(9) \qquad g(e^{Lt}y) = e^{Lt}g(y).$$

This formulation for the normal form does not require that L be in diagonal form (L must be diagonalizable!). This is the classical normal form as found in Diliberto (1961) et al. For example if $n = 3$ and

(10)
$$L = \begin{pmatrix} 0 & -1 & 0 \\ 1 & 0 & 0 \\ 0 & 0 & -1 \end{pmatrix}$$

so L has eigenvalues -1, and $\pm i$ then the normal form is

$$\dot{u} = -v - v\, a(u^2 + v^2) + u\, b(u^2 + v^2)$$

(11)
$$\dot{v} = u + u\, a(u^2 + v^2) + v\, b(u^2 + v^2)$$

$$\dot{w} = -w + w\, c(u^2 + v^2),$$

where the a, b and c are arbitrary series. This normal form yields the so called center manifold since the plane $w = 0$ is invariant and the equations on this center manifold are in Poincare's normal form for a center.

Invariant tori. Consider a system of coupled van der Pol equations written in polar coordinates. Or more generally a system of the form

(12)
$$\dot{r} = R_*(r, \theta, \varepsilon) = \sum_{i=0}^{\infty} \left(\frac{\varepsilon^i}{i!} \right) R_i^0(r, \theta)$$

$$\dot{\Theta} = \theta_*(r, \theta, \varepsilon) = \sum_{i=0}^{\infty} \left(\frac{\varepsilon^i}{i!} \right) \Theta_i^0(r, \theta)$$

where r is a m-vector, θ is an n-vector of angles, R_i^0 and Θ_i^0 have finite Fourier series in the θ's with coefficients which are polynomials in the r variables. Let $\mathcal{P}_i$ be the space of all such functions.

Assume that $\Theta_0^0 = \omega$ is a constant vector, $R_0^0 = P(r)$ and that there exists a constant vector r_0 such that $P(r_0) = 0$ and $\dfrac{\partial P}{\partial r}(r_0)$ has no eigenvalue with zero real part. Then there is a formal change of variables $(r, \theta) \to (\rho, \phi)$ such that the equations (11) are of the form

(13)
$$\dot{\rho} = R^*(\rho, \phi, \varepsilon) = \sum_{i=1}^{\infty} \left(\frac{\varepsilon^i}{i!} \right) R_0^i(\rho, \phi)$$

$$\dot{\phi} = \Phi^*(\rho, \phi, \varepsilon) = \sum_{i=1}^{\infty} \left(\frac{\varepsilon^i}{i!} \right) \Phi_0^i(\rho, \phi)$$

where R^* and Φ^* are like R_* and Θ_* but have the additional property that

(14)
$$R^*(r_0, \phi, \varepsilon) \equiv 0 \quad \text{and} \quad \Phi^*(r_0, \phi + \omega t, \varepsilon) \equiv 0.$$

The first condition in (14) says that $r = r_0$ is an invariant torus for the equations (13) and the second condition says that the equations on the invariant torus are in normal form. If there are no resonances among the frequencies ω then $\Phi^*(r_0, \phi, \varepsilon) \equiv 0$.

Here

$$Z_0^0 = \begin{pmatrix} P(r) \\ \omega \end{pmatrix}, \quad C = \begin{pmatrix} U \\ V \end{pmatrix} = \begin{pmatrix} u(r)e^{ik\theta} \\ v(r)e^{ik\theta} \end{pmatrix}$$

(15)
$$A = \begin{pmatrix} a(r)e^{ik\theta} \\ \alpha(r)e^{ik\theta} \end{pmatrix}, \quad B = \begin{pmatrix} b(r)e^{ik\theta} \\ \beta(r)e^{ik\theta} \end{pmatrix}$$

$$k\theta = k_1\theta_1 + \cdots + k_n\theta_n$$

then

(16)
$$[Z_0^0, C] = \begin{pmatrix} \dfrac{\partial P}{\partial r} & 0 \\ 0 & 0 \end{pmatrix} \begin{pmatrix} U \\ V \end{pmatrix} - \begin{pmatrix} \dfrac{\partial U}{\partial r} & \dfrac{\partial U}{\partial \theta} \\ \dfrac{\partial V}{\partial r} & \dfrac{\partial V}{\partial \theta} \end{pmatrix} \begin{pmatrix} P \\ \omega \end{pmatrix} =$$

$$\begin{pmatrix} \dfrac{\partial P}{\partial r} u e^{ik\theta} & -\dfrac{\partial u}{\partial r} e^{ik\theta} P - iuk\omega e^{ik\theta} \\ & -\dfrac{\partial v}{\partial r} e^{ik\theta} P - ivk\omega e^{ik\theta} \end{pmatrix} = \begin{pmatrix} (b-a)e^{ik\theta} \\ (\beta - \alpha)e^{ik\theta} \end{pmatrix}.$$

To solve the second set of equations take

(17)
$$v = \dfrac{\alpha}{k\omega} \qquad \beta = \dfrac{d\alpha}{dr}\dfrac{P}{k\omega} \qquad \text{when} \quad k\omega \neq 0$$

$$v = 0 \qquad \beta = \alpha \qquad \text{when} \quad k\omega = 0.$$

For the first equation in (16) first let $D = \dfrac{\partial P}{\partial r}(r_0)$ and note that $u = (D - ik\omega I)^{-1}a$ solves $Du - ik\omega\, u = -a$ for all k since D has no eigenvalue with zero real part by assumption. So we take

(18)
$$u = (D - ik\omega)^{-1}a$$

$$b = \left(\dfrac{\partial P}{\partial r}(r) - \dfrac{\partial P}{\partial r}(r_0) \right) u - \dfrac{\partial u}{\partial r} P(r).$$

This formulas satisfy the equations and clearly $b(r_0) = 0$. The space $\mathcal{J}_i$ is the span of all the solutions given for B and the space $\mathcal{R}_i$ is the span of all the solutions given for C above. Thus we have verified the conditions of the Theorem 1. This was the procedure used in Meyer and Schmidt (1977) to calculate the regions in parameter space where two coupled van der Pol oscillators had frequencies that were locked in. The so called entrainment domains.

V. Non-Autonomous Differential Equations. In many applications the differential equations involve time explicitly so one must consider equations of the form $\dot{x} = f(t, x)$. In this case one would allow the transformation generated by W to depend on t also. But this case can be reduced to the previous case by replacing the original system with the equivalent autonomous system $\dot{x} = f(\tau, x)$, $\dot{\tau} = 1$ where τ is a new variable.

Consider the system

$$
\text{(1)} \qquad \dot{x} = Z_*(t, x, \varepsilon) = \sum_{j=0}^{n} \left(\frac{\varepsilon^j}{j!} \right) Z_j^0(t, x),
$$

and the near identify transformation

$$
\text{(2)} \qquad x = x(t, \xi, \varepsilon) = \xi + \cdots
$$

generated as a solution of the equation

$$
\text{(3)} \qquad \frac{dx}{d\varepsilon} = W(t, x, \varepsilon) = \sum_{j=0}^{n} \left(\frac{\varepsilon^j}{j!} \right) W_{j+1}(t, x), \qquad x(0) = \xi
$$

which transforms (1) to

$$
\text{(4)} \qquad \dot{\xi} = Z^*(t, \xi, \varepsilon) = \sum_{j=0}^{n} \left(\frac{\varepsilon^j}{j!} \right) Z_0^j(t, \xi).
$$

The translation of the Theorem 1 to the non-autonomous case goes as follows.

THEOREM 2. *Let $\mathcal{P}_j$ ($\mathcal{R}_j$ respectively) be linear spaces of smooth time dependent tensor (respectively vector) fields defined for $j = 1, 2, \ldots, x \in D \subset \mathbf{R}^n$ and $t \in \mathbf{R}$ and let $\mathcal{I}_j$ be a subspace of $\mathcal{P}_j$. If i) $Z_j^0 \in \mathcal{P}_j$ for $j = 0, 1, 2, \ldots$ ii) $[\mathcal{P}_i, \mathcal{R}_j] \subset \mathcal{P}_{i+j}$, $i, j = 0, 1, 2, \ldots$ iii) for any $i = 1, 2, 3, \ldots$ and any $A \in \mathcal{P}_i$ there exist $B \in \mathcal{I}_i$ and $C \in \mathcal{R}_i$ such that*

$$
\text{(5)} \qquad B = A + [Z_0^0, C] - \dot{C},
$$

then one can construct W as in (3) with $W_i \in \mathcal{R}_i$ which generates a transformation (2) which takes (1) to (4) with $Z_0^i \in \mathcal{I}_i$.

The method of averaging. The method of averaging is a special case of the normal form theorem given above. The method of averaging deals with a periodic system of the form (1) where $Z_0^0 = 0$, i.e. $\dot{x} = \varepsilon Z_1^0(t, x) + \cdots$. One seeks a periodic change of variables, so the function W must be periodic in t also. Equation (5) reduces $B = A - \dot{C}$. Given a periodic A in order to have a periodic C it is necessary and sufficient that we take B as the average over a period of A, so B is independent of t, and C as any indefinite integral of $A - B$. This shows that the normalized equation (4) are autonomous, i.e. Z_0^i is independent of t. The name comes from the fact that Z_0^1 is the time average of Z_1^0.

The Floquet exponents and the Liapunov transformation. A classical problem is to compute the characteristic exponents of Mathieu's equation $\ddot{x} + (a + b\cos 2\pi t)x = 0$ or other similar linear periodic systems. Assume that $Z_0^0(t, x) = Lx$ where L is diagonal matrix $L = \text{diag}(\lambda_i, \ldots, \lambda_n)$ and $Z_i^0(t, x) = A_i(t)x$ where $A_i(t)$ in an $n \times n$ 2π-periodic matrix, so let $\mathcal{P}_i$ be the space of all linear 2π-periodic systems. Seek a linear 2π-periodic change of variables, so seek $W_i(t, x) = C_i(t)x$ where $C_i(t)$ is to be 2π-periodic also and take $\mathcal{R}_i$ be the space $\mathcal{P}_i$. Equation (5) becomes

$$(6) \qquad B(t) = A(t) + C(t)L - LC(t) - \dot{C}(t)$$

where A, B and C are matrices. The equation for the ij^{th} component is

$$(7) \qquad b_{ij} = a_{ij} + (\lambda_i - \lambda_j)c_{ij} - \dot{c}_{ij}.$$

This is a linear first order differential equation in c_{ij}. Let $\lambda_i - \lambda_j \neq n\sqrt{-1}$ for $i \neq j$. Then when $i \neq j$ take $b_{ij} = 0$ and c_{ij} as the unique 2π-periodic solution of (7). When $i = j$ take b_{ii} as the average of a_{ii} and c_{ii} as any indefinite integral of $(-a_{ii} + b_{ii})$. Thus the space $\mathcal{J}_i$ is all linear systems with constant diagonal coefficient matrices. Thus we can compute a linear periodic change of coordinates which reduces the linear periodic system (1) to the linear diagonal constant system (4), this transformation is known as the Liapunov transformation. The entries on the diagonal are the Floquet exponents. The equation (6) has been studied in the more general case when L is not necessarily diagonal. The presentation given here is merely a simple example.

A very similar problem is to calculate the series expansion of a solution of a linear differential equation at a regular singular point.

VI. The Computational Darboux Theorem. To our knowledge the method of Lie transforms has not been used on tensor fields more complicated than vector fields. Here we will give a somewhat frivolous example to illustrate the generality of the method. In order to avoid the notational overload found in modern treatises like Kobayashi and Nomizu (1963) or Abraham and Marsden (1978), we shall use classical tensor notation. Thus repeated indices are summed over. Since the problem is a computational one we must use coordinates in the end anyway. Flanders (1963) is a highly recommended introduction to differential forms. The fundamental geometry of Hamiltonian mechanics is embodied in a *symplectic structure*, Ω, i.e. a closed, non-degenerate 2-form. In a neighborhood of the origin in $\mathbf{R}^{2n}$

$$(1) \qquad \Omega = \Omega_{ij}(x) \, dx^i \wedge \, dx^j$$

where we have used the summation convention, $\Omega_{ij} = -\Omega_{ji}$, and the $\Omega_{ij}(x)$ are the real analytic in x. $\{\Omega_{ij}\}$ is a 2-covariant tensor, so if you change coordinates by $x = x(y)$ then the tensor in the y coordinates is

$$(2) \qquad \Omega(y) = \Omega_{ij}(x(y))\frac{\partial x^i}{\partial y^m} \frac{\partial x^j}{\partial y^n} \, dy^m \wedge \, dy^n.$$

Sometimes we will think of $\Omega(x)$ as the skew-symmetric matrix $(\Omega_{ij}(x))$, the coefficient matrix of the form (1). Ω is non-degenerate means that the matrix $\Omega(x)$ is nonsingular for all x. (1) means that the matrix Ω transforms by

$$(3) \qquad \qquad \Omega \to \frac{\partial x^T}{\partial y} \, \Omega \, \frac{\partial x}{\partial y} \, .$$

Ω is closed means that

$$(4) \qquad \qquad d\Omega = \frac{\partial \Omega_{ij}}{\partial x_k} \, dx^i \wedge dx^j \wedge dx^k = 0.$$

Since we are working locally, a closed form is exactly by Poincare's lemma so there is a one form $\alpha(x) = \alpha_i(x)dx^i$ such that $\Omega = d\alpha$.

This matrix $\Omega(0)$ is nonsingular and skew symmetric so there is a nonsingular matrix P such that

$$(5) \qquad \qquad P^T \Omega(0) P = J = \begin{pmatrix} 0 & I \\ -I & 0 \end{pmatrix},$$

which means that after a linear change of coordinates the coefficient matrix of $\Omega(0)$ is J. Darboux's theorem says there is a nonlinear change of coordinates defined in a neighborhood of the origin in $\mathbf{R}^{2n}$ so that in the new coordinates the coefficient matrix of Ω is identically J in the whole neighborhood. Our computational procedure follows the proof given by Weinstein (1971).

Assume that the linear change of variables has been made so that $\Omega(0) = J$ and scale by $x \to \varepsilon x$, $\Omega \to \varepsilon^{-1}\Omega$ so that

$$(6) \qquad \qquad \Omega = \sum_{s=0}^{\infty} \left(\frac{\varepsilon^s}{s!} \right) \omega_s^0 \, ,$$

where ω_s^0 is closed 2-form with coefficients that are homogeneous polynomials in x of degree s. Let $\mathcal{P}_s$ be the vector space of such forms and $\mathcal{I}_s = \{0\}$. Let $A \in \mathcal{P}_s$, $B = 0 \in \mathcal{I}_s$, and $C \in \mathcal{R}_s$, where $\mathcal{R}_s$ is the vector space of vector fields which are homogeneous polynomials of degree $s + 1$. In coordinates, equation (II.18) for this problem is

$$(7) \qquad \qquad 0 = A_{sm} + J_{im} \frac{\partial C^i}{\partial x^s} + J_{sj} \frac{\partial C^j}{\partial x^m} \, .$$

(In general there would be a term $+\dfrac{\partial J_{sm}}{\partial x^i} C^i$ in (7) but this term is zero since J is constant.)

Since A is a closed two form there is a one form α such that $A = d\alpha$ so (7) becomes

$$(8) \qquad \qquad 0 = \frac{\partial \alpha_s}{\partial x^m} - \frac{\partial \alpha_m}{\partial x^s} + J_{im} \frac{\partial C^i}{\partial x^s} + J_{sj} \frac{\partial C^j}{\partial x^m}.$$

This equation has a solution $C^i = \alpha_{i+n}$ for $1 \le i \le n$, $C^i = -\alpha_{i-n}$ for $n \le i \le 2n$, or $C = J\alpha$. Thus there is a solution of (II.18) and so the coordinate change given by Darboux's theorem can be computed by Lie transforms.

VII. Hamiltonian Systems. For Hamiltonian systems the Lie bracket is replaced by the Poisson bracket. Let F, G and H be smooth real valued functions defined in an open set in $\mathbf{R}^{2n}$, the Poisson bracket of F and G is the smooth function $\{F, G\}$ defined by

$$(1) \qquad \{F, G\} = \frac{\partial F^T}{\partial x} J \frac{\partial G}{\partial x}$$

where J is as in (VI.5) the usual $2n \times 2n$ skew symmetric matrix of Hamiltonian mechanics. A Hamiltonian differential equation (generated by the Hamiltonian H) is

$$(2) \qquad \dot{x} = J\frac{\partial H}{\partial x} \,.$$

The Poisson bracket and the Lie bracket are related by

$$(3) \qquad J\frac{\partial}{\partial x}\, \{F, G\} = \left[J\frac{\partial F}{\partial x},\ J\frac{\partial G}{\partial x} \right]$$

so the Hamiltonian vector field generated by $\{F, G\}$ is the Lie bracket of the Hamiltonian vector fields generated by G and F, see Abraham and Marsden (1978).

Consider a Hamiltonian perturbation problem given by the Hamiltonian

$$(4) \qquad H_*(x, \varepsilon) = \sum_{j=0}^{n} \left(\frac{\varepsilon^j}{j!} \right) H_j^0(x).$$

A near identity symplectic change of coordinates $x = \phi(\xi, \varepsilon) = \xi + \cdots$ can be generated as the solution of the Hamiltonian differential equations

$$(5) \qquad \frac{dx}{d\varepsilon} = J\frac{\partial W}{\partial x}(x, \varepsilon), \quad x(0) = \varepsilon, \quad W(x, \varepsilon) = \sum_{j=0}^{n} \left(\frac{\varepsilon^j}{j!} \right) W_{j+1}(x).$$

It transforms (4) to

$$(6) \qquad H^*(x, \varepsilon) = \sum_{j=0}^{n} \left(\frac{\varepsilon^j}{j!} \right) H_0^j(x).$$

THEOREM 3. *Let $\mathcal{P}_j, \mathcal{I}_j$, and $\mathcal{R}_j$ be vector spaces of smooth Hamiltonians on D with $\mathcal{I}_j \subset \mathcal{P}_j$. Assume that i) $Z_j^0 \in \mathcal{P}_j$ for $j = 1, 2, 3\ldots$ ii) $\{\mathcal{P}_i, \mathcal{R}_j\} \subset \mathcal{P}_{i+j}$ for $i, j = 1, 2, 3, \ldots$ iii) for any j and any $A \in \mathcal{P}_j$ there exist $B \in \mathcal{I}_j$ and $C \in \mathcal{R}_j$ such that*

$$(7) \qquad B = A + \{H_0^0, C\}.$$

Then one can compute a formal expansion for W in (5) with $W_j \in \mathcal{R}_j$ for all j which transforms (4) to (6) where $H_0^j \in \mathcal{I}_j$ for all j.

The classical Birkhoff normal form for a Hamiltonian system near an equilibrium point is as follows. Assume that the Hamiltonian (4) came from scaling a system about an equilibrium point at the origin. That is, $H_0^0(x)$ is a quadratic form and H_j^0 is a homogeneous polynomial of degree $j + 2$. Assume that the linear Hamiltonian system

$$\text{(8)} \qquad \dot{x} = J\frac{\partial H_0^0}{\partial x} = Ax$$

is such that A is diagonalizable. Then one can compute a symplectic change of variables generated by (5) which transforms (4) to (6) with

$$\text{(9)} \qquad H^*(e^{At}x, \varepsilon) = H^*(x, \varepsilon).$$

For a Lie transform proof see Meyer (1974).

Kummer (1976) has shown that Lie algebra theory is useful in studying normal forms in some special cases in celestial mechanics. Taking this lead Cushman, Deprit and Mosak (1983) have used results from representation theory to give a complete description of the normal forms for Hamiltonian systems without the diagonalizable assumption.

VIII. The General Lie Transform Algorithm. In this section we will give a proof of the main algorithm of Deprit, Theorem 4, and the main perturbation algorithm, Theorem 5, for general tensor fields. Theorem 5 is a slight extension of Theorem 1. A general reference for the tensor analysis and notation used here is Abraham and Marsden (1978).

Let $\mathsf{E}, \mathsf{F}, \mathsf{G}$ and $\mathsf{E}_1, \ldots, \mathsf{E}_k$ be vector spaces over K where K is the real numbers R or the complex numbers C; $L(\mathsf{E}; \mathsf{F})$ be the space of bounded linear functions from E to F; $\mathsf{E}^* = L(\mathsf{E}, \mathsf{K})$ be the dual space of E; and $L^k(\mathsf{E}_1, \ldots, \mathsf{E}_k; \mathsf{K})$ be the space of bounded multilinear maps from $E_1 \times \cdots \times E_k$ into K. Define $T_s^r(\mathsf{E}) = L^{r+s}(\mathsf{E}^*, \ldots, \mathsf{E}^*, \mathsf{E}, \ldots, \mathsf{E}; \mathsf{K})$ – r copies of E^* and s copies of E, so if $Z \in T_s^r(\mathsf{E})$ then $Z : \mathsf{E}^* \times \cdots \times \mathsf{E} \times \cdots \times \mathsf{E} \to \mathsf{K}$ is linear in each argument. The elements, $Z \in T_s^r(\mathsf{E})$ are called r-contravariant, s-covariant tensors or simply (r, s)-tensors. In the case $r = s = 0$ we define $T_0^0(\mathsf{E}) = \mathsf{K}$. If $A : \mathsf{E} \to \mathsf{E}$ is an invertible linear map and $A^* : \mathsf{E}^* \to \mathsf{E}^*$ is the dual map, then $A_s^r : T_s^r(\mathsf{E}) \to T^r(\mathsf{E})$ is the invertible linear map defined by $(A_s^r Z)(\alpha^1, \ldots, \alpha^r, \beta_1, \ldots, \beta_s) = Z(A^*\alpha^1, \ldots, A^*\alpha^r, A^{-1}\beta_1, \ldots, A^{-1}\beta_s)$.

Let M be a smooth manifold modeled on a vector space E and $p \in M$ any point. In the classical and still most important case M is simply an open set D in R^m and E is R^m itself. The tangent space to M at p, denoted by T_pM is isomorphic to E itself; the cotangent space to M at p, denoted by T_p^*M, is the dual of T_pM; and the space of r-contravariant, s-covariant tensors at p is $T_s^r(T_pM)$. The vector bundles built on T_pM, T_p^*M, and $T_s^r(T_pM)$ are respectively: TM, the tangent bundle; T^*M, the cotangent bundle; and T_s^rM, the (r, s)-tensor bundle. Smooth sections in these bundles are called respectively: vector fields (or contravariant vector fields or ordinary differential equations); covector fields (or one forms); and (r, s)-tensor fields. Let $\mathfrak{T}(M)$ be the space of smooth vector fields, $\mathfrak{T}^*(M)$ the space of smooth one-forms, and $\mathfrak{T}_s^r(M)$ in the space of smooth (r, s)-tensors. Let $V : M \to M$ be a diffeomorphism, $p \in M, q = V(p)$ and $DV(p) : TM \to T_1M$ be the derivative of

V at p then $DV_s^r(p): T_s^r(T_pM) \to T_s^r(T_qM)$. The results of this section are quite general so M could be a Banach manifold modeled on a reflexive Banach space E, but the author has no examples which require this level of generality.

Consider the case where M is an open set in $\mathbf{R}^m$ with coordinates $(x^1, \ldots, x^m)$. A (0,0)- tensor field is simply a smooth function $Z: M \to \mathbf{K}$. A vector field, Z, is given by

$$(1) \qquad Z = Z^1(x)\frac{\partial}{\partial x^1} + \cdots + Z^m(x)\frac{\partial}{\partial x^m},$$

where $Z^1, \ldots, Z^m$ are smooth real valued functions on M. The vector field Z is the same as the differential equation

$$(2) \qquad \dot{x} = Z(x) \quad (\text{or } \dot{x}^i = Z^i(x), \ i = 1, \ldots, m).$$

A covector field, Z, is given by

$$(3) \qquad Z = Z_1(x)dx^1 + \cdots + Z_m(x)dx^m,$$

where again $Z_1, \ldots, Z_m$ are smooth functions.

Let U be a smooth vector field (autonomous differential equation) on M and let $X(\tau, y)$ be the general solution of the differential equation

$$(4) \qquad x' = \frac{dx}{d\tau} = U(x)$$

which satisfies $X(0, y) = y$. That is, $X'(\tau, y) = U(X(\tau, y))$. Assume that there is an $\tau_0 > 0$ such that $X: (-\tau_0, \tau_0) \times M \to M$ is defined and smooth. X is a function of two arguments and let $'$ denote the partial derivative with respect to the first argument, $' = \partial/\partial\tau$, and let D denote the partial derivative with respect to the second argument, $D = \partial/\partial y$, thus $DX(\tau, p): T_pM \to T_qM$, $q = X(\tau, p)$ and $DX_s^r(\tau, p): T_s^r(T_pM) \to T_s^r(T_qM)$. Let $Z: M \to \mathfrak{T}_s^r(M)$ be a smooth (r, s)-tensor field on M, $p \in M$, $q = X(\tau, p)$. Then $Z(p) \in T_s^r(T_pM)$, $Z(X(\tau, p)) \in T_s^r(T_qM)$, and $A(\tau) = DX_s^r(\tau, p)^{-1}Z(X(\tau, p)) \in T_s^r(T_pM)$, so $A(\tau)$ is a smooth curve of (r, s)-tensors in the fixed tensor space $T_s^r(T_pM)$. The Lie derivative of Z in the direction of U (or along U) is denoted by $[Z, U]$ and is defined as

$$(5) \qquad [Z, U](p) = \frac{\partial}{\partial\tau}A(\tau)\bigg|_{\tau=0} = \frac{\partial}{\partial\tau}DX_s^r(\tau, p)^{-1}Z(X(\tau, p))\bigg|_{\tau=0}$$

Since $A(\tau) \in T_s^r(T_pM)$ for all τ its derivative is in $T_s^r(T_pM)$ so $[Z, U](p) \in T_s^r(T_pM)$ and $[Z, U]$ is a smooth (r, s)-tensor field also and $[\cdot, \cdot]: \mathfrak{T}_s^r(M) \times \mathfrak{T}(M) \to \mathfrak{T}_s^r(M)$ is bilinear. $[\cdot, \cdot]$ is called the Lie bracket.

If M is an open set in $\mathbf{R}^m$ and $Z: M \to \mathbf{R}$ is a smooth function $((0,0)$-tensor field) then in classical notation

$$(6) \qquad [Z, U](x) = \nabla Z(x) \cdot U(x)$$

so $[Z, U]$ is the directional derivative of Z in the direction U. If Z is a smooth vector field (ordinary differential equation) as in (2) then

(7)
$$[Z, U](x) = \frac{\partial Z}{\partial x}(x)U(x) - \frac{\partial U}{\partial x}(x)Z(x)$$

where z and U are column vectors. If Z is a one form though of as a column vector then

(8)
$$[Z, U](x) = \frac{\partial Z}{\partial x}(x)^T U(x) + \frac{\partial U}{\partial x}(x)^T Z(x).$$

Suppose that the perturbation problem is given as an (r, s)-tensor field $Z = Z_*$ on M which has a formal expansion in a small parameter ε. Consider

(9)
$$Z(\varepsilon, x) = Z_*(\varepsilon, x) = \sum_{j=0}^{\infty} \left(\frac{\varepsilon^j}{j!} \right) Z_j^0(x)$$

where each $Z_j^0 : M \to T_s^r M$ is an (r, s)-tensor field.

To simplify the perturbation problem given by Z_* in (9) we seek a near identity change of coordinates of the form

(10)
$$x = X(\varepsilon, y) = y + \cdots$$

where $X(\varepsilon, y)$ is constructed as a formal solution of the nonautonomous system of differential equations

(11)
$$\frac{dx}{d\varepsilon} = W(x, \varepsilon) = \sum_{j=0}^{\infty} \left(\frac{\varepsilon^j}{j!} \right) W_{j+1}(x),$$

satisfying the initial condition

(12)
$$x(0) = y$$

where each $W_j : M \to TM$ is a smooth vector field.

The Lie transform of $Z(= Z_*)$ by W, denoted by $\mathcal{L}(W)Z$ or Z^* for short, is the tensor field Z_* expressed in the new coordinates and so is an (r, s)-tensor field depending on the parameter ε also. Specifically,

(13)
$$Z^*(\varepsilon, y) = \mathcal{L}(W)Z(\varepsilon, y) = DX_s^r(\varepsilon, y)^{-1} Z_*(\varepsilon, X(\varepsilon, y))$$

In the new coordinates y the tensor $Z_*(x, \varepsilon)$ becomes

(14)
$$Z^*(\varepsilon, y) = \mathcal{L}(W)Z(\varepsilon, y) = \sum_{j=0}^{\infty} \left(\frac{\varepsilon^j}{j!} \right) Z_0^j(y).$$

We say (10) or (11) transforms (9) into (14). The method of Lie transforms introduces a double indexed array of tensor fields $\{Z_j^i\}$, $i, j = 0, 1, \ldots$ which agree with the definitions given in (9) and (14) when either i or j is zero. The other terms are intermediary terms introduced to facilitate the computation. The main theorem on Lie transforms by Deprit (1969) in this general context is the following.

THEOREM 4. *Using the notation given above, the tensor fields* $\{Z^i_j\}, i = 1, 2, \ldots, j = 0, 1, \ldots$ *satisfy the recursive identities*

(15)
$$Z^i_j = Z^{i-1}_{j+1} + \sum_{k=0}^{j} \binom{j}{k} [Z^{i-1}_{j-k}, W_{k+1}].$$

REMARKS. The above formulas contain the standard binomial coefficient $\binom{j}{k} = \dfrac{j!}{k!(j-k)!}$. Note that since the transformation generated by W is a near identity transformation the first term in Z_* and Z^* are the same, namely Z^0_0. Also note that the first term in the expansion for W starts with W_1. This convention imparts some nice properties to the formulas in (15). Each term in (15) has indices summing to $i + j$ and each term on the right hand side has upper index $i - 1$.

The interdependent of the $\{Z^i_j\}$ can easily be understood by considering the Lie triangle

(16)
$$
\begin{array}{ccccc}
Z^0_0 & & & \\
\downarrow & & & \\
Z^0_1 & \!\!\longrightarrow\!\! & Z^1_0 & \\
\downarrow & & \downarrow & \\
Z^0_2 & \!\!\longrightarrow\!\! & Z^1_1 & \!\!\longrightarrow\!\! & Z^2_0 \;. \\
\downarrow & & \downarrow & & \downarrow
\end{array}
$$

The coefficients of the expansion of the old tensor field Z_* are in the left column and those of the new tensor field Z^* are on the diagonal. The formula (15) says that to calculate any element in the Lie triangle you need the entries in the column one step to the left and up.

Proof of Theorem 4. Let $Y(\varepsilon, x)$ be the inverse of $X(\varepsilon, y)$ so $Y(\varepsilon, X(\varepsilon, y)) \equiv y$, $X(\varepsilon, Y(\varepsilon, x)) \equiv x$, $DX(\varepsilon, y)^{-1} = DY(\varepsilon, X(\varepsilon, y))$, and $DX^r_s(\varepsilon, y)^{-1} = DY^r_s(\varepsilon, X(\varepsilon, y))$. Thus (13) becomes $\mathcal{L}(W)Z(\varepsilon, y) = DY^r_s(\varepsilon, X(\varepsilon, y))Z_*(\varepsilon, X(\varepsilon, y))$.

Define the differential operator $\mathcal{D} = \mathcal{D}_w$ acting on (r, s)-tensor fields depending on a parameter ε by

(17)
$$\mathcal{D}K(\varepsilon, x) = \frac{\partial K}{\partial \varepsilon}(\varepsilon, x) + [K, W](\varepsilon, x).$$

In computing the Lie bracket in (17) the ε is simply a parameter and so held fixed during any differentiation. With this notation we have

(18)
$$\frac{d}{d\varepsilon}\left\{ DY^r_s(\varepsilon, x)K(\varepsilon, x)\Big|_{x=X(\varepsilon, y)} \right\} = DY^r_s(\varepsilon, x)\mathcal{D}K(\varepsilon, x)\Big|_{x=X(\varepsilon, y)}.$$

Define new functions by $Z^0 = Z, Z^i = \mathcal{D}Z^{i-1}, i \geq 1$. Let these functions have series expansions

(19)
$$Z^i(\varepsilon, x) = \sum_{k=0}^{\infty} \left(\frac{\varepsilon^k}{k!}\right) Z_k^i(x),$$

so

(20)

$$Z^i(\varepsilon, x) = \mathcal{D} \sum_{k=0}^{\infty} \left(\frac{\varepsilon^k}{k!}\right) Z_k^{i-1}(x) =$$

$$\sum_{k=0}^{\infty} \left(\frac{\varepsilon^{k-1}}{(k-1)!}\right) Z_k^{i-1}(x) + \sum_{k=}^{\infty} \left[\left(\frac{\varepsilon^k}{k!}\right) Z_k^{i-1}(x), \sum_{s=0}^{\infty} \left(\frac{\varepsilon^s}{s!}\right) W_{s+1}(x) \right] =$$

$$\sum_{j=0}^{\infty} \left(\frac{\varepsilon^j}{j!}\right) \left(Z_{j+1}^{i-1} + \sum_{k=0}^{i} \binom{j}{k} [Z_{j-k}^{i-1}, W_{k+1}] \right).$$

So the functions Z_j^i are related by (15). It remains to show that $Z_* = G$ has the expansion (14). By Taylor's theorem and (18)

(21)
$$G(\varepsilon, y) = \sum_{n=0}^{\infty} \left(\frac{\varepsilon^n}{n!}\right) \frac{d^n}{d\varepsilon^n} G(\varepsilon, y) \bigg|_{\varepsilon=0} =$$

$$\sum_{n=0}^{\infty} \left(\frac{\varepsilon^n}{n!}\right) \frac{d^n}{d\varepsilon^n} \left(DY_s^r(\varepsilon, x) Z(\varepsilon, x) \bigg|_{x=X(\varepsilon,y)} \right) \bigg|_{\varepsilon=0} =$$

$$\sum_{n=0}^{\infty} \left(\frac{\varepsilon^n}{n!}\right) \left(DY_s^r(\varepsilon, x) \mathcal{D}^n Z(\varepsilon, x) \bigg|_{x=X(\varepsilon,y)} \right) \bigg|_{\varepsilon=0} =$$

$$\sum_{n=0}^{\infty} \left(\frac{\varepsilon^n}{n!}\right) Z_0^n(x).$$

In the cases of interest the tensor field is given and the change of variables is sought to simplify it. When the field is sufficiently simple it is said to be in 'normal form'. The main Lie transform algorithm starts with a given field which depends on a small parameter, ε, and constructs a change of variables so that the field in the new variables is simple. The algorithm is built around the following observation.

Consider the series (9) as given so all the Z_i^0 are known. Assume that all the entries in the Lie triangle are known down to the N row, so the Z_j^i are known for $i + j \leq N$ and assume the W_i are known for $i \leq N$. Let $\underline{Z}_j^i$ be computed from the same differential equation, so $\underline{Z}_i^0 = Z_i^0$ for all i, and with $\underline{W}_1, \ldots, \underline{W}_N$ where $\underline{W}_i = W_i$ for $i = 1, 2, \ldots, N - 1$ but $\underline{W}_N = 0$. Then

(22)
$$Z_j^i = \underline{Z}_j^i \quad \text{for} \quad i + j < N$$
$$Z_j^i = \underline{Z}_j^i + [Z_0^0, W_N] \quad \text{for} \quad i + j = N.$$

This is easily seen from the recursive formulas in Theorem 4. Recall the remark that the sum of all the indices must add to the row number, so W_N does not effect the terms in the first $N - 1$ rows. The second equation in (22) follows from a simple induction across the Nth row. The algorithm can be used to prove a general theorem which includes almost all applications, see Meyer and Schmidth (1977).

THEOREM 5. *Let* $\{\mathcal{P}_i\}_{i=0}^\infty$, $\{\mathcal{Q}_i\}_{i=1}^\infty$ *and* $\{\mathcal{R}_i\}_{i=1}^\infty$ *be sequences of linear spaces of smooth fields defined on a manifold* M *where* $\{\mathcal{P}_i\}_{i=0}^\infty$ *and* $\{\mathcal{Q}_i\}_{i=1}^\infty$ *are* (r,s)*-tensor fields and* $\{R_i\}_{i=1}^\infty$ *are a vector fields. Assume:*

i) $\mathcal{Q}_i \subset \mathcal{P}_i$, $i = 1, 2, \ldots$

ii) $Z_i^0 \in \mathcal{P}_i$, $i = 0, 1, 2, \ldots$

iii) $[\mathcal{P}_i, \mathcal{R}_j] \subset \mathcal{P}_{i+j}$, $i, j = 0, 1, 2, \ldots$

iv) *for any* $A \in \mathcal{P}_i$, $i = 1, 2 \ldots$ *there exists* $B \in \mathcal{Q}_i$ *and* $C \in \mathcal{R}_i$ *such that*

$$(23) \qquad\qquad B = A + [Z_0^0, \ C].$$

Then there exists a W *with a formal expansion of the form (11) with* $W_i \in \mathcal{R}_i$, $i = 1, 2, \ldots$, *which transforms the tensor field* Z_* *with the formal series expansion given in (9) to the field* Z^* *with the formal series expansion given by (14) with* $Z_0^i \in \mathcal{Q}_i$, $i = 1, 2, \ldots$

Proof. Use induction on the rows of the Lie triangle. Induction Hypothesis I_n: Let $Z_j^i \in \mathcal{P}_{i+j}$ for $0 \leq i + j \leq n$ and $W_i \in \mathcal{R}_i$, $Z_0^i \in \mathcal{Q}_i$ for $1 \leq i \leq n$.

I_0 is true by assumption and so assume I_{n-1}. By (15)

$$(24) \qquad Z_{n-1}^1 = Z_n^0 + \sum_{k=0}^{n-2} \binom{n-1}{k} [W_{k+1}, Z_{n-1-k}^0] + [W_n, Z_0^0].$$

The last term is singled out because it is the only term that contains an element, W_n, which is not covered either by the induction hypothesis or the hypothesis of the theorem. All the other terms are in $\mathcal{P}_n$ by I_{n-1} and iii). Thus

$$(25) \qquad\qquad Z_{n-1}^1 = K^1 + [W_n, Z_0^0]$$

where $K^1 \in \mathcal{P}_n$ is known. A simple induction on the columns of the Lie triangle using (15) shows that

$$(26) \qquad\qquad Z_{n-s}^s = K^s + [W_n, Z_0^0]$$

where $K^s \in \mathcal{P}_n$ for $s = 1, 2, \ldots, n$ and so

$$(27) \qquad\qquad Z_n^0 = K^n + [W_n, Z_0^0].$$

By iv) solve (27) for $W_n \in \mathcal{R}_n$ and $Z_0^n \in \mathcal{Q}_i$. Thus I_n is true. □

The theorem given above is formal in the sense that the convergence of the various series is not discussed. In interesting case the series diverge, but useful information can be obtained in the first few terms of the normal form. One can stop the process at any order, N, to obtain a W which is a polynomial in ε and so converges. From the proof given above it is clear that the terms in series for Z^* up to order N are unaffected by the termination.

REFERENCES

ABRAHAM, R. AND MARSDEN J.E. 1978:, *Foundations of Mechanics*, Benjamin/Cummings Publ. Co., Reading, Mass..

BIRKHOFF, G.D. 1927:, *Dynamical Systems*, Am. Math. Soc., Providence, R.I..

CUSHMAN, R., DEPRIT, A. AND MOSAK, R. 1983:, *Normal forms and representation theory*, J. Math. Phy. 24 (8), 2102–2116.

CUSHMAN, R., AND SANDERS, J. 1986:, *Nilpotent normal forms and representation theory of* $sl(2, \mathbf{R})$, Contemporary Mathematics 56, Amer. Math. Soc., Providence, R.I., 31–51.

DEPRIT, A. 1969:, *Canonical transformation depending on a small parameter*, Celestial Mechanics 72, 173–79.

DILIBERTO, S.P. 1961:, *Perturbation theorems for periodic systems*, Circ. Mat. Palermo 9 (2), 265–299, 10 (2), 111–112.

DILIBERTO, S.P. 1967:, *New results on periodic surfaces and the averaging principle*, Differential and Integral Equations, Benjamin, New York, 49–87.

FLANDERS, H. 1963:, *Differential Forms*, Academic Press, New York.

GANTMACHER, F.R. 1960:, *The Theory of Matrices*, Chelsea Publ., New York.

HENRARD J. 1970:, *On a perturbation theory using Lie transforms*, Celestial Mech. 3, 107–120.

HORI, G. 1966:, *Theory of general perturbations with unspecified canonical variables*, Publ. Astron. Soc. Japan 18 (4), 287–296.

KAMEL, A. 1970:, *Perturbation method in the theory of nonlinear oscillations*, Celestial Mechanics 3, 90–99.

KOBAYASHI, S. AND NOMIZU 1969:, K, *Foundations of Differential Geometry*, Interscience, New York.

KUMMER 1976:, M., *On resonant non-linear coupled oscillators with two equal frequencies*, Comm. Math. Phy. 48, 137–139.

MEYER, K.R. 1974:, K.R., *Normal forms for Hamiltonian systems*, Celestrial Mech. 9, 517–522.

MEYER, K.R. 1984:, *Normal forms for the general equilibrium*, Funkcialaj Ekv. 27 (2), 261–271.

MEYER, K.R. 1990:, *A Lie transform tutorial*, to appear in the Proceedings of the Conferences on Symbolic Computations at IBM, Yorktown Heights.

MEYER, K.R. AND SCHMIDT, D.S. 1977:, *Entrainment domains*, Funkcialaj Ekvacioj 20 (2), 171–92.

MEYER, K. R. AND SCHMIDT, D.S 1988:, *Bifurcations of relative equilibria in the N-body and Kirchhoff problems*, SIAM J. Math Anal. 19 (6), 1295–1313.

POSTON, T. AND STEWART, I. 1978:, *Catastrophe Theory and its Applications*, Pitman, Boston.

WEINSTEIN, A. 1971:, *Symplectic manifolds and their Lagrangian submanifolds*, Adv. Math. 6, 329–346.

INTERVAL TOOLS FOR COMPUTER AIDED PROOFS IN ANALYSIS

RAMON E. MOORE*

Abstract. A brief survey of theory and software implementations of interval and related techniques for computing with machine representable sets is presented with applications to computer aided proofs in analysis. Recent work on variable precision software is discussed.

1. Introduction. During the past three decades, a body of theory and computer software has evolved enabling computations with finitely representable sets. In the references are listed some 24 books dealing exclusively or partly with these techniques. There are a few other books in English, which have been omitted, with only brief discussions. New books are in progress and there are some in other languages (German, Russian, and Chinese, for example), also omitted. Only a few papers are listed, but bibliographies containing thousands of others are included.

New developments involving variable precision computation are discussed.

Fritz Krückeberg [27] has presented an excellent research plan for the development of a "three-layered methodology":

- computer algebra procedures
- numerical algorithms
- an interval arithmetic with variable and controllable word length.

He envisions feedback from levels 2 and 3 to earlier levels. To do this I would like to add the remark that, with interactive programming, a user can and will often want to get involved in such feedback. I would like to see the explicit development of more tools within programming languages to facilitate interactive computing. Automation is powerful and has its place, but it is, after all, only a tool to assist the human brain. A computer can be much more powerful than otherwise if the brain of a mathematician is allowed as one of the "peripherals". It can be argued that such a practice will slow things down intolerably. To this it may be countered that it is absurd to allow an expensive automated tool to run away at breakneck speed in some useless direction. Especially in exploratory computations such as those involved in preliminary designs with somewhat conflicting constraints or desiderata, it makes sense to invite the human designer to participate in decisions regarding reasonable compromises. The same goes for many "real time computations". In spite of airplane accidents often being attributed to "pilot error", I would have strong misgivings about embarking on a flight that was going to be on auto-pilot the whole trip including take-off and landing.

With a pair of machine numbers, we can represent an **interval** on the real line. With vectors of these, we can represent sets in finite-dimensional spaces, and with unions, more complicated sets. Using polynomials with interval coefficients, we can

---

*Department of Computer and Information Science, Ohio State University

represent on computers, sets in function spaces. Again we can deal with vectors of such sets. We can program a test for empty intersection of two such sets and and for **inclusion** of one such set in another. When two such sets have non-empty intersection, we can find the intersection.

We can extend the arithmetic operations of the real number field to arithmetic operations on intervals, using elementary properties of inequalities: $a \leq x \leq b$ and $c \leq y \leq d$ imply $a + c \leq x + y \leq b + d$, etc. See [5], [15], [17] for discussions of properties of **interval arithmetic**. It turns out that interval numbers no longer form a field; many of the nice algebraic properties of real numbers are lost in the extension. For example, distributivity, $x(y + z) = xy + xz$, is replaced by **sub-distributivity**, $X(Y + Z)$ is **contained** in $XY + XZ$. On the other hand, it turns out that we can compute, in a finite number of machine operations, intervals containing the **range of values** of programmable functions in this way, whether or not the functions are monotone. It is efficient, of course, to make use of monotonicity whenever possible.

In the complex plane, we can represent sets conveniently either by rectangles (vectors of intervals) [5] or by discs, or circular rings [22], with appropriate extensions of complex arithmetic for such sets.

Using directed rounding [5], [10], [15], [17], [24], we can be certain that machine-computed intervals really do contain the sets of all possible results of real (infinite-precision) arithmetic operations. Thus, we have a means of rigorously bounding the effects of rounding error in any machine computation.

We can extend integration to interval-valued functions [14], [15], [17], [28]. By these procedures the classical inequalities generalize to **inclusion** relations. Recall the completely general "sub-set property": for an **arbitrary** mapping, $f : X \to Y$, where X and Y are arbitrary sets, we have

$$Z \underline{C} X \quad \text{implies} \quad f(Z) \underline{C} f(x).$$

This property is enjoyed also by natural interval extensions of programmable functions and by their outwardly rounded computer implementations. It is also preserved by integration. It follows that the classical integral (and differential) inequalities can be generalized to inclusion relations for set-valued mappings with **no** assumptions about monotonicity of any kind for the operators [13], [17] and [28].

In keeping with Krückeberg's program, we should definitely include, among the **computer algebra** routines to made available in connection with interval algorithms: pre-conditioning transformations for linear systems [4], [5], [8], [17]; recursive automatic differentiation [15], [17], [23], [29]; special transformation (centered forms, etc.) for sharper bounds on ranges of values [8], [15], [17], and especially [25]; coordinate transformations to fight the effect of "wrapping error" in initial value problems in ordinary differential equations [7], [8], [15], [17], [20], [21], [22], [24]; in addition to any algebraic routines available for carrying out factorizations or symbolic cancellations to avoid or reduce loss of significant digits, e.g. through subtraction of nearly equal numbers in the computer.

2. Tools for what? What kind of computer aided proofs in analysis can we do with the sorts of interval tools mentioned in the introduction? These include at least the following.

2.1 Existence. We can program computer verification of sufficient conditions for existence theorems, for example via fixed point theorems involving continuous mappings of compact convex sets into themselves. Since we can compute, using interval methods, sets (such as multi-dimensional intervals) containing ranges of values of mappings, we can test whether such sets are contained in a given set of arguments. If so, we have a computer aided existence proof; for this and many other techniques for proving existence, see e.g. [3], [4], [5], [9], [11], [12], [13], [14], [15], [16], [17], [18], [19], [20], [21], [22], [24], [26], [27], [28].

2.2 Non-existence. By testing for empty intersections, for instance, a computer can determine that no solutions exist in a certain set. See the same set of references as in 2.1 above. As a simple example, if the computer finds, using interval computation or otherwise, that $f(X)$ does not contain 0, then f has no zero in the set X. As another example, interval Newton methods for nonlinear operator equations in finite or infinite dimensional spaces (see e.g. [4], [17], [19]) involve some mapping, say P, with the property that if X contains a solution then so does $P(X)$. If the computer finds that X has empty intersection with $P(X)$, then it has proved that there is no solution in X. This procedure is used very effectively in interval methods for **global optimization** [19], [26] and in other application areas as well. In optimization problems, tests for feasibility follow along similar lines, based on set inclusions verified by the computer.

2.3 Uniqueness and convergence. Computationally verifiable conditions for uniqueness and for convergence of iterative methods are also discussed extensively in the references given.

2.4 Finite termination. A general theory of finite termination of computer programs of certain types is discussed in [17]. Many iterative interval methods have the form $X_{n+1} = P(X_n)$, with a mapping P with the property that $P(X)$ still contains a fixed point x whenever X does. If we implement an outwardly rounded evaluation in fixed precision of P, say $\mathbf{P}$, then we can compute the sequence

$$\mathbf{X}_{n+1} = \mathbf{P}(\mathbf{X}_n) \ \Omega \ \mathbf{X}_n, \quad \text{(where } \Omega \text{ means intersection)}.$$

As long as we know by construction that $\mathbf{P}$ has the subset property (inclusion isotonicity: X contained in Y implies $\mathbf{P}(X)$ contained in $\mathbf{P}(Y)$), then, if we can find an $\mathbf{X}_0$ such that $\mathbf{P}(\mathbf{X}_0) \underline{C} \mathbf{X}_0$, it follows that the sequence $\{\mathbf{X}_n\}$ is nested and converges in a finite number of steps to an interval (or interval vector) which satisfies the computationally verifiable test: $\mathbf{X}_{n+1} = \mathbf{X}_n$. This is so because there is only a finite set of different machine numbers in fixed precision.

Search procedures for finding such an $\mathbf{X}_0$ are also discussed in [17], [26] based in part on exclusion tests for non-existence. We can start with a large region and throw away parts shown not to contain a solution or a fixed point or a global minimum as the case may be. It can be shown that such procedures will terminate in a finite number of steps with the information that the initial region contains no solution or else that there are solutions in certain computed sub-regions and perhaps still others in regions to be further analyzed with higher precision.

2.5 Computable rigorous bounds. Interval algorithms discussed in the references given can provide machine computable sets guaranteed to contain solutions for standard classes of problems: roots of polynomials, zeros of functions, solutions of linear systems and non-linear systems, values of definite integrals, solutions of initial and two-point boundary value problems for linear and nonlinear ordinary differential equations, certain types of problems in partial differential equations (much remains to be done in this area!), global optimization (see the recent excellent book [26]), and much else.

3. Dichotomy vs. trichotomy. When testing a relation such as: $[x \leq y?]$, when x and y are computed numbers, a simple yes or no answer may not be appropriate. If the algorithm involved supposes x and y to be produced by certain operations and the computer instead only approximates those operations, then a wrong logical path may be followed because of approximation error in x or y. Thus, for example, even the seemingly safe bisection method for finding real zeros of continuous functions can go wrong on a computer because of approximation error (for example round-off error) in evaluating a function. Such flaws in standard methods of computing can be avoided easily with outwardly rounded interval computation and three-valued logic.

If we compute intervals X and Y known to contain the (unknown) exact values of x and y, then we can test: $[X \leq Y?]$. There are now **three** possible outcomes. Suppose $X = [a, b]$ and $Y = [c, d]$. (1) If $b \leq c$, then certainly $[x \leq y]$ is true; (2) If $d < a$, then certainly $[x \leq y]$ is false; (3) Otherwise we **don't know** whether $[x \leq y]$ is true or false, so we must take other action depending on the situation–perhaps stop or perhaps repeat part or all of the computation with higher precision. In this way we can, for instance, make the bisection method completely rigorous.

Similarly, if for intervals or other sets, $X \underline{C} Y$ implies existence say, and an empty intersection of X and Y implies non-existence in a certain situation, then there is still a third possibility, namely that X and Y overlap without satisfying either of the first two conditions. In that case, we draw no conclusion but proceed to appropriate further analysis.

4. Fixed precision interval software. While a number of computer programs and software systems employing interval computation were written as long ago as 1959 or earlier, it is not until very recently that readily available portable software packages have appeared. We now have ARITH, a package of FORTRAN subroutines, and FORTRAN-SC (for Scientific Computation) [19] which run on IBM mainframes of recent vintage. FORTRAN-SC is a very high level language with many features intended to make it user friendly such as allowing vectors and matrices to be defined as data types and even user defined operators and data types. We also have PASCAL-SC [6] which runs on IBM and other PC's and is similarly powerful. Automatic differentiation routines are also available in PASCAL-SC.

These systems enjoy an additional keen advantage. They provide maximally accurate inner products at very high efficiency by use of long virtual accumulators and an algorithm of Bohlender [6], [10], [19]. This enables such software to obtain very high accuracy in the solution of even ill-conditioned linear systems for example.

These are very powerful programming languages for implementing the kinds of interval algorithms discussed in this paper and deserve extensive use.

5. Variable precision interval software. To my knowledge the first reported variable precision interval programs were those of F. Krückeberg [27], pages 95–101, in 1985, mentioned in the introduction of this paper. He reported results of guaranteed accuracy for a system of three differential equations, for example, to 50 and 70 decimal digits.

More recently, there has appeared the remarkable book of Oliver Aberth [3]. It comes complete with floppy disks for IBM-PC's. It allows up to 122 decimal digits of precision. The precision, number of digits carried, can be varied during the course of a computation. Aberth's implementation of interval arithmetic is aimed mainly at problems with precise inputs, so he used a form of interval arithmetic he calls "range arithmetic", carrying the midpoint of an interval represented by up to 122 decimal digits and a "range" represented by two decimal digits and indicating ± that amount in the last two digits carried.

Aberth's work [3] also includes exact computation with rational numbers with application to exact solution of linear systems with rational coefficients.

In range arithmetic with variable precision, he provides, in addition to the general programming language PBASIC, some problem solving routines all of which provide answers correct to the last decimal place displayed. These include: function evaluation, zeros of functions, solving systems of linear equations, finding real and complex zeros of polynomials, finding eigenvectors of symmetric square matrices, derivatives of functions, definite integrals, linear differential equations with constant coefficients, first and second order nonlinear differential equations.

The idea in all this variable precision work is to allow a user to specify the accuracy he wants and the computer gets it. And guarantees it. The work of Aberth is a giant step in that direction. To some extent he has already done it. I think what remains is to extend problem domains, and improve portability and efficiency and "user-friendliness". Of course portability and efficiency are somewhat conflicting goals. The closer to the hardware we get in our implementations the more efficient but the less portable and vice versa. Perhaps the best approach at the moment seems to be two versions: one very portable and the other very efficient (machine dependent, largely in assembly language or even microprogrammed).

Variable precision interval software [3] offers important new tools for those interested in computer aided proofs in analysis.

REFERENCES

[1]   G.F. CORLISS ET AL., *Bibliography on interval methods for ODEs*, Marquette University, Dept. of Math., Stat. & C.S., Tech. Rept. no. 289 (September, 1988).

[2] J. GARLOFF ET AL., *Interval Mathematics-a bibliography*, Freiburger Intervall-Berichte 85/6 and 87/2 (331 pp.).

[3] O. ABERTH, *Precise Numerical Analysis*, Wm. C. Brown Publ. (with variable-precision software on IBM-PC disks) (1988).

[4] G. ALEFELD AND R.D. GRIGORIEFF (EDS.), *Fundamentals of Numerical Computation (Computer-oriented Numerical Analysis)*, Computing Supplementum 2, Springer (1980).

[5] G. ALEFELD AND J. HERZBERGER, *Introduction to Interval Computations*, Academic Press (1983).

[6] G. BOHLENDER, C. ULLRICH, J. WOLFF VON GUDENBERG, AND L.B. RALL, *Pascal-SC. A Computer Language for Scientific Computation*, Academic Press (1987).

[7] J.W. DANIEL AND R.E. MOORE, *Computation and Theory in Ordinary Differential Equations*, Freeman (1970).

[8] E.R. HANSEN (ED.), *Topics in Interval Analysis*, Oxford U. Press (1969).

[9] E.W. KAUCHER AND W.L. MIRANKER, *Self-validating Numerics for Function Space Problems*, Academic Press (1984).

[10] U.W. KULISCH AND W.L. MIRANKER, *Computer Arithmetic in Theory and Practice*, Academic Press (1981).

[11] ——————————, *A New Approach to Scientific Computation*, Academic Press (1983).

[12] U.W. KULISCH AND H.J. SETTER (EDS.), *Scientific Computation with Automatic Result Verification*, Springer (1988).

[13] V. LAKSHMIKANTHAM, S. LEELA, AND A.A. MARTYNYUK, *Stability Analysis of Nonlinear Systems*, Marcel Dekker, Inc. (1988).

[14] R.E. MOORE, *Interval Arithmetic and Automatic Error Analysis in Digital Computing*, Applied Math. and Stat. Lab. Report No. 25 (1965).

[15] ——————————, *Interval Analysis*, Prentice-Hall (1966).

[16] ——————————, *Mathematical Elements of Scientific Computing*, Holt, Rinehart and Winston (1975).

[17] ——————————, *Methods and Applications of Interval Analysis*, SIAM Studies in Applied Mathematics (1979).

[18] ——————————, *Computational Functional Analysis*, Ellis Horwood and John Wiley (1985).

[19] —————————— (EDS), *Reliability in Computing. The Role of Interval Methods in Scientific Computing*, Academic Press (1988).

[20] K. NICKEL (ED.), *Interval Mathematics*, Lecture Notes in Computer Science, No. 29, Springer (1975).

[21] ——————————, *Interval Mathematics 1980*, Academic Press (1980).

[22] ——————————, *Interval Mathematics 1985*, Lect. Notes in C.S., Springer, No. 212 (1985).

[23] L.B. RALL, *Automatic Differentiation*, Lecture Notes in Computer Science, No. 120, Springer (1981).

[24] —————————— (EDS.), *Error in Digital Computation, Vol. I and II*, Wiley (1965).

[25] H. RATSCHEK AND J. ROKNE, *Computer Methods for the Range of Functions*, Ellis Horwood and John Wiley (1984).

[26] ——————————, *New Computer Methods for Global Optimization*, Ellis Horwood and John Wiley (1988).

[27] F. KRÜCKEBERG, *Arbitrary accuracy with variable precision arithmetic*, In K.Nickel (1985), pp. 95–101.

[28] R.E. MOORE, *Set-valued extensions of integral inequalities*, Journal of Integral Equations 5 (1983), pp. 187–198.

[29] R.D. NEIDINGER, *An efficient method for the numerical evaluation of partial derivatives of arbitrary order*, personal communication (March, 1989).

TOOLS FOR MATHEMATICAL COMPUTATION

L. B. RALL*

Abstract. Methodology for the validation of computation of values of functions using floating-point arithmetic is discussed and illustrated by an example.

1. Mathematical Computation. The term mathematical computation will be used here to denote areas of scientific computation in which goals include obtaining assertions about the validity of results in addition to the results themselves. This includes computations which are inherently exact, such as those with logical variables or integers (provided sufficient precision is available), and also symbolic computation. In the absence of hardware or software bugs, which usually make themselves manifest in one way or another, one can depend on the result of this kind of computation in much the same way as on the result of a carefully proved theorem. However, in the case of computations based on real numbers or other objects without finite representations, one is forced to work with approximations, and assertions of validity of results may not be easy to obtain. Attention will be devoted here to some helpful computational tools to assist with this problem.

Close examination of the problem of validation of numerical calculations shows that it goes beyond classical error analysis. Such analysis is tedious and usually has to consider worst cases, so gross overestimates of actual error are common. A better approach is to design the computation itself to produce the desired guarantees of validity of its results [7]. Progress in this direction turns out to involve interaction between computer arithmetics, programming languages, and mathematical algorithms. This is a wide-ranging topic, since computer arithmetics are intimately related to computer hardware itself, and mathematical algorithms suitable for computational validation of results are still in the research stage in certain areas. Programming languages occupy a central position in this scheme. On one hand, it should be easy to program the chosen algorithm correctly in a way accessible to others. On the other, the programming language should permit direct access to features of the computer arithmetic necessary for validation (such as directed rounding, high-precision scalar products and standard functions, for example). While considerable progress has been made to date, there is room for advancement on all fronts. Some directions will be indicated below.

2. Maximum quality computer arithmetic. In the past, analysis and implementations of computer arithmetics has tended to focus on the details of representation of floating-point numbers, such as their radix, precision (number of digits), and exponent range. A breakthrough was made in this area by an axiomatic formulation of computer arithmetic which is independent of these details [6],[8]. If the arithmetic unit of the computer is built in conformance with these axioms, then

*Department of Mathematics, University of Wisconsin-Madison, Madison, Wisconsin 53706

its behavior can be deduced. In the case of real arithmetic, these axioms yield a floating-point computer arithmetic of maximum quality, as defined below.

Given the real numbers $\mathbf{R}$ and a finite set of floating-point numbers $\mathbf{F}$, a floating-point number $\xi \in \mathbf{F}$ is said to be an approximation of *maximum quality* to a real number $x \in \mathbf{R}$, if there is no floating-point number η between ξ and x. The concept of maximum quality is independent of the actual construction of the floating-point numbers being used, while the more usual idea of the *accuracy* of ξ as an approximation to x is determined by the spacing of the floating-point grid. Thus, while real numbers cannot in general be represented exactly on a computer, there always exists a floating-point approximation of maximum quality to a given real number. Furthermore, real numbers between the least and greatest elements of $\mathbf{F}$ (assumed to be the case in what follows) always can be enclosed by an interval with floating-point endpoints. Such an interval enclosure X of a real number $x \in X$ is said to be of *maximum quality* if the interior of X contains at most one floating-point number. If $\text{int}(X) \cap \mathbf{F} = \emptyset$, then each endpoint of X is an approximation of maximum quality to x, otherwise, $\eta \in \text{int}(X)$ is an approximation of maximum quality to x.

The problem of validation would be solved if it were possible to compute a maximum quality floating-point approximation to the exact answer for each problem, or an interval inclusion of maximum quality. While this is probably not a reasonable goal for most problems, it certainly can and should be a minimum standard for floating-point computer arithmetics, which will be considered to consist of arithmetic operations and the set Φ of standard functions defined in the programming language used. For example, in Pascal and Pascal-SC [16], one has arithmetic operators $+, -, *, /$, and the set of standard functions

$$(2.1) \qquad \Phi = \{\text{abs,sqr,sqrt,exp,ln,sin,cos,arctan}\}.$$

Of course, these sets can be augmented by additional useful operators and functions in actual implementations or other languages.

The axiomatic approach to computer arithmetic is based on requirements which a mapping $\square : \mathbf{R} \to \mathbf{F}$ (naturally called a *rounding*) must satisfy. It follows from

$$(1) \quad \square\xi = \xi \qquad \text{and} \qquad (2) \quad x \leq y \Rightarrow \square x \leq \square y$$

for all $\xi \in \mathbf{F}$ and $x, y \in \mathbf{R}$ that the rounding $\square$ is of *maximum quality*, that is, there is no floating-point number η between the real number x and its rounded value $\xi = \square x$. Maximum quality floating-point arithmetic operations $\boxed{\circ}$ are defined in $\mathbf{F}$ by

$$(3) \quad \xi \boxed{\circ} \eta = \square(\xi \circ \eta), \qquad \circ \in \{+, -, *, /\},$$

division by 0 of course being excluded, where $\xi \circ \eta$ denotes the *exact* result in $\mathbf{R}$. In actual practice, operations overflowing the range of real values represented in $\mathbf{F}$ are also undefined. The value of $\xi \circ \eta$ does not have to be computed exactly to implement this axiom, all that is required is an approximation which can be guaranteed to be of maximum quality, or a maximum quality interval inclusion of the exact result. If the final axiom

$$(4) \quad \square(-x) = -(\square x)$$

is satisfied for all $x \in \mathbf{R}$, then the mapping $\square$ is said to be a *semimorphism*. Axiom (4) also defines the unary *negation operator* $\boxminus$ in $\mathbf{F}$. (It is assumed that $\mathbf{F}$ contains $-\xi$ for all $\xi \in \mathbf{F}$.)

Computer arithmetics satisfying these axioms are not hard to implement in hardware (or emulate in software with less efficiency). For example, rounding to the closest floating-point number toward or away from 0 is satisfactory, as is rounding to the closest floating-point number if (4) is used to break ties. Implementations of standard functions should also be of maximum quality, which means that the floating-point standard functions $\square \phi$ should satisfy

$$(2.2) \qquad (\square \phi)(\xi) = \square \left(\phi(\xi) \right)$$

for all $\phi \in \Phi$ and $\xi \in \mathbf{F}$, where $\phi(\xi)$ again denotes the exact value in $\mathbf{R}$. Pascal-SC [2] requires maximum quality arithmetic operations and standard functions, but most current languages require neither. An exception is Ada, which requires maximum quality only for arithmetic operations. In order to provide users with consistent, dependable numerical results, standards for programming languages should include requirements of maximum quality for arithmetic operations and standard functions. Furthermore, the same holds for implementations of language extensions which include additional arithmetic operations or standard functions.

3. Directed rounding and interval arithmetic. One way to validate a numerical computation is to obtain a floating-point interval which can be guaranteed to contain the exact result. Even if the inclusion is of not of maximum quality, the interval may be narrow enough to show that some floating-point number which it contains approximates the exact result with sufficient accuracy. On the other hand, an interval inclusion may be too wide initially to be useful. In this case, it may be possible to reduce its width by subsequent calculations. Since its introduction by R. E. Moore for this purpose [9], [10], interval arithmetic has been one of the fundamental tools of validation of numerical computation (see also [1]).

The concept of maximum quality inclusion of a real interval $X = [x, y] \in \mathbf{R}$ by a floating-point interval $\Xi = [\xi, \eta] \in \mathbf{F}$ is similar to the idea developed for the real case. Here, ξ is the approximation of maximum quality to x such that $\xi \leq x$, and η is the maximum quality approximation to y such that $y \leq \eta$. Hence, there are no floating-point numbers between ξ and x, and none between y and η. In other words, Ξ is the smallest floating-point interval which contains X or, stated differently, Ξ is the intersection of all floating-point intervals Υ such that $X \subseteq \Upsilon$.

Maximum quality floating-point interval arithmetic can be implemented using real floating-point arithmetic with *directed roundings* [10]. These roundings are denoted respectively by ∇ (downward) with $\nabla x \leq x$ and Δ (upward), for which $x \leq \Delta x$. If (1) and (2) are satisfied, then ∇, Δ will be of maximum quality. The corresponding arithmetic operations $\triangledown, \blacktriangle$ with directed rounding are defined by (3) for $\circ \in \{+, -, *, /\}$. Instead of (4), one has

$$(3.1) \qquad \nabla(-x) = -(\Delta x), \qquad \Delta(-x) = -(\nabla x),$$

for all $x \in \mathbf{R}$.

In addition to forming the basis for floating-point interval arithmetic, directed rounding can be used to compute guaranteed lower or upper bounds for quantities of interest. This could be crucial to the verification of hypotheses of theorems, for example. However, implementation of directed rounding in the computer arithmetic is not sufficient for these purposes unless the programming language permits easy access to this capability. This is another failure of most current languages. Pascal-SC, however, provides the operator symbols $+ <, - <, * <, / <$ for the corresponding downwardly rounded arithmetic operations, and $+ >, - >, * >, / >$ for upward rounding, with the ordinary symbols $+, -, *, /$ representing rounding to nearest. Thus, programming languages should have an adequate set of operator symbols and provision for introducing others and "overloading" the meanings of existing symbols. For example, it should be possible to use "+" to denote addition of integers, floating-point real and complex numbers, intervals, vectors, matrices, and so on. Pascal-SC and Ada are examples of languages in which introduction of operators and overloading of operator symbols are possible. Rounding to nearest, downward, and upward actually yield twelve different arithmetic operations on floating-point numbers. If maximum quality rounding toward and away from 0 is also needed, then the number of distinct arithmetic operations increases to twenty. The important point is that the programmer should have convenient access to whatever provided operations are significant to the computation being done.

A useful operation based on directed rounding is the *outward rounding* $\Diamond$ of real numbers and intervals to floating-point intervals defined respectively by

(3.2) $$\Diamond x = \Diamond[x, x] = [\nabla x, \Delta x], \qquad \Diamond[x, y] = [\nabla x, \Delta y].$$

4. Maximum quality scalar products. A cornerstone of the theory of computer arithmetic developed by Kulisch and Miranker [6], [8] is the maximum quality scalar product

(4.1) $$\mathbf{u} \boxdot \mathbf{v} = \Box(\mathbf{u} \cdot \mathbf{v}) = \Box\left(\sum_{i=1}^{n} u_i * v_i\right)$$

of floating point vectors $\mathbf{u} = (u_1, u_2, \dots, u_n)$ and $\mathbf{v} = (v_1, v_2, \dots, v_n)$ of arbitrary length. As before, $(\mathbf{u} \cdot \mathbf{v})$ denotes the *exact* real value of the product. There are several ways to implement this evaluation, the simplest apparently being the provision of an accumulator sufficiently long to hold all digits of the sum of products in (4.1) [8]. The actual length of this accumulator depends on the number of digits in the mantissa of the floating-point numbers used, and on the exponent range, but is not excessively large. For example, about 400 digits suffice for the twelve decimal digit floating-point numbers and the exponent range of -99 to $+99$ used in the PC implementation of Pascal-SC [4].

The scalar product (4.1), introduced to permit maximum quality vector and matrix floating-point arithmetic, has many other uses. For example, addition of a sequence of floating-point numbers $v_1, v_2, \dots, v_n$ can be done with maximum quality by computing the scalar product of the floating-point vector $\mathbf{v}$ with the summands as components and the vector $\mathbf{u} = (1, 1, \ldots, 1)$. Ordinarily, floating-point addition is notoriously nonassociative, for example, one can rearrange the order of addition of $-10^{50}, 1, 2, 3, 4, 10^{50}$ to obtain any integer from 0 to 10. The use of the maximum scalar product gives the correct answer in this case without having to order the components of $\mathbf{v}$.

Another use of the maximum quality scalar product is the accurate solution of linear systems of equations $Ax = b$ by the method of *iterative refinement*. Supposing that x_n is an approximate solution of this system and R is an approximation to A^{-1} (assumed to exist), then the calculation

$$(4.2) \qquad x_{n+1} - x_n = R(b - Ax_n)$$

gives a better approximation if R is sufficiently good. In ordinary floating-point arithmetic, a difficulty in application of this method is loss of accuracy in the calculation of the *residual* $r_n = b - Ax_n$, which does not occur if the maximum quality scalar product (4.1) is used.

For the purpose of validation (that is to say, obtaining interval inclusions of the exact results), the maximum quality scalar products with directed rounding,

$$(4.3)$$

$$\mathbf{u} \bigtriangledown \mathbf{v} = \nabla(\mathbf{u} \cdot \mathbf{v}) = \nabla\left(\sum_{i=1}^{n} u_i * v_i\right), \qquad \mathbf{u} \bigtriangleup \mathbf{v} = \Delta(\mathbf{u} \cdot \mathbf{v}) = \Delta\left(\sum_{i=1}^{n} u_i * v_i\right),$$

are powerful tools for linear and nonlinear problems ([17], also [7], pp. 51–120). The operators (4.3) can be used to construct floating-point interval matrix-vector arithmetic of maximum quality. As an application, suppose that x_n is an approximation to the solution x of a linear system $Ax = b$, and Γ is an interval matrix which contains A^{-1}. It follows that

$$(4.4) \qquad x \in x_n + \Gamma\Diamond(b - Ax_n).$$

As well as validating the accuracy of an approximation x_n to x, interval methods can be used to verify the existence of A^{-1} computationally, which in turn implies the existence of a unique solution x of the linear system ([7], pp. 51–120). This in turn applies to computational problems which can be reduced to the solution of linear systems, such as the evaluation of polynomials. In most cases, inclusions of maximum quality can be obtained by use of interval iteration. Utility subroutines for problems of this kind are included in Pascal-SC software [4], [5].

5. Code list representation of functions. In this section, representation of functions by code lists [12] will be introduced for computational purposes including the validation of function evaluation. For example, the function $f(x, y)$ represented by the formula

$$(5.1) \qquad f = (xy + \sin x + 4)(3y^2 + 6)$$

is also represented by the code list

$$
\begin{aligned}
t_1 &= x * y, \\
t_2 &= \sin(x), \\
t_3 &= t_1 + t_2, \\
t_4 &= t_3 + 4, \\
t_5 &= \mathrm{sqr}(y), \\
t_6 &= 3 * t_5, \\
t_7 &= t_6 + 6, \\
t_8 &= t_4 * t_7.
\end{aligned}
$$

(5.2)

Neither representation is unique.

Code lists can be used to represent functions in a general computer arithmetic $\mathcal{A}$ based on a set of elements E, unary operations U, and binary operations B. Given a set of *inputs* $I = \{i_1, i_2, \dots, i_\delta\} \subset E$, a *code list* $t = (t_1, t_2, \dots, t_n)$ is a finite sequence of *terms* t_k, each of which is of the form

(5.3) $\qquad t_k = u(a), \quad u \in U, \quad a \in I \cup \{t_1, t_2, \dots, t_{k-1}\},$

or

(5.4) $\qquad t_k = a_1 \circ a_2, \quad \circ \in B, \quad a_1, a_2 \in I \cup \{t_1, t_2 \dots, t_{k-1}\}.$

In other words, each argument of the operation which defines t_k is required to be an input or a previous term $t_1, t_2, \dots, t_{k-1}$ of the code list.

If all terms of the code list t are defined, then its final term t_n is said to be the *value* of the function $f : I \to E$ represented by the code list t, that is,

(5.5) $\qquad t_n = f(i_1, i_2, \dots, i_\gamma).$

This value will also be called the *output* of the code list.

The set I of inputs is often taken to consist of *variables* $V = \{x_1, x_2, \dots, x_\alpha\}$, *constants* $C = \{c_1, c_2, \dots, c_\beta\}$, and *parameters* $P = \{p_1, p_2, \dots, p_\gamma\}$ with $\alpha + \beta + \gamma = \delta$. In this context, the dependence of the function f on constants and parameters is usually suppressed, and one writes $t_n = f(x_1, x_2, \dots, x_\alpha)$.

For the present purpose, the U of unary operations will consist of unary $+, -$ and the set (2.1) of standard functions ϕ, and the binary operations will be the arithmetic operations $+, -, *, /$. The code list can be evaluated in any computer arithmetic $\mathcal{A}$ in which these operations are defined. For example, (5.2) could be evaluated in complex or interval arithmetic just as well as in real arithmetic. The result would be the value of the complex extension or an interval inclusion of the corresponding real function, assuming of course that all terms of the code list remain defined. The same observation applies to the corresponding floating-point arithmetics. The result of evaluation of a code list thus depends on the *type* of elements E on which the arithmetic is based, and the definitions of the operations involved.

In real floating-point arithmetic, the quality of the output of a code list is generally unknown, even if each operation and function is evaluated with maximum quality. Thus, even maximum quality floating-point arithmetic is inadequate for the purpose of validation. A simple and direct approach to determination of the quality of the output is to evaluate the code list in floating-point interval arithmetic, with the inputs replaced by the corresponding point intervals, for example $X = [x, x]$ and $Y = [y, y]$ in the case of (5.2). The results of real and interval floating-point evaluation of (5.2) for $x = 1.556$, $y = 9.87654321098$ are given below:

Term	Computed Value	Interval Inclusion
t_1	$1.53679012363E + 01$	$[1.53679012362E + 01, 1.53679012363E + 01]$
t_2	$9.99890536354E - 01$	$[9.99890536353E - 01, 9.99890536354E - 01]$
t_3	$1.63677917727E + 01$	$[1.63677917725E + 01, 1.63677917727E + 01]$
t_4	$2.03677917727E + 01$	$[2.03677917725E + 01, 2.03677917727E + 01]$
t_5	$9.75461057984E + 01$	$[9.75461057983E + 01, 9.75461057984E + 01]$
t_6	$2.92638317395E + 02$	$[2.92638317394E + 02, 2.92638317396E + 02]$
t_7	$2.98638317395E + 02$	$[2.98638317394E + 02, 2.98638317396E + 02]$
t_8	$6.08260306405E + 03$	$[6.08260306397E + 03, 6.08260306408E + 03]$

This computation, which was done with the PC implementation of Pascal-SC [4] shows that the approximation 6082.60306405 to $f(x, y)$ is of very high quality, since there are only 10 floating-point numbers in the interior of the interval inclusion $[6082.0306397, 6082.0306408]$ of the exact value of $f(x, y)$, and at most 7 between the computed and exact values. While this quality would ordinarily be considered satisfactory for practical purposes, it is not maximum. Furthermore, straightforward interval evaluation of code lists in other cases may yield interval inclusions of the exact values which are too wide to be useful. Thus, a method for the reduction of width of these interval inclusion to maximum or at least sufficiently high quality is needed.

6. Validation of code list evaluation. The improvement of the quality of interval inclusions of functions represented by code lists is based on an idea similar to the iterative refinement method (4.2) for linear systems. The code list (5.2) can be rewritten as the nonlinear system of equations

$$
\begin{aligned}
t_1 - x * y &= 0, \\
t_2 - \sin(x) &= 0, \\
t_3 - (t_1 + t_2) &= 0, \\
t_4 - (t_3 + 4) &= 0, \\
t_5 - \mathrm{sqr}(y) &= 0, \\
t_6 - 3 * t_5 &= 0, \\
t_7 - (t_6 + 6) &= 0, \\
t_8 - t_4 * t_7 &= 0,
\end{aligned}
$$

(6.1)

for $t_1, t_2, \ldots, t_8$. Similarly, a general code list can be transformed into a nonlinear system $\theta(t) = 0$ for $t = (t_1, t_2, \ldots, t_n)$. This suggests the use of Newton's method

$$(6.2) \qquad t^{(k+1)} - t^{(k)} = -\theta'(t^{(k)})^{-1}\theta(t^{(k)})$$

to obtain improved approximations $t^{(1)}, t^{(2)}, \ldots$ to the exact solution t of $\theta(t) = 0$, starting from the floating-point evaluation of the code list as the initial approximation $t^{(0)}$ [12].

Because of the structure of a code list, the application of Newton's method to this problem is much easier than in the general case. First of all, the Jacobian matrix $\theta'(t)$ is of the form $\theta'(t) = I - L(t)$, where $L(t)$ is strictly lower-triangular. Hence if $L(t)$ is defined, then

$$(6.3) \qquad \theta'(t)^{-1} = I + L(t) + \cdots + L^{n-1}(t)$$

exists, and any solution of $\theta(t) = 0$ is unique, as one expects. The matrix $L(t)$ is also *sparse*, with no more than two nonzero entries in any row, and can be computed automatically because the partial derivatives of the standard functions and arithmetic operations are known [13]. The Jacobian matrix of the system (6.1), for example, is

$$(6.4) \qquad \theta'(t) = \begin{bmatrix} 1 & 0 & 0 & 0 & 0 & 0 & 0 & 0 \\ 0 & 1 & 0 & 0 & 0 & 0 & 0 & 0 \\ -1 & -1 & 1 & 0 & 0 & 0 & 0 & 0 \\ 0 & 0 & -1 & 1 & 0 & 0 & 0 & 0 \\ 0 & 0 & 0 & 0 & 1 & 0 & 0 & 0 \\ 0 & 0 & 0 & 0 & -3 & 1 & 0 & 0 \\ 0 & 0 & 0 & 0 & 0 & -1 & 1 & 0 \\ 0 & 0 & 0 & -t_7 & 0 & 0 & -t_4 & 1 \end{bmatrix},$$

which is very easy to compute. Furthermore, as in the example above, the computed value $t^{(0)}$ of t is often a good approximation from which to start the iteration. However, as in the case of iterative refinement for linear systems, the difficulty in the application of (6.2) lies in the sufficiently accurate computation of the residual, which in this case is the value of $\theta(t^{(k)})$. The accuracy of the approximation $t^{(k+1)}$ thus cannot be validated without further information.

The interval evaluation of the code list does produce an initial interval $T^{(0)}$ which is guaranteed to contain the exact value of t. By the interval mean-value theorem [14], if $t \in T$, and $\bar{t} \in T$ is an approximation to t, then

$$(6.5) \qquad t - \bar{t} \in \theta'(\bar{t})^{-1}\{-\theta(\bar{t}) + (\theta'(\bar{t}) - \theta'(T))(T - \bar{t})\}.$$

It follows that if $\theta'(t^{(k)})^{-1} \in \Gamma_k$, $\theta(t^{(k)}) \in \Theta(t^{(k)})$, $\theta'(t^{(k)}) \in \Theta'(t^{(k)})$, and $\theta(T^{(k)}) \subseteq \Theta(T^{(k)})$, then

$$(6.6) \qquad t \in t^{(k)} + \Gamma_k\{-\Theta(t^{(k)}) + (\Theta'(t^{(k)}) - \Theta'(T^{(k)}))(T^{(k)} - t^{(k)})\},$$

which provides validation for the accuracy of $t^{(k)}$ as an approximation to t. If the width of the interval $T^{(k)}$ is small, then the width of the interval on the right of (6.6) will depend crucially on the width of the interval inclusion $\Theta(t^{(k)})$ of the residual $\theta(t^{(k)})$, which should be as small as possible [18]. For this purpose, a new computer arithmetic called interval residual arithmetic will be introduced.

7. Interval residual arithmetic. A computer arithmetic will now be defined for the set of elements E which consists of ordered pairs $\rho = (r, R)$, where r is a real floating-point number, and R is a real floating-point interval. The arithmetic operations are defined by

$$(7.1) \qquad \rho \circ \sigma = (r, R) \circ (s, S) = \left(r \boxdot s, \Diamond(r \boxdot s - r \circ s) \right), \qquad \circ \in \{+, -, *, /\},$$

where $r \boxdot s$ is the floating-point and $r \circ s$ the exact result of the indicated operation, with the outward rounding $\Diamond$ to whatever precision is being used. Similary, for standard functions $\phi \in \Phi$,

$$(7.2) \qquad \phi(\rho) = \phi((r, R)) = \left((\Box \phi)(r), \Diamond((\Box \phi(r) - \phi(r)) \right).$$

Unary $+, -$ (and hence the function abs) are exact for floating-point arguments, $+\rho = +(r, R) = (r, [0, 0])$ and $-\rho = -(r, R) = (-r, [0, 0])$. This arithmetic will be called *interval residual arithmetic*. Note that the results of operations and functions in this arithmetic depend only on the first components of the arguments.

It follows from the definition of this arithmetic that the evaluation of a function defined by a code list results in the ordered pairs $\left(\tau_i, \Diamond(\tau_i - t_i) \right) = (\tau_i, T_i)$ for $i = 1, 2, \ldots, n$, where τ_i is the computed floating-point value of the exact result t_i, and T_i is thus an interval enclosure of the residual for each term. In the notation of §6, one has

$$(7.3) \qquad \theta(t) \in \Theta(t) = (T_1, T_2, \ldots, T_n),$$

which provides an interval inclusion of the residual.

For example, evaluation of the code list (5.2) in interval residual arithmetic with the arguments $x = 1.556$, $y = 9.87654321098$ of §5 gives

Term	Computed Value τ_i	Interval Residual $\Diamond(\tau_i - t_i)$
t_1	$1.53679012363E + 01$	$[\ 1.51200000000E - 11,\ \ 1.51200000000E - 11]$
t_2	$9.99890536354E - 01$	$[\ 2.04084613230E - 13,\ \ 2.04084613250E - 13]$
t_3	$1.63677917727E + 01$	$[\ 4.60000000000E - 11,\ \ 4.60000000000E - 11]$
t_4	$2.03677917727E + 01$	$[\ 0.00000000000E + 00,\ \ 0.00000000000E + 00]$
t_5	$9.75461057984E + 01$	$[\ 4.48712074396E - 11,\ \ 4.48712074396E - 11]$
t_6	$2.92638317395E + 02$	$[-2.00000000000E - 10, -2.00000000000E - 10]$
t_7	$2.98638317395E + 02$	$[\ 0.00000000000E + 00,\ \ 0.00000000000E + 00]$
t_8	$6.08260306405E + 03$	$[-8.52296116500E - 10, -8.52296116500E - 10]$

Evaluation of (6.6) using these values and the ones in §5 as initial values gives the following maximum quality interval inclusions of the exact values of the terms of the code list:

Term	Interval Inclusion
t_1	$[1.53679012362E+01, 1.53679012363E+01]$
t_2	$[9.99890536353E-01, 9.99890536354E-01]$
t_3	$[1.63677917726E+01, 1.63677917727E+01]$
t_4	$[2.03677917726E+01, 2.03677917727E+01]$
t_5	$[9.75461057983E+01, 9.75461057984E+01]$
t_6	$[2.92638317395E+02, 2.92638317396E+02]$
t_7	$[2.98638317395E+02, 2.98638317396E+02]$
t_8	$[6.08260306403E+03, 6.08260306404E+03]$

Thus, the maximum quality interval inclusion

$$(7.4) \qquad f(1.556, 9.87654321098) \in [6082.60306403, 6082.60306404]$$

in twelve-digit decimal floating-point arithmetic was obtained in one step. The computation was simple in this case, since the interval matrix

$$(7.5) \qquad M = \Theta'(t^{(0)}) - \Theta'(T^{(0)})$$

has only two nonzero components, $M_{84} = [-10^{-9}, 10^{-9}]$ and $M_{87} = [-2 \times 10^{-10}, 0]$, and the interval vector $Z = M(T^{(0)} - t^{(0)})$ has only one nonzero component, $Z_8 = [-4 \times 10^{-19}, 4 \times 10^{-19}]$. Again, the PC implementation of Pascal-SC was used. Of course, if the first application of (6.6) does not yield maximum quality, then interval iteration can be used to reduce the width of this interval to its minimum value, which will give an inclusion of maximum or at least best possible quality [10].

The maximum quality scalar product with directed rounding can be used to implement the arithmetic operations of interval residual arithmetic, and the standard functions sqr,sqrt. For example, for $\rho * \sigma$, one has

$$(7.6) \qquad \Diamond(r \boxtimes s - r * s) \subseteq \left[\mathbf{u} \triangledown \mathbf{v}, \mathbf{u} \triangle \mathbf{v} \right],$$

where $\mathbf{u} = (r \boxtimes s, r)$ and $\mathbf{v} = (1, -s)$. Implementation of the transcendental standard functions exp,ln,cos,sin,arctan requires higher precision versions of the corresponding real functions. These are not provided in the PC implementation of Pascal-SC, so a value of $\sin(1.556)$ calculated to 23 significant digits was used in the above example. The ST implementation of Pascal-SC [5] does provide standard functions with slightly more precision on demand.

8. Conclusions. From the above, it appears that validation of computations done with floating-point approximations requires a number of capabilities. First of all, maximum quality floating-point arithmetic with directed rounding, and hence maximum quality interval arithmetic is needed, as well as maximum quality scalar products and standard functions with directed rounding. In addition, standard functions should be available on demand with specified levels of higher precision. For greatest efficiency, the corresponding computer arithmetics should be implemented as much as possible in the hardware of the computer.

Programming languages should have function and operator notation which allow programs to be easily written and understood. For example, it should be possible to evaluate an expression in different arithmetics by changing the types of the arguments instead of rewriting it as a sequence of subroutine calls. Pascal-SC has this capability, but Pascal does not [16]. Furthermore, one should be able to introduce more or less exotic new arithmetics such as interval residual arithmetic or differentiation arithmetics [15] (see also [7], pp. 291–309) by defining the corresponding data types and overloading the symbols for arithmetic operations and standard function identifiers. If this is to be done with maximum quality, then the programming language has to provide access to the needed features of the computer arithmetic, such as directed rounding, scalar products, and high-precision standard functions.

Mathematical algorithms should be chosen which are capable of producing guaranteed interval inclusions of results, with the possibility of improvement by iteration or other additional computation. An example is the interval mean-value theorem (6.6) used above. Another is the method of self-validating numerical integration which uses interval remainder terms [3]. There are already a number of such algorithms for important numerical problems, such as optimization [11], but others need to be developed.

REFERENCES

[1] G. ALEFELD AND J. HERZBERGER, *Introduction to Interval Computations*, Academic Press, 1983.

[2] G. BOHLENDER, C. ULLRICH, J. WOLFF VON GUDENBERG, AND L. B. RALL, *Pascal-SC: A Computer Language for Scientific Computation*, Academic Press, 1987.

[3] GEORGE F. CORLISS AND L. B. RALL, *Adaptive, self-validating numerical quadrature*, SIAM J. Scientific and Statistical Computing, 8 (1987), 831–847 ().

[4] U. KULISCH (ED.), *Pascal-SC for the IBM PC*, B. G. Teubner, 1987.

[5] U. KULISCH (ED.), *Pascal-SC for the Atari ST*, B. G. Teubner, 1987.

[6] U. W. KULISCH AND W. L. MIRANKER, *Computer Arithmetic in Theory and Practice*, Academic Press, 1981.

[7] U. W. KULISCH AND W. L. MIRANKER (EDS.), *A New Approach to Scientific Computation*, Academic Press, 1983.

[8] U. W. KULISCH AND W. L. MIRANKER, *The arithmetic of the digital computer: A new approach*, SIAM Review, 28, (1986), 1–40.

[9] R. E. MOORE, *Interval Analysis*, Prentice-Hall, 1966.

[10] R. E. MOORE, *Methods and Applications of Interval Analysis*, Society for Industrial and Applied Mathematics, 1979.

[11] R. E. MOORE (ED.), *Reliability in Computing*, Academic Press, 1988.

[12] L. B. RALL, *Computational Solution of Nonlinear Operator Equations*, Wiley, 1969.

[13] L. B. RALL, *Automatic Differentiation: Techniques and Applications*, Lecture Notes in Computer Science No. 120, Springer, 1981.

[14] L. B. RALL, *Mean value and Taylor forms in interval analysis*, SIAM J. Math. Anal., 14 (1983) 223–238.

[15] L. B. RALL, *The arithmetic of differentiation*, Math. Mag., 59, (1986), 275–282.

[16] L. B. RALL, *Pascal and Pascal-SC*, in *Encyclopedia of Physical Science and Technology*, Vol. 10, pp. 183–209, Academic Press, 1987.

[17] S. M. RUMP, *Solution of linear and nonlinear algebraic problems with sharp, guaranteed bounds*, Computing, Suppl. 5 (1984), 147–168.

[18] H. J. STETTER, *Intervals revisited*, Herrn Professor Dr. Karl Nickel zum 60. Geburtstag gewidmet, Vol. 2, pp. 519–538, University of Freiburg i. Br., 1984.

SHADOWING TRAJECTORIES OF DYNAMICAL SYSTEMS

TIM SAUER* AND JAMES A. YORKE†

Abstract. Computer simulation of the trajectories of deterministic systems involve truncation and rounding errors. A theorem is presented which describes a computer-assisted method for checking whether there is a true trajectory near the computer-generated one.

1. Introduction. One of the characteristic properties of a chaotic dynamical system is the existence of trajectories which are sensitive to initial conditions. This means that nearby initial conditions generate trajectories that exponentially diverge from each other. There is a resulting serious effect on the computer simulation of the dynamical system. If a small error is made in the initial condition of the trajectory to be computed, or at any step during the computation, the error will tend to be magnified by future evolution of the system.

This leads to a significant problem of information loss in computer simulations. If a finite-precision computer produces a trajectory from a deterministic model, making small rounding and truncation errors at each step, is there any assurance that the computed trajectory will have any relation to a true trajectory of the model?

It turns out that the question can be answered affirmatively in many cases of interest. We state a theorem which says that if certain quantities evaluated at points of the computer-generated trajectory, called a *pseudo-trajectory* or *pseudo-orbit*, are not too large, then there exists a true trajectory near the computer-generated one. Rigorous upper bounds for these quantities can be generated by the computer as it produces the pseudo-trajectory. If these quantities satisfy the hypotheses of the theorem, which again can be rigorously checked by the computer, the result is a computer-assisted proof of the existence of a true trajectory near the computer-generated pseudo-trajectory.

A true orbit that stays near the pseudo-orbit is said to *shadow* the pseudo-orbit. Bowen [1] proved a shadowing result for hyperbolic maps on a differential manifold. That result says that given any prescribed shadowing distance ε (between the pseudo-orbit and true orbit) there exists a $\delta > 0$ so that any δ-pseudo-orbit can be ε-shadowed by a true orbit.

There are two factors that make Bowen's proof impractical for use in computer experiments. First, the δ that is produced can be orders of magnitude smaller than the machine epsilon of existing digital computers. Second, many interesting dynamical systems currently being studied are not hyperbolic.

In ([2], [3]) a method is developed which uses a type of interval arithmetic to create computer-assisted proofs of the existence of finite-length shadowing orbits

---

*Department of Mathematical Sciences, George Mason University, Fairfax, VA 22030.

†Institute for Physical Sciences and Technology, University of Maryland, College Park, MD 20742.

on a case–by–case basis. They apply the method to one-dimensional maps and the two-dimensional Henon and Ikeda maps, none of which are hyperbolic. A key technique of that paper is a method for refining pseudo-orbits.

We use that technique in the present approach, but instead of using it explicitly on the computer-generated pseudo-orbit, we use it in the proof of the main shadowing theorem. The hypotheses of the theorem guarantee that iterative refinement of the pseudo-orbits would converge to a true orbit nearby. Bowen's shadowing theorem for hyperbolic maps follows as a corollary.

It is not necessary to actually carry out any refinements on the computer, although this may be done if desired, to the extent that higher precision is available on the computer. It was the lack of such extra precision that motivated the sublimation of the explicit refinement process into the proof of the theorem.

2. Shadowing theorem. The theorem can be used to shadow orbits of diffeomorphisms or trajectories of differential equations. In the latter case, the flow of the differential equation maps a point in phase space at time t to a point at time $t + h$. This map, called the time-h map of the flow, can be approximated by a one-step ODE solver in a computer program, resulting in a pseudo-orbit of the map. Thus the flow case reduces to the diffeomorphism case.

In the case of an autonomous differential equation, the induced time-h map will be the same for all t. On the other hand, if the differential equation is non-autonomous, the time-h map will depend on t. The following definition of an orbit of a dynamical system is made to encompass both the autonomous and non-autonomous cases.

DEFINITION 2.1. For each $0 \leq n < N$, let $f_n : R^m \to R^m$ be a C^2-diffeomorphism. The finite sequence $\{y_n\}, n = 0, \dots, N$ of points in R^m is called an *orbit of the dynamical system* $\{f_n\}, n = 0, \dots, N-1$ if $f_n(y_n) = y_{n+1}$ for $n = 0, \dots, N-1$. The finite sequence $\{x_n\}$ is called a δ-*pseudo-orbit* of $\{f_n\}$ if $|f_n(x_n) - x_{n+1}| < \delta$ for $n = 0, \dots, N-1$. The δ-pseudo-orbit $\{x_n\}$ is ε-*shadowed* by the orbit $\{y_n\}$ of the dynamical system $\{f_n\}$ if $|x_n - y_n| < \varepsilon$ for $n = 0, \dots, N$.

Here, as below, we use the l^2 norm:

$$|v| = \left(\sum_{i=1}^{m} v_i^2 \right)^{1/2}$$

for a vector $v = (v_1, \dots, v_m)$, and

$$|A| = \max_{v \in R^m, |v|=1} |Av|$$

for an $m \times m$ matrix A.

THEOREM 2.2. Let $f_n : R^m \to R^m$ be a C^2-diffeomorphism for $n = 0, \dots, N-1$. Let $\{x_n\}, n = 0, \dots, N$ be a δ-pseudo-orbit for the dynamical system $\{f_n\}$. For a fixed integer k, for $n = 0, \dots, N$, let S_n and U_n be subspaces of R^m of dimensions

k and $m - k$, respectively, such that $\mathcal{S}_n + \mathcal{U}_n = R^m$. Let S_n and U_n be the (unique) projections from R^m onto $\mathcal{S}_n$ and $\mathcal{U}_n$, respectively, which satisfy $S_n + U_n = I$.

Assume that $B, p, r_0, \dots, r_N$ and $t_0, \dots, t_N$ are constants satisfying

(1) $|\dfrac{\partial^2 f_n(x)}{\partial x_i \partial x_j}| \leq B, |\dfrac{\partial^2 f_n^{-1}(x)}{\partial x_i \partial x_j}| \leq B$, for all i, j, n, and for x on a compact set
    containing an open $4\delta^{1/2}$-neighborhood of each x_n.

(2) $\delta^{-2p} \geq 2(B+1)(m+4)m^{3/2}$

(3) $r_n \geq \max_{v \in \mathcal{S}_n, |v|=1} |Df_n(x_n)(v)|$

(4) $t_n \geq \max_{v \in \mathcal{U}_{n+1}, |v|=1} |Df_n^{-1}(x_{n+1})(v)|$

Define $C_0 = 0$; $C_n = |S_n| + (r_{n-1} + \delta + \delta^{1/2-p}|S_n|)C_{n-1}$ for $n > 0$.

Define $D_N = 0$; $D_n = |U_n| + (t_n + \delta + \delta^{1/2-p}|U_n|)D_{n+1}$ for $n < N$.

If $C_n \leq \delta^{p-1/2}$ and $D_n \leq \delta^{p-1/2}$ for $n = 0, \dots, N$, then there exists an orbit of the dynamical system $\{f_n\}$ which shadows $\{x_n\}$ within $4\delta^{1/2}$.

The proof is given in [4]. The key technique is the refinement process of [3]. Define the maps $F_s, F_u : (R^m)^N \to (R^m)^N$ by

$$F_s(x_1, \dots, x_N) = (f(x_0) - x_1, \dots, f(x_{N-1}) - x_N)$$

$$F_u(x_0, \dots, x_{N-1}) = (f(x_0) - x_1, \dots, f(x_{N-1}) - x_N).$$

Let G_s (respectively G_u) be the Newton's method map for F_s (respectively F_u). If the pseudo-orbit distance is small, the iteration of G_s, using the pseudo-orbit as the starting point, would converge to the root $(f(x_0), f^2(x_0), \dots, f^{N-1}(x_0))$. On the other hand, iteration of G_u would converge to $(f^{-N}(x_N), \dots, f^{-1}(x_N))$. Neither of these is desired. Although these are true orbits, in general neither of them is near the original pseudo-orbit. Instead, our refinement process uses G_s and G_u in a slightly different way.

To refine the orbit, we move the point $x_n^0 := x_n$ in the pseudo-orbit to a new location x_n^1. The point x_n^1 is defined by adding two contributions to x_n^0, one in the $\mathcal{S}_n$ direction and one in the $\mathcal{U}_n$ direction. The stable contribution is the projection onto $\mathcal{S}_n$ of component i of $(G_s - I)(x^0)$, and the unstable contribution is the projection onto $\mathcal{U}_n$ of component n of $(G_u - I)(x^0)$. The new pseudo-orbit $\{x_n^1\}$ is a refined pseudo-orbit. It is proved in [4] that under the hypotheses of the Theorem 2.2, iteration of this refinement process results in a true orbit $\{x_n^\infty\}$ near the original pseudo-orbit $\{x_n^0\}$.

Theorem 2.2 gives an alternative approach to Bowen's shadowing lemma [1]. Let $f : R^m \to R^m$ be a C^2-diffeomorphism. A compact invariant set Λ is called *hyperbolic* if there is a continuous splitting of the tangent space $T_x R^m = E_x^s \oplus E_x^u$ for $x \in \Lambda$, and positive constants $\lambda < 1, C$ such that

(1) $Df(x)(E_x^s) = E_{f(x)}^s$

(2) $Df(x)(E_x^u) = E_{f(x)}^u$

(3) $|Df^n(x)(v)| \leq C\lambda^{-n}|v|$ for $v \in E_x^s$

(4) $|Df^{-n}(x)(v)| \leq C\lambda^{-n}|v|$ for $v \in E_x^u$

for all $x \in \Lambda$ and for all $n \geq 0$.

THEOREM 2.3. (Bowen) *Suppose* Λ *is a hyperbolic set for* f. *For each* $\varepsilon > 0$ *there is a* $\delta > 0$ *so that every* δ-*pseudo-orbit in* Λ *can be* ε-*shadowed.*

Proof. We remark, as Bowen did, that it suffices to prove the conclusion for f^k, where k is a positive integer. In fact, assume we have done so and let c be such that $|f^i(x) - f^i(y)| \le c|x - y|$ for $0 \le i < k$ on Λ. Given $\varepsilon > 0$, choose δ small enough so that given a δ-pseudo-orbit of f, there exists a true orbit $f^{ik}(y_0)$ of f^k shadowing the pseudo-orbit x_{ik} of f^k within $\varepsilon/2c$. This is possible because a δ-pseudo-orbit for f is a δkc-pseudo-orbit for f^k. Further choose $\delta < \varepsilon/2kc$. If K is an integer multiple of k and $0 \le i < k$, then

$$|f^{K+i}(y_0) - x_{K+i}| \le |f^i(f^K(y_0)) - f^i(x_K)|$$
$$+ \sum_{j=0}^{i-1} |f^{i-j-1}(f(x_{K+j})) - f^{i-j-1}(x_{K+j+1})|$$
$$\le c\frac{\varepsilon}{2c} + kc\delta$$
$$< \frac{\varepsilon}{2} + \frac{\varepsilon}{2} = \varepsilon$$

It remains to prove the conclusion of the theorem for $g = f^k$, where k is large enough so that $|Df^k(x)(v)| \le \alpha|v|$, and $|Df^{-k}(x)(v)| \le \alpha|v|$, for $\alpha < 1$, for all $x \in \Lambda$. For an arbitrary orbit length N, let $S_n = E^s_{x_{nk}}$ and $U_n = E^u_{x_{nk}}$ for $n = 0, \dots, N$ as in Theorem 2.2. Since the splitting is continuous on the closed set Λ, there is an upper bound A on $|S_n| + 1$ and $|U_n| + 1$. Choose B to satisfy (1) of Theorem 2.2 and set $p = 1/4$. Choose $\delta > 0$ small enough so that (2) of Theorem 2.2 is satisfied, $\delta^{1/2-p} \le (1 - \alpha)/2A$, and $\delta < \varepsilon^2/16$.

Then

$$C_n < A + (\alpha + A\delta^{1/2-p})C_{n-1}$$
$$\le A + (\alpha + \frac{1-\alpha}{2})C_{n-1}$$
$$\le A + \frac{\alpha+1}{2}C_{n-1}$$

It follows easily from this recurrence that

$$C_n \le \frac{2A}{1-\alpha} \le \delta^{p-1/2}$$

for all $n \le N$. The same argument holds for D_n. The hypotheses of Theorem 2.2 are satisfied, thus there is a true orbit within $4\delta^{1/2} < \varepsilon$ of any δ-pseudo-orbit.

There now exist shadowing orbits $\{f^i(y_N)\}, i = -N, \dots, N$ of arbitrary length. Letting y be an accumulation point of $\{y_N\}$, we get $|f^i(y) - x_i| \le \varepsilon$ for $-\infty < i < \infty$ as claimed.

3. Computer-assisted shadowing. In this section we describe a computer algorithm which uses the above Theorem 2.2 to verify the existence of true orbits of a dynamical system near the pseudo-orbit determined by a numerical computation. Along with the pseudo-orbit being computed, there are some auxiliary calculations to be made to check that the hypotheses of the Theorem are satisfied. We next describe these auxiliary calculations, which if successful provide a computer-assisted proof of the existence of a true orbit.

The most useful choice of the subspaces S_n and $\mathcal{U}_n$ is to choose them to approximate the stable and unstable directions, respectively, for the dynamical system $\{f_n\}$ at the particular map f_n. One way to accomplish this is as follows. Begin with an orthonormal set $\{u_{01}, \dots, u_{0l}\}$ of vectors in R^m chosen arbitrarily. Inductively define the orthonormal set $\{u_{n+1,1}, \dots, u_{n+1,l}\}$ to be the computed results of applying the technique of Gram-Schmidt orthogonalization, followed by normalization, to the set $\{Df_n(x_n)u_{n1}, \dots, Df_n(x_n)u_{nl}\}$. Because of computer round-off, these computations will be only approximate, which is not important. The subspace $\mathcal{U}_n$ is defined to be the span of the computer memory values of $\{u_{n1}, \dots, u_{nl}\}$.

The subspace S_n is defined analogously. Begin with an arbitrary orthonormal set $\{s_{N1}, \dots, s_{Nk}\}$ from R^m. Inductively, given $\{s_{n1}, \dots, s_{nk}\}$, apply Gram-Schmidt to the set $\{Df_{n-1}^{-1}(x_n)s_{n1}, \dots, Df_{n-1}^{-1}(x_n)s_{nk}\}$. The definition of the set $\{s_{n-1,1}, \dots, s_{n-1,k}\}$, is the set of stored values of the resulting finite-precision computation, and S_n is defined to be the span of these k vectors.

The projection matrices S_n and U_n are then found as follows. Let A_n be the $m \times m$ matrix whose columns are $\{s_{n1}, \dots, s_{nk}, u_{n1}, \dots, u_{nl}\}$, and let $B_n = A_n^{-1}$. Let B_n^s be the $m \times m$ matrix whose top k rows are the same as those of B_n and whose bottom l rows are filled with zeros. Let B_n^u be the $m \times m$ matrix whose top k rows are the filled with zeros and whose bottom l rows are the same as those of B_n. Note that $B_n = B_n^s + B_n^u$.

Now define $S_n = A_n B_n^s$ and $U_n = A_n B_n^u$. It is clear that S_n and U_n are projections onto S_n and $\mathcal{U}_n$, respectively, and that

$$S_n + U_n = A_n(B_n^s + B_n^u) = I.$$

Further, S_n and U_n are the unique $m \times m$ matrices with these properties.

We turn now to computer verification of the hypotheses of Theorem 2.2. The number δ is the truncation error of the process being approximated. Also necessary is an a priori upper bound B on the second partial derivatives. Given B, a constant p is chosen just large enough to satisfy the second hypothesis. Finally, we describe a method for finding values of the r_n and t_n which satisfy hypotheses (3) and (4) of the Theorem. We will need the following definition and lemma.

DEFINITION 3.1. If A is a real symmetric matrix, define $\sigma(A)$ to be the largest eigenvalue of A.

LEMMA 3.2. Let A be an $m \times m$ matrix and W a subspace of R^m with basis $\{w_1, \dots, w_k\}$. Let W be the $m \times k$ matrix with columns $\{w_1, \dots, w_k\}$, and $B = AW$. Then

$$\max_{v \in W, |v|=1} |Av| \leq \left(\frac{\sigma(B^T B)}{1 - |W^T W - I|} \right)^{1/2}.$$

234

The lemma is applied for each n with $\mathcal{W} = \mathcal{S}_n, w_i = s_{ni}, i = 1, \ldots, k$, and $A = Df_n(x_n)$. Since $\{s_{n1}, \ldots, s_{nk}\}$ is near to being an orthonormal set, an upper bound for $|W^T W - I|$ can be computed that is much smaller than 1. An upper bound for $\sigma(B^T B)$ can also be computed. (In both instances, it is necessary, and straightforward, to allow for numerical roundoff in assigning the upper bound.) Then the Lemma gives an upper bound r_n for the right-hand-side of assumption (3) of the Theorem. The Lemma is also applied with $\mathcal{W} = \mathcal{U}_n$ and $A = Df_n^{-1}(x_n)$ in an analogous way to find suitable t_n for assumption (4) of the Theorem.

4. Examples.

As a first example, consider the Henon map

$$f(x, y) = (A - x^2 - By, x)$$

of the plane. For parameter values $A = 1.4$, $B = 0.3$, this map has an apparently chaotic orbit. Using the method described above, a computer-generated δ-pseudo-orbit with initial condition $(0, 0)$ and $\delta = 10^{-14}$ was found to have a true orbit within $\epsilon = 4 \times 10^{-7}$ for over one million iterates. Similar statements apply for other initial conditions.

This map was originally shadowed in [2], and similar results were reported. In that paper, a different approach was taken, which uses higher precision arithmetic (machine-epsilon $= 10^{-28}$) to verify shadowing of a δ-pseudo-orbit with $\delta = 10^{-14}$. The method of the present paper does not require such higher precision.

This point becomes especially relevant when systems are studied that are inherently more difficult to shadow. Consider the forced damped pendulum, which satisfies the differential equation

$$y'' + Ay' + \sin y = B \cos t.$$

To achieve good shadowing results for this differential equation we needed to generate a δ-pseudo-trajectory with $\delta = 10^{-18}$. We accomplish this by using a seventh order ODE-solver method with an explicit truncation error formula, and using a step size of $h = \pi/1000$.

For the forced damped pendulum with parameters $A = 0.2$ and $B = 2.4$, there is an apparently chaotic trajectory beginning at $(0, 0)$ at time $t = 0$. Using the techniques described above, we were able to prove the existence of a true trajectory within 4×10^{-9} for the computer-generated trajectory for time t ranging from 0 to $10^4 \pi$. Again, there are similar results for other initial conditions, and other values of A and B.

REFERENCES

[1] R. BOWEN, ω-limit sets for Axiom A diffeomorphisms, J. of Differential Equations, 18 (1975), pp. 333–339.

[2] C. GREBOGI, S. HAMMEL, AND J. YORKE, Do numerical orbits of chaotic dynamical processes represent true orbits?, J. Complexity, 3 (1987), pp. 136–145.

[3] —————, Numerical orbits of chaotic processes represent true orbits, Bulletin A.M.S., 19 (1988), pp. 465–470.

[4] T. SAUER AND J. YORKE, A shadowing theorem for differential equations.

TRANSFORMATION TO VERSAL NORMAL FORM

1. Introduction. Often the first step in analyzing a given problem requires the transformation of the linearized system into its Jordan canonical form. If the given system depends on parameters, say ε this reduction to Jordan canonical form can be an unstable operation, since the normal form and the transformation itself can depend in a discontinuous way on these parameters. The difficulty to which we elude occurs when several eigenvalues of the linearized system coincide, say for $\varepsilon = 0$. In the generic case the matrix will be non-semi-simple, i.e. not diagonalizable for $\varepsilon = 0$.

In order to overcome this difficulty Arnol'd [1] introduced a new normal form for matrices which depend on parameters. It differs from the usual Jordan form by having some elements functions of ε. These functions vanish for $\varepsilon = 0$, where the matrix takes on the standard Jordan canonical form. Arnol'd has called this the versal normal form of a matrix. It distinguishes itself from other normal forms as the smallest number of entries in the matrix are nonzero and at the same time the transformation matrix remains continuous.

Despite the advantages of versal normal forms they have not yet been used extensively in applications, as they have a reputation of being difficult to work with. This reputation may have originated from the fact that versal normal forms may not be unique[1] or from the fact that Arnol'd did not write down an explicit algorithm on how to construct the transformation. Finally the reputation may also have arisen, because versal normal are typically more applicable for problems in higher dimensions where calculations by hand are more tedious anyway.

A computer algebra system like Macsyma can handle such cases easily. Based on our limited experience we found that the calculations for the versal normal form are not significantly more time consuming than those for the Jordan canonical form. Furthermore, it turned out that our approach did fit well into the framework of Hamiltonian matrices where a straight forward method allows us to insure that the transformation is symplectic.

Although the work which is described here is applicable to a wider class of problems it was motivated and carried out in order to prove theorems about the motion near the Lagrangian point $\mathcal{L}_4$ in the restricted problem of three bodies. The

*Department of Computer Science, University of Cincinnati, Cincinnati, Ohio 45221-0008. Supported by a grant from ACMP of DARPA administered by NIST
[1]Arnol'd derived the term versal from the word universal by dropping its prefix which indicates uniqueness

problem has been studied by Meyer and Schmidt [4], Schmidt [5], van der Meer [6] and many others, but it is often done with adhoc methods.

It is hoped that the versal normal form allows a uniform treatment of this and similar problems. The price to be paid for is the more complicated linear transformation, but with the help of computer algebra this work is managable.

The versal normal form for the problem at $\mathcal{L}_4$ has already been determined by Cushman et al.[2]. Unfortunately they start with a system that is nearly in normal form. It means that the elements in their transformation matrix have removable singularites at $\varepsilon = 0$. It appears that the authors took advantage of this fact when they constructed their transformation but it makes it very difficult to compare their work with ours except via series expansion.

2. Definition of Versal Normal Form. Arnol'd gives a geometric definition for the versal normal form for matrices but it can also be understood in the context of normal forms of systems of differential equations.

Definition 1 : *The system of differential equations $y' = Cx + f(x)$ is in normal form if $f(e^{C^T t}x) = e^{C^T t}f(x)$, that is, $f(x)$ evaluated at the solution of the adjoint linear system is the same as following the solutions of the adjoint system starting at $f(x)$.*

Typically C is already in Jordan canonical form and $f(x)$ represents nonlinear terms. In our case we take C to be A_0 the Jordan canonical form of our system for $\varepsilon = 0$ and we set $f(x) = Bx$. We then try to determine those terms in B which may exist for $\varepsilon \neq 0$. From the definition it follows that $Be^{A_0^T t}x = e^{A_0^T t}Bx$ has to hold for all t and x. Therefore A_0^T and B have to commute.

When A_0 is given in Jordan canonical form the solution to the equation $A_0^T B = B A_0^T$ can be found in [3, p.234] and also in the paper of Arnol'd [1]. It shows which terms in B can be nonzero for $\varepsilon \neq 0$. The values of these terms can be found by comparing the characteristic polynomial for $A(\varepsilon)$ with the one for $A_0 + B$.

The following example illustrates this approach. Assume that $A(\varepsilon)$ has two pairs of purely imaginary eigenvalues for $\varepsilon = 0$, and that the Jordan canonical form for A_0 is

$$A_0 = \begin{pmatrix} i\omega & 1 & 0 & 0 \\ 0 & i\omega & 0 & 0 \\ 0 & 0 & -i\omega & 1 \\ 0 & 0 & 0 & -i\omega \end{pmatrix}.$$

The form for B is then

$$B = \begin{pmatrix} v_1 & 0 & 0 & 0 \\ v_2 & v_1 & 0 & 0 \\ 0 & 0 & v_3 & 0 \\ 0 & 0 & v_4 & v_3 \end{pmatrix}.$$

If $A(\varepsilon)$ is real the following reality conditions hold $\bar{v}_1 = -v_3, v_2 = v_4 = \bar{v}_2$, so that we can write for the versal normal form for $A(\varepsilon)$

$$\Lambda = A_0 + B = \begin{pmatrix} i(\omega + v_1) & 1 & 0 & 0 \\ v_2 & i(\omega + v_1) & 0 & 0 \\ 0 & 0 & -i(\omega + v_1) & 1 \\ 0 & 0 & v_2 & -i(\omega + v_1) \end{pmatrix}$$

where the new quantities v_1 and v_2 are now purely real. Their values can be determined by comparing the characteristic polynomials for $A(\varepsilon)$ and the one for Λ.

The eigenvalues for Λ are $\pm i(\omega + v_1) \pm \sqrt{v_2}$. From this it is clear that the versal normal form captures the generic behavior of $A(\varepsilon)$ as ε passes through 0. For $v_2 \leq 0$ the eigenvalues are purely imaginary whereas for $\varepsilon > 0$ they lie in the complex plane.

The problem to be discussed here is how to calculate the transformation matrix T such that $T^{-1}AT = \Lambda$. It follows from the work of Arnol'd that T is continuous in ε.

3. Finding the Transformation Matrix.

The transformation matrix T in the example of the previous section can be calculated easily with the following method.

Set $T = (\alpha_1, \alpha_2, \alpha_3, \alpha_4)$ where $\alpha_j, j = 1, \ldots, 4$ are complex valued column vectors with $\alpha_3 = \overline{\alpha_1}, \alpha_4 = \overline{\alpha_2}$. The condition $AT = T\Lambda$ for this example gives

(1)
$$(A - i(\omega + v_1)I)\alpha_1 = \alpha_2$$

and
(2)
$$(A - i(\omega + v_1)I)\alpha_2 = v_2\alpha_2.$$

Since v_2 can be zero we eliminate α_2 and obtain

(3)
$$((A - i(\omega + v_1)I)^2 - v_2I)\alpha_1 = 0.$$

The rank of this coefficient matrix has been reduced by 2 so that the general solution is $\alpha_1 = r_1\beta_1 + r_2\beta_2$ with β_1 and β_2 two linearly independent solutions of (3) and r_1, r_2 arbitrary scalar factors. The value of α_2 can then be found with the help of (1) using only matrix multiplications.

The above method is similar to the one which can be used to find generalized eigenvectors of a matrix. In numerical computations this approach is not used because it is numerical unstable. Therefore the standard method is to find the eigenvectors first and only then the generalized eigenvectors. Since all our calculations are carried out exactly numerical instabilty is of no concern to us. Futhermore, the method has the advantage that it will give at once the most general transformation matrix T. This is helpful when T has to be restricted further. Such a case will be discussed in the next section, where we deal with a Hamiltonian matrix $A(\varepsilon)$ and the matrix T has to be symplectic, that is, it has to satisfy $T^T J T = J$, where J is the usual matrix of Hamiltonian mechanics.

4. The Lagrangian Point $\mathcal{L}_4$.

The method of the previous section will be applied to the linearized system near the Lagrangian point $\mathcal{L}_4$ in the restricted problem of three bodies. The parameter μ represents the mass ratio of the two primaries. The matrix of the linearized system is

$$
(4) \qquad A(\mu) = \begin{pmatrix} 0 & 1 & 1 & 0 \\ -1 & 0 & 0 & 1 \\ \frac{-1}{4} & \frac{3\sqrt{3}}{4}(1-2\mu) & 0 & 1 \\ \frac{3\sqrt{3}}{4}(1-2\mu) & \frac{5}{4} & -1 & 0 \end{pmatrix}.
$$

It's characteristic equation is

$$
(5) \qquad \lambda^4 + \lambda^2 + \frac{27}{4}\mu(1-\mu) = 0.
$$

The eigenvalues lie on the imaginary axis for $0 < \mu \leq \mu_1$ and they move into the complex plane for $\mu > \mu_1$. The repeated eigenvalues are $\pm i\frac{\sqrt{2}}{2}$ and they occur for the value of the mass ratio $\mu_1 = \frac{1}{2}(1 - \frac{1}{9}\sqrt{69})$, which is known as Routh's critical mass ratio.

The situation near μ_1 is therefore similar to the case described in the previous section with the exception that (4) comes from a Hamiltonian sytem and its normal form has to be adjusted so that it remains a Hamiltonian matrix. Furthermore, the transformation matrix has to be symplectic and although we write the normal form in complex coordinates, we have to make sure that the implied transformation to the real normal form can be carried out in real coordinates. For this reason the versal normal form for (4) near μ_1 is

$$
\Lambda = \begin{pmatrix} -i(\frac{\sqrt{2}}{2} + v_1) & 0 & 0 & v_2 \\ 0 & i(\frac{\sqrt{2}}{2} + v_1) & v_2 & 0 \\ 0 & -1 & i(\frac{\sqrt{2}}{2} + v_1) & 0 \\ -1 & 0 & 0 & -i(\frac{\sqrt{2}}{2} + v_1) \end{pmatrix}.
$$

By comparing the characteristic equation of Λ with (5) we find

$$
v_1 = -\frac{\sqrt{2}}{2} + \frac{1}{2}\sqrt{1 + \sqrt{27\mu(1-\mu)}}
$$

$$
v_2 = \frac{1}{4}(1 - \sqrt{27\mu(1-\mu)})
$$

The choice of the parameter has a significant effect on the amount of calculations which has to be done by Macsyma. We found that by setting $\sigma = \sqrt{27\mu(1-\mu)}$ the computations could be kept at a minimum. Routh's critical mass ratio μ_1

corresponds to $\sigma = 1$. The versal normal form of (4) is therefore

$$\Lambda(\varepsilon) = \begin{pmatrix} -iw & 0 & 0 & v \\ 0 & iw & v & 0 \\ 0 & -1 & iw & 0 \\ -1 & 0 & 0 & -iw \end{pmatrix}$$

with $w = \frac{1}{2}\sqrt{1+\sigma}$ and $v = \frac{1}{4}(1-\sigma)$.

It would have been natural to set $v = \varepsilon$ in which case we would have $w = \sqrt{\frac{1}{2} - \varepsilon}$ but this choice resulted in increased computing time. On the other hand the final result can easily be expressed in ε since $\sigma = 1 - 4\varepsilon$.

The transformation matrix $T = (\alpha_1, \alpha_2, \alpha_3, \alpha_4)$ has to satisfy now the reality conditions $\alpha_2 = \bar{\alpha}_1$ and $\alpha_3 = \bar{\alpha}_4$. From $T\Lambda = AT$ we obtain

$$(A - iwI)\alpha_1 = \alpha_4$$
$$(A - iwI)\alpha_4 = \varepsilon\alpha_1$$

and solve first for α_1 to get

$$\alpha_1 = r_1 \begin{bmatrix} 1 \\[6pt] 0 \\[6pt] \dfrac{2\sqrt{27-4\sigma^2}-i(3-2\sigma)\sqrt{1+\sigma}}{12-4\sigma} \\[10pt] \dfrac{6-i\sqrt{(1+\sigma)(27-4\sigma^2)}}{12-4\sigma} \end{bmatrix} + r_2 \begin{bmatrix} 0 \\[6pt] 1 \\[6pt] \dfrac{6-i\sqrt{(1+\sigma)(27-4\sigma^2)}}{12-4\sigma} \\[10pt] \dfrac{-2\sqrt{27-4\sigma^2}-i(9-2\sigma)\sqrt{1+\sigma}}{12-4\sigma} \end{bmatrix}$$

and then from the first equation

$$\alpha_4 = r_1 \begin{bmatrix} \dfrac{-2\sqrt{27-4\sigma^2}-3i\sqrt{1+\sigma}}{12-4\sigma} \\[10pt] \dfrac{6-4\sigma+i\sqrt{(1+\sigma)(27-4\sigma^2)}}{12-4\sigma} \\[10pt] -\dfrac{3+2\sigma}{8} \\[10pt] -\dfrac{\sqrt{27-4\sigma^2}+4i\sqrt{1+\sigma}}{8} \end{bmatrix} + r_2 \begin{bmatrix} \dfrac{-18+4\sigma+i\sqrt{(1+\sigma)(27-4\sigma^2)}}{12-4\sigma} \\[10pt] \dfrac{2\sqrt{27-4\sigma^2}+3i\sqrt{1+\sigma}}{12-4\sigma} \\[10pt] \dfrac{-\sqrt{27-4\sigma^2}+4i\sqrt{1+\sigma}}{8} \\[10pt] -\dfrac{9+2\sigma}{8} \end{bmatrix} .$$

Since r_1 and r_2 are arbitrary we can construct from these vectors the most general transformation matrix T which puts (4) into its versal normal form. We want T to be symplectic, that is, it has to satisfy $T^T J T = J$. At Routh's critical mass ratio μ_1, i.e. at $\sigma = 1$ it is seen that r_1 and r_2 have to be real. The condition for T to be symplectic gives then rise to the following two equations

$$(6) \qquad (3 - 2\sigma)r_1^2 + 2\sqrt{27 - 4\sigma^2}\,r_1 r_2 + (9 - 3\sigma)r_2^2 = 0$$

$$(-3 + 5\sigma + 2\sigma^2)r_1^2 + 2(1 + \sigma)\sqrt{27 - 4\sigma^2}\,r_1 r_2 + (9 + 11\sigma + 2\sigma^2)r_2^2 = 4(3 - \sigma)$$

The second equation can be simplified with the help of the first one to read

$$r_1^2 + r_2^2 = \frac{3 - \sigma}{\sigma(1 + \sigma)}.$$

Therefore we have to find the intersection of a pair of lines through the origin with a circle. A solution is

$$r_1 = -\frac{1}{2}\frac{\sqrt{(9 + 2\sigma)(3 - \sigma) + \sqrt{2\sigma(3 - \sigma)(27 - 4\sigma^2)}}}{\sqrt{\sigma(1 + \sigma)(3 + \sigma)}}$$

$$r_2 = \frac{\sqrt{\sigma(27 - 4\sigma^2)} - 2\sigma\sqrt{2(3 - \sigma)}}{2\sigma(9 - 2\sigma)\sqrt{(1 + \sigma)(3 + \sigma)}} \times \sqrt{(9 + 2\sigma)(3 - \sigma) + \sqrt{2\sigma(3 - \sigma)(27 - 4\sigma^2)}}.$$

In closing we would like to remark that if the transformation is done in real coordinates, then

$$T = (\mathrm{Re}\,\alpha_1, -\mathrm{Im}\,\alpha_1, \mathrm{Re}\,\alpha_4, -\mathrm{Im}\,\alpha_4)$$

and the real normal versal form for (4) is

$$\begin{pmatrix} 0 & -w & v & 0 \\ w & 0 & 0 & v \\ -1 & 0 & 0 & -w \\ 0 & -1 & w & 0 \end{pmatrix}.$$

REFERENCES

[1] V. ARNOLD, *On matrices depending on parameters*, Russian Math. Surveys, 26:29–43, 1971.

[2] R. CUSHMAN, A. KELLEY, AND H. KOCAK, *Versal normal form at the Lagrangian equilibrium $\mathcal{L}_4$*, J. Differential Equations, 64:340–374, 1986.

[3] F. R. GANTMACHER, *Matrizentheorie*, Springer, 1986.

[4] K. R. MEYER AND D. S. SCHMIDT, *Periodic orbits near $\mathcal{L}_4$ for mass ratio's near the critical mass ratio of Routh*, Celestial Mech., 4:99–109, 1971.

[5] D. S. SCHMIDT, *Periodic solutions near a resonant equilibrium of a Hamiltonian system*, Celestial Mech., 9:81–103, 1974.

[6] J. C. VAN DER MEER, *The Hamiltonian Hopf Bifurcation*, Springer, 1986.

COMPUTER ASSISTED LOWER BOUNDS
FOR ATOMIC ENERGIES

LUIS A. SECO*

Abstract. For an atom of nuclear charge Z, the ground state energy is defined to be the lowest possible value of the energy Hamiltonian. We describe an algorithm to produce rigorous lower bounds for the ground state energy of atoms as well as its implementation.

0. Introduction. The Hamiltonian for an atom of charge Z is

$$H = \sum_{i=1}^{Z}[(-\tfrac{1}{2}\Delta_{x_i}) - \frac{Z}{|x_i|}] + \frac{1}{2}\sum_{i \neq j}\frac{1}{|x_i - x_j|}$$

acting on

(0.1)
$$\mathcal{H} = \bigwedge_{i=1}^{Z}\left(L^2\left(\mathbf{R}^3\right) \otimes \mathbf{C}^2\right)$$

The antisymmetric tensor product "$\wedge$" has to do with Pauli's exclusion principle, and $L^2\left(\mathbf{R}^3\right) \otimes \mathbf{C}^2$ is the set of states of one electron with two spins. We will refer to $\mathcal{H}$ as the space of antisymmetric wave functions. It is an important problem in Quantum Mechanics to compute good bounds for the ground state energy of the atom,

$$E = \inf_{\lambda \in Spec(H)} \lambda = \inf_{\substack{\psi \in \mathcal{H} \\ \|\psi\|_2 = 1}} \langle H\psi, \psi \rangle$$

Upper bounds to E can be obtained by restricting the infimum in the definition to a smaller class of functions. In Hartree-Fock method, for example, the infimum is taken over functions that can be written as antisymmetric products of one–electron functions. See [FF] for more information on upper bounds.

The problem of obtaining lower bounds is considerably more complicated, and was treated numerically before by Hertel, Lieb and Thirring in [HLT]. In this paper we will present very general ideas to improve the lower bounds they obtained that will moreover produce rigorous results. For complete information and refinements of this method, see [Se].

The basic idea of the method is to construct a radial function $V(|x|)$ and a constant C such that

(0.2)
$$\frac{1}{2}\sum_{i \neq j}\frac{1}{|x_i - x_j|} \geq \sum_{i=1}^{Z}V(|x_i|) - C.$$

---
*Department of Mathematics, Princeton University. Supported by a Sloan Foundation Dissertation Fellowship

which provides the operator inequality

$$H \geq \sum_{i=1}^{Z} \left(-\tfrac{1}{2}\Delta_{x_i} - \frac{Z}{|x_i|} + V(|x_i|) \right) - C$$

This implies that, if $\lambda_1 < \cdots < \lambda_n < \cdots < 0$ are the negative eigenvalues of

$$-\tfrac{1}{2}\Delta - \frac{Z}{|x|} + V(|x|) \overset{\text{def}}{=} H^{1 \text{ electron}}$$

then

$$E \geq \begin{cases} 2\sum_{i=1}^{Z/2} \lambda_i - C & Z \text{ even} \\[2ex] 2\sum_{i=1}^{(Z-1)/2} \lambda_i + \lambda_{(Z+1)/2} - C & Z \text{ odd} \end{cases}$$

The factor 2 appears because, by (0.1), eigenvalues come with multiplicity 2: one for each spin.

Separation of variables tells us then that the negative eigenvalues of $H^{1 \text{ electron}}$ are the same as the negative eigenvalues of the ODE operator

(0.3)
$$-\tfrac{1}{2}u'' - \left(\frac{Z}{r} - \frac{l(l+1)}{2r^2} - V(r) \right) u$$

acting on

$$\mathcal{H}_{\text{ODE}} = \{ f : r^{-1} \cdot f(r) \in L^2[0,\infty) \}$$

for $l = 0, 1, 2, \ldots$; every eigenvalue has multiplicity $2l+1$. For our eigenvalue problem, this space is equivalent to $L^2(0,\infty)$ with Dirichlet boundary conditions at 0 and ∞.

This ODE in general cannot be solved explicitly, but we will still be able to estimate its eigenvalues. For this, we will use computer assisted techniques, as will be explained in section 3.

1. The Potential. In order to obtain V and C in (0.2), we use an idea introduced in [FL1]; they wrote

$$\frac{1}{|x-y|} = \frac{1}{\pi} \int_{R>0} \int_{z \in \mathbb{R}^3} \chi_{B(z,R)}(x) \cdot \chi_{B(z,R)}(y) \, \frac{dz \, dR}{R^5}$$

where $\chi_{B(z,R)}$ is the characteristic function of the ball $B(z,R)$. This implies that

$$\frac{1}{2}\sum_{i \neq j} \frac{1}{|x_i - x_j|} = \frac{1}{2\pi} \int_{R>0} \int_{z \in \mathbb{R}^3} N(N-1) \frac{dz dR}{R^5}$$

where

$$N = N(x_1, \ldots, x_Z; z, R) = \sum_{i=1}^{Z} \chi_{B(z,R)}(x_i)$$

is the number of x_i that belong to $B(z,R)$.

Observe that given *any* function $\bar{N} = \bar{N}(z,R)$ defined in the space of all balls in $\mathbf{R}^3$ that takes values $k + \frac{1}{2}$ with k a nonnegative integer, we have

$$N(N-1) = (N - \bar{N})^2 + (2\bar{N} - 1)N - \bar{N}^2 \geq \frac{1}{4} + (2\bar{N} - 1)N - \bar{N}^2$$

therefore,

$$\frac{1}{2} \sum_{i \neq j} \frac{1}{|x_i - x_j|} \geq \frac{1}{2\pi} \sum_{i=1}^{Z} \iint (2\bar{N} - 1)\chi_{B(z,R)}(x_i) \frac{dz\,dR}{R^5} - \frac{1}{2\pi} \iint (\bar{N}^2 - \frac{1}{4}) \frac{dz\,dR}{R^5}$$

$$= \sum_{i=1}^{Z} V(|x_i|) - C.$$

with

$$V(|x|) = \frac{1}{2\pi} \iint_{\substack{|z - x| < R \\ R > 0}} 2(\bar{N}(z,R) - 1) \frac{dz\,dR}{R^5}$$

$$C = \frac{1}{2\pi} \iint (\bar{N}^2(z,R) - \frac{1}{4}) \frac{dz\,dR}{R^5}.$$

Note that whatever our choice of $\bar{N}(z,R)$ we obtain a lower bound; however, different choices of $\bar{N}$ will give different results, and it is important to make a good choice for $\bar{N}$. The way $\bar{N}$ is chosen is by selecting a charge density $\rho(x) \geq 0$, with $\int_{\mathbf{R}^3} \rho(x)dx = Z$ that we believe (but need not prove) is close to the real one, and then choose $\bar{N}$ according to the following rule:

Define functions $R_i(z)$, $1 \leq i \leq Z - 1$, in such a way that

$$\int_{B(z, R_i(z))} \rho(x)dx = i.$$

Then, set

$$\bar{N}(z,R) = \begin{cases} \frac{1}{2} & \text{if } R < R_1(z) \\ i + \frac{1}{2} & \text{if } R_i(z) < R < R_{i+1}(z) \\ Z - \frac{1}{2} & \text{if } R > R_{Z-1}(z) \end{cases}$$

The freedom in choosing $\bar{N}$ then translates in the freedom to choose the R_i. This has as a consequence that we can make our potential V have the following properties:

1. If we take the R_i to be piecewise-linear functions in $|z|$, we can write the integrals over $\mathbf{R}_+^4$ in elementary terms that involve only sums of rational expressions of degree at most 5.

    Since we can approximate any such R_i by piecewise linear functions, this is not a severe restriction.

2. The potential is piecewise analytic, i. e., there exist finitely many points

$$\infty > x_0 > x_1 > \cdots > x_n > 0$$

such that V has a power series expansion around each x_i convergent in a disk that contains both x_{i-1} and x_{i+1}, and V agrees with this power series *to the left* of x_i. Globally, V has in general only 1 continuous derivative, (except at x_0 and x_n where it is merely continuous) and all discontinuities happen at the x_i. Moreover, the partition $\{x_i\}$ can be refined as needed. This will be useful, for example, to obtain small steps for the ODE solver.

3. V is constant around 0, i. e.

$$(1.1) \qquad V(x) = \text{constant} \qquad 0 \le x \le x_n$$

4. Around ∞, V can be taken to have the special form

$$(1.2) \qquad V(x) = \frac{2\lambda k}{x} - \frac{l(l+1)}{x^2} \qquad x \ge x_0$$

where λ is any positive number and k is a positive integer that depends on λ.

2. The Functional Analysis. The underlying Banach space in this theory is the space of piecewise analytic functions, with a lower bound on the size of the domains of analyticity. The purpose of this section is to formalize definitions and set up the framework for computer assisted analysis in function space. For similar and more detailed analysis, see [Ra], [EKW], [EW], [Mo] and [KM], for example.

Consider the Banach space

$$H^1 = \left\{ f(z) \;\middle|\; f(z) = \sum_{n=0}^{\infty} a_n z^n, \; \sum_{n=0}^{\infty} a_n < \infty \right\}$$

with norm

$$\|f\| = \sum_{n=0}^{\infty} |a_n|$$

This is a subspace of the set of analytic functions in the unit disk, that becomes a Banach Algebra with $\|\ \|$.

We consider a neighborhood basis consisting of sets of the form

$$(2.1)$$
$$\mathcal{U}(I_1, \dots, I_N; C) = \left\{ f(z) = \sum_{n=0}^{\infty} a_n z^n \;\middle|\; a_n \in I_n, \quad 0 \le n \le N, \; \sum_{n=N+1}^{\infty} |a_n| \le C \right\}$$

where C is a positive real number and I_n are intervals in the real line. For the computer implementation, C will run over $\mathcal{R}$, the set of computer-representable numbers, and the intervals will be those with representable endpoints; we call denote as $\mathcal{I}$ the set of this intervals.

The reason why this is a convenient space to work in is because elementary operations, such as addition, product, integration, differentiation (composed with a slightly contracting dilation), evaluation at points in the domain of analyticity and integration of initial value problems in ordinary differential equations are expressible by elementary formulas in terms of this set of neighborhoods. The question is now how to perform these elementary computations in an exact way using a computer. For this we use interval arithmetic. Although this point is something that several articles in these Proceedings are going to discuss, here is a very brief account of the technique:

Let $\mathcal{R}$ be the set of computer representable numbers. Given any real number r, the idea is to work with an interval in which r is contained, $[r_1, r_2]$ with $r_1, r_2 \in \mathcal{R}$, and translate in terms of these intervals whatever manipulation we intend to do with real numbers. In this way, matters are reduced to obtaining upper and lower bounds for manipulations of representable numbers in terms of representable numbers. This is possible using the capabilities of a computer. Standard references for these ideas are [KM] and [Mo] .

Observe that if we have a function $f(z)$ which is analytic in some disk, $|z - z_0| < r$, then, for any $\tilde{r} < r$, if we define $\tilde{f}(\tilde{z}) = f(\frac{\tilde{z}-z_0}{\tilde{r}})$, then $\tilde{f} \in H^1$. This allows to translate analytic functions into functions in H^1 of the unit disk.

In the real analytic case, $H^1[a, b]$ will denote H^1 of the disc with center a and radius $|b - a|$.

In the previous section, we saw that we will have to deal with functions with are sums of rational expressions. It is immediate to produce neighborhoods of type (2.1) that contain these functions locally.

3. The ODE. In this section we will discuss how our ODE problem (0.3) can be dealt with using the functional analysis introduced in the previous section. This presentation is taylored to deal with our special problem, but it can be modified trivially, at the expanse of complication, to deal with more general problems.

We consider first the solution of initial value problems. Lemma 3.1 below takes care of the solutions of an IVP with analytic coefficients. Lemmas 3.2 and 3.3 take care of the expansion of the solutions at the singularities of the ODE, around 0 and ∞. All three lemmas can be proved by matching coefficients.

LEMMA 3.1: *Consider the ODE:*

$$\left.\begin{array}{r} u'' + qu = 0 \\ u(0) = u_0 \\ u'(0) = u_1 \end{array}\right\}$$

where $q(x) \in \mathcal{U}(q_0, \cdots, q_N; \delta)$. *Then,* $u \in \mathcal{U}(u_0, \cdots, u_{N+2}; C)$ *where*

$$u_{n+2} = -\frac{1}{(n+2)(n+1)} \sum_{i=0}^{n} u_i q_{n-i} \qquad 0 \le n \le N$$

and

$$C \leq \frac{\sum_{i=0}^{N+2}\left\{\sum_{k=N+1-i}^{N} \frac{|q_k|}{(k+i+2)(k+i+1)} + \frac{\delta}{(N+3+i)(N+2+i)}\right\}|u_i|}{1 - \left(\sum_{k=0}^{N} \frac{|q_k|}{(k+N+5)(k+N+4)} + \frac{\delta}{(2N+6)(2N+5)}\right)}$$

And this scales trivially to deal with $q \in H^1[a,b]$ for any a and b.

LEMMA 3.2: Consider the ODE

$$u'' + \left(\frac{2\lambda k}{r} - \lambda^2\right)u = 0$$

for k a positive integer. Then, the only solution of the ODE that vanishes at ∞ is

$$u = e^{-\lambda r}\sum_{n=0}^{k} a_n r^n$$

where a_k is an arbitrary constant, and

$$a_n = a_{n+1}\frac{n(n+1)}{2\lambda(n-k)} \qquad n \leq k-1$$

LEMMA 3.3: Consider the ODE

$$u'' + \left(a + \frac{b}{r} - \frac{n(n+1)}{r^2}\right)u = 0$$

for n a positive integer. Then, the only solution of the ODE that vanishes at 0 is

$$u = \sum_{k=n+1}^{\infty} u_k r^k$$

where u_{n+1} is an arbitrary constant, and

$$u_{n+2} = -\frac{b \cdot u_{n+1}}{(n+2)(n+1) - n(n+1)}$$

$$u_{k+2} = -\frac{b \cdot u_{k+1} - a \cdot u_k}{(k+2)(k+1) - n(n+1)} \qquad k \geq n+1$$

$$\sum_{k>N+2} |u_k| \leq \frac{\frac{|b \cdot u_{N+2}| + |a \cdot u_{N+1}|}{(N+3)(N+2) - n(n+1)} + \frac{|a \cdot u_{N+2}|}{(N+4)(N+3) - n(n+1)}}{1 - \frac{|b|}{(N+4)(N+3) - n(n+1)} - \frac{|a|}{(N+5)(N+4) - n(n+1)}}$$

We now pass to discuss how to use these lemmas to the problem of the localization of eigenvalues.

A crucial device in the study of eigenvalue problems is the "*match*" function, $M(\lambda)$ associated with the ODE operator $-u'' - q \cdot u$ acting on $\mathcal{H}_{\text{ODE}}$. It is defined as follows:

given $\lambda > 0$, $-\lambda^2$ is a negative eigenvalue iff

$$u'' + (q - \lambda^2) \cdot u = 0$$

has a solution in $\mathcal{H}_{\text{ODE}}$. Take any point y, and consider u_0, a solution of the ODE which vanishes at 0, and u_∞, a solution that vanishes at ∞. Then, define

$$M(\lambda) = \frac{u_0'(y) \cdot u_\infty(y) - u_0(y) \cdot u_\infty'(y)}{\sqrt{u_0^2(y) + u_0'^2(y)} \cdot \sqrt{u_\infty^2(y) + u_\infty'^2(y)}}.$$

Then, $-\lambda^2$ is an eigenvalue iff $M(\lambda) = 0$. The eigenvalues of the ODE thus correspond to the zeroes of M.

For our analysis we will use a computer bound for M, $\mathcal{M} : \mathcal{R} \to \mathcal{J}$, that will satisfy the property that for any representable r,

(3.1) $$M(r) \in \mathcal{M}(r).$$

One possible way of implementing $\mathcal{M}$ is as follows: First, define the "phase" of a function u to be the point in the unit circle given by

$$\Phi_u(x) = \frac{(u'(x), u(x))}{\sqrt{u(x)^2 + u'(x)^2}}.$$

Note that the phase is invariant under multiplication of u by a nonzero constant, and it is only defined for functions that do not vanish to order two: since the functions we will be working with will be nonzero solutions of an ODE problem, their phase is defined. Note also that

$$M(\lambda) = \det\left(\Phi_{u_0}(y), \Phi_{u_\infty}(y)\right)$$

For positive $\lambda \in \mathcal{R}$ the eigenvalue problem we need to solve is

$$\left.\begin{array}{c} \frac{1}{2}u'' + p(x)u = 0 \\ u(0) = u(\infty) = 0 \end{array}\right\}$$

where

(3.2) $$p(x) = \frac{Z}{x} - \frac{l(l+1)}{2x^2} - V(x) - \lambda^2$$

From section 1, we have a finite set of real numbers, $x_0 > x_1 > \cdots > x_n > 0$ such that the coefficients of the ODE are in $H^1[x_i, x_{i+1}]$ for $0 \leq i \leq n-1$, and (1.1) and (1.2) say that the ODE takes the special form dealt with in lemmas 3.2 and 3.3 around 0 and ∞. Elsewhere, it takes the form dealt with in Lemma 3.1.

With the aid of Lemma 3.2, we can determine a neighborhood of type (2.1) of $H^1[x_0, \infty]$ that contains u_∞. This allows us to obtain intervals that contain $u_\infty(x_0)$ and $u'_\infty(x_0)$ and we can therefore give bounds for $\Phi_{u_\infty}(x_0)$.

With the aid of Lemma 3.1, we can solve the initial value problem at x_0, thus obtaining another neighborhood of type (2.1) of $H^1[x_0, x_1]$ that contains u_∞, and again obtain intervals that contain $\Phi_{u_\infty}(x_1)$.

Repeating this argument, we can obtain bounds for $\Phi_{u_\infty}(x_n)$.

With the aid of Lemma 3.3, we can determine bounds for $\Phi_{u_0}(x_n)$.

Then, we define
$$\mathcal{M}(\lambda) = \det\left(\Phi_{u_0}(x_n), \Phi_{u_\infty}(x_n)\right).$$

where the determinant is taken in the interval arithmetic sense. It is just clear that $M(\lambda) \in \mathcal{M}(\lambda)$ for $y = x_n$.

The idea now is to create heuristic (e.g. using numerical analysis) representable numbers
$$\lambda_1^{\mathrm{up}} > \lambda_1^{\mathrm{dn}} > \lambda_2^{\mathrm{up}} > \lambda_2^{\mathrm{dn}} > \cdots \lambda_k^{\mathrm{up}} > \lambda_k^{\mathrm{dn}} > 0$$

and then compute $\mathcal{M}(\lambda_i^{\mathrm{up,\, dn}})$.

If we can prove that

(3.3)
$$\begin{aligned} \mathcal{M}(\lambda_i^{\mathrm{up}}) > 0 \quad \mathcal{M}(\lambda_i^{\mathrm{dn}}) < 0 \quad i = 1, 3, \ldots \\ \mathcal{M}(\lambda_i^{\mathrm{up}}) < 0 \quad \mathcal{M}(\lambda_i^{\mathrm{dn}}) > 0 \quad i = 2, 4, \ldots \end{aligned}$$

using (3.1) and the fact that M is continuous we would have proved that each interval $\lambda_i^* = (\lambda_i^{\mathrm{dn}}, \lambda_i^{\mathrm{dn}})$ contains at least one eigenvalue.

This previous procedure does not work if we substitute the phase by simply the vector $(u(x), u'(x))$; the reason is that interval arithmetic estimates are far to conservative and the bounds we obtain are very bad after a few steps. The reason why the previous algorithm proves good bounds for the eigenvalues has to do with the fact that the solution of this particular ODE is of the form $e^{-\lambda x}$ with a factor with only polynomial growth. As a consequence, the normalizing factor in the phase has a contractive effect that makes the bounds more stable. In other words, if you look at the time flow
$$T_t\left(\Phi_u(x)\right) = \Phi_u(x+t)$$

for the particular ODE we are working with, the phase of the eigenfunctions are almost always in the unstable manifold, and expansion of the phase backward in time gives good answers. If the phase were in the stable manifold, then this procedure would be stable expanding forward in time. In general, however, one sided shooting is not enough, and you need to consider the projection into the stable and unstable manifolds separately. For this important generalization see [LR].

The fact that (3.3) holds thus tells us that there are negative eigenvalues $-\lambda^2$, with $\lambda_i \in \lambda_i^*$, $i = 1, \cdots, k$, with eigenfunctions u_k, but we still have to show that we didn't miss any eigenvalue, that is, that if $-\lambda^2$ is a negative eigenvalue and $-\lambda^2 \leq -\lambda_k^2$ then $\lambda = \lambda_i$ for some i; in other words, λ_k really is the k'th eigenvalue. In order to do this, we use the fact that ours is a Sturm-Liouville problem. Therefore if a certain eigenfunction u_k has $k - 1$ zeroes, then its eigenvalue is the k'th one. Hence, there are no other eigenvalues between $-\lambda_1^2$ and $-\lambda_k^2$.

In order to check that u_k has $k-1$ zeroes, note first of all the following corollary to comparison theorems for Sturm-Liouville problems:

Lemma 3.4: *Let u be any solution of*

$$\tfrac{1}{2}u'' - \left(V(x) + \frac{l(l+1)}{2x^2} - \frac{Z}{x} + \lambda^2\right) u = 0$$

on $[a, b]$, with $0 < a < b < \infty$, with V decreasing. Then, if

$$b - a \leq \pi \left(\frac{2Z}{b} - 2V(a) - \frac{l(l+1)}{a^2} - 2\lambda^2\right)_+^{-\frac{1}{2}}$$

then u cannot have two zeros in $[a, b]$.

Since our potential (3.2) is bounded above, we can arrange from the start that our partition $x_0, \cdots, x_n$ is fine enough so every interval (x_i, x_{i-1}) satisfies the hypothesis of this lemma. Also, for $l \neq 0$, since p is negative around 0 and ∞, we can take x_0 sufficiently big and x_n sufficiently close to 0 so $p(x_0)$ and $p(x_n)$ are both negative, and therefore u_k has no zeros in $(0, x_n)$ and (x_0, ∞); the special case $l = 0$ can be dealt with a trivial refinement of Lemma 3.4.

So, everything is reduced to counting the zeros of u_k in (x_n, x_0). For this, consider u_∞^{up} and u_∞^{dn}, the solutions of the ODE with parameters λ_k^{up} and λ_k^{dn} respectively which are 0 at ∞. Again, comparison theorems tell us that the number of zeros of u_∞ in (x_n, x_0) is bounded between the number of zeros in (x_n, x_0) of u_∞^{up} and u_∞^{dn}, and thus, it suffices to check that they both have $k - 1$ zeros. In order to check this, note that Lemma 3.4 guarantees that the number of zeros of u_∞^{up} and u_∞^{dn} in (x_n, x_0) are (essentially) the same as the number of sign changes in the sequences $\{u_\infty^{\text{up}}(x_i)\}_{i=0}^n$ and $\{u_\infty^{\text{dn}}(x_i)\}_{i=0}^n$ respectively. From the ODE solver, we have good bounds for these sequences, hence the number of sign changes can be just counted, and if in both cases it is $k - 1$, there are no more eigenvalues.

It is clear that with the control we have over the solutions of ODE's it is possible to bound the number of zeros without using Sturm-Liouville theory, but comparison theorems simplify the algorithm enormously.

4. The Results. The previous algorithm was carried out using as charge density guess an approximation to Thomas-Fermi density introduced in [Ti]. Once we solve the ODE's, the solutions give us another charge density in a trivial way, and this provides an iterative procedure that produces lower bounds for the ground state energy of atoms.

The following is a sample of results that can be obtained with this method for some values of the atomic charge Z: E_{lb} stands for the lower bounds obtained by the previous method; E_{ub} stands for known numerical —non rigorous— upper bounds to the energy obtained using Hartree-Fock's method (see [FF]).

Z	E_{lb}	E_{ub}	error (%)
10	−138.90	−128.54	7.74
20	−707.75	−676.75	4.47
30	−1,853.60	−1,777.84	4.17
40	−3,661.60	−3,538.99	3.40
50	−6,217.81	−6,022.93	3.18
60	−9,560.54	−9,283.88	2.98
70	−13,719.54	−13,391.45	2.45
80	−18,812.97	−18,408.99	2.20

The computer programs were written in **C**, and run on a SUN 3/60 workstation. The interval arithmetic package was supplied to me by D. Rana, and is the one he used in [Ra]. The program was divided into parts with a more general scope in mind than this particular problem, and most of them apply to more general situations. Execution time for the heuristics is of about two days for the largest atom, and the rigorous part takes about a month. A feature of the program is that most of the memory allocation is done during execution, and disposed of when no longer needed. This is done for two reasons: one, is to allow the degree of the Taylor expansions to be chosen adaptively, and second, because it requires too much memory otherwise. Freeing memory, the program uses about 3Mb of memory, and without freeing memory it cannot run after a few hours.

I wish to express my gratitude to Charles Fefferman and Rafael de la Llave for introducing me to these problems and for expert advise. I am grateful to D. Rana for providing me with his interval arithmetic package. I thank also P. Solovej and T. Spencer for useful conversations.

REFERENCES

[EKW] ECKMANN, J. P., KOCH, H. AND WITTWER, P., *A computer Assisted Proof of Universality in Area Preserving Maps*, Memoirs, A.M.S., Vol 289 (1984).

[EW] ECKMANN, J. P. AND WITTWER, P., *Computer Methods and Borel Summability Applied to Feigenbaum's equation*, Lecture Notes in Mathematics, Springer Verlag (1985).

[FL1] FEFFERMAN, C. AND LLAVE, R., *Relativistic Stability of Matter, I*, Revista Matemática Iberoamericana, Vol 2 no.1&2, pp. 119-213 (1986).

[FF]   FROESE-FISHER, C., *The Hartree-Fock Method for Atoms*, Wiley, New York (1977).

[HLT] HERTEL, P., LIEB, E. AND THIRRING, W., *Lower Bound to the Energy of Complex Atoms*, Journal of Chemical Physics, Vol. 62 no.8, p. 3355 (1975).

[KM] KAUCHER, E. W. AND MIRANKER, W. L., *Self-validating Numerics for Function Space Problems*, Academic Press, New York (1984).

[LR]   LLAVE, R. AND RANA, D., *Algorithms for the Rigorous Proof of Existence of Special Orbits*, *(to appear)*.

[Mo] MOORE, R. E., *Methods and Applications of Interval Analysis*, S.I.A.M., Philadelphia (1979).

[Ra] RANA , D., *Proof of Accurate Upper and Lower Bounds for Stability Domains in Denominator Problems*, Thesis, Princeton University (1987).

[Se]   SECO, L., *Lower Bounds for the Ground State Energy of Atoms*, Thesis, Princeton University (1989).

[Ti]   TIETZ, T., *Atomic Energy Levels for the Thomas-Fermi Potential*, J. of Chem. Phys., Vol 25, p. 787 (1956).